An Introduction to
Infrared Spectroscopy

An Introduction to

INFRARED SPECTROSCOPY

DR WERNER BRÜGEL
Physicist of the Badische Anilin- and Soda-Fabrik AG.
Ludwigshafen a. Rh.

Translated from the German original

by A. R. KATRITZKY
Fellow of Churchill College, Cambridge

and A. J. D. KATRITZKY

LONDON: METHUEN & CO LTD
NEW YORK: JOHN WILEY & SONS INC

CHEMISTRY

Einführung in die Ultrarotspektroskopie
was first published in 1957
by Dr Dietrich Steinkopff, Darmstadt
This translation first published 1962
© 1962 by Methuen & Co Ltd
Printed in Great Britain
by Spottiswoode Ballantyne & Co Ltd
London and Colchester
Cat. No. (Methuen) 2/6420/11

Contents

Part I. Fundamentals of the Theory of Infrared Spectra in the Gaseous State

vii

Contents

Contents

Contents

From the Preface to
the German Edition

A comprehensive treatment of infrared spectroscopy including all the theoretical and experimental details would amount to many times the size of the present volume, especially as it would necessarily include Raman and microwave spectroscopy. The purpose of this book is not to provide such a comprehensive treatment but to help the numerous scientists using infrared spectroscopy to overcome difficulties connected with commencing in a new field. The selection and treatment of the material handled inevitably depends to a certain extent on the particular interests and experience of the author and this should be borne in mind by the reader.

The treatment in this book emphasizes the experimental side of the subject. However, a theoretical section has been included which, although brief, contains all the essentials. At least a rough knowledge of the theoretical relationships is needed to understand and interpret infrared spectra. In the theoretical section, much use has been made of Professor Herzberg's excellent treatment, and his terminology has been followed for the most part.

The fact that a second edition was quickly needed showed that the book filled a gap in the literature. In the second edition details of individual smaller molecules have been omitted. It was possible to do this because certain specialist works have appeared which deal only with such applications. Also, the documentation of molecular spectroscopy has now begun in earnest.

It has been especially difficult to take into account all the work that has appeared since 1954. The volume of literature dealing with experimental techniques is beginning to decrease markedly, and work of this type up to the end of 1956 and in part to the spring 1957 could be included almost completely. However, the ever-increasing literature on applications has almost completely been left to the documentation system as fits the character of this book.

W. BRÜGEL

xiii

Translators' Preface

This book is a translation of the second German edition of Dr W. Brügel's *Einführung in die Ultrarotspektroskopie*, which appeared in 1957. Quite extensive additions, which Dr Brügel provided, have been incorporated. These will bring the book up to the stage of the third German edition which is due to appear in 1961. In this work we have attempted to reproduce faithfully Dr Brügel's account of the subject and selection of subject matter made and the opinions advanced are not necessarily the same as those of the translators.

We should like to thank Dr Brügel for all the help which he has given to us and in particular for accepting the suggestion that one of us made for amending the section dealing with aromatic and heterocyclic compounds. Our thanks are also due to Messrs. R. A. Jones, A. J. Boulton, and R. A. Y. Jones for reading the manuscript and for making many helpful suggestions.

A. R. KATRITZKY

A. J. D. KATRITZKY

Cambridge

Introduction

The region of the electromagnetic spectrum known as the infrared is next to
the visible region of the spectrum, i.e. that to which the human eye is sensitive (Fig. 1). Infrared waves are often known as heat waves because they
cause a warm sensation to the human skin, and because the most important
sources of these waves are usually heated solids. However, the description
'heat waves' is less appropriate and less exact than 'infrared'. This region
of the spectrum is of great theoretical and practical importance, and is of
especial interest to the physicist and the chemist because it gives particularly definite information about molecular structure. In this spectral region
lie the rotational and vibrational spectra of molecules, i.e. the manifestations of the changes in the molecular rotational and vibrational energy that

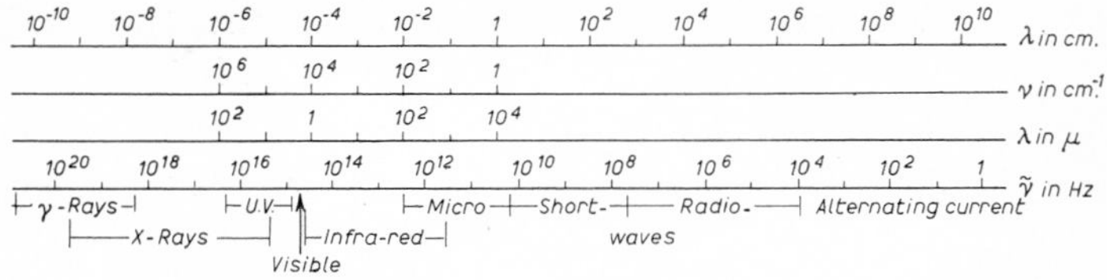

FIG. 1. *Position of the infrared region in the electromagnetic
spectrum.*

can occur under certain conditions by the interaction of infrared radiation
with matter.

The rotational and vibrational frequencies which thus appear, and
various other experimentally observed quantities, are directly dependent
on certain molecular constants, for example, the force constants between
atoms and the moments of inertia about certain axes. These molecular
constants can thus be calculated on the basis of quite simple molecular
models. It is easy to understand that infrared spectroscopy, i.e. that part of
the science of optics concerned with the infrared region, has become today
an indispensable auxiliary technique for chemical structural analysis. In
addition to these theoretical applications, infrared spectroscopy has practical uses which are no less valuable. These are fundamentally based on the

B

1

same experimental data and laws. Techniques of qualitative and quantitative analysis have been developed which are already quite refined but which are certainly not to be taken as final. This book is concerned both with the investigation of molecular structure by means of infrared spectra, and with the techniques of infrared spectral analysis, in so far as they can be considered within the framework of an introductory treatise.

The physicist and the technologist characterize infrared radiation, just as that of other spectral regions, by its wavelength. The micron (μ) is generally used as a unit and the connection between this and several other units of wavelength is the following:

$$1\mu = 10^{-3}\,\text{mm} = 10^{-4}\,\text{cm} = 10^{-6}\,\text{m} = 10^3\,\text{m}\mu = 10^4\,\text{Å}$$

Chemists, especially those concerned with the investigation of molecular structure, prefer another convention. The radiation is characterized by stating the number of waves which are contained in a length of 1 cm. This number, the reciprocal of the wavelength in centimetre units, is called the *wave number*; its unit is the reciprocal centimetre (cm^{-1}).†

Up to date no binding agreement over the exclusive use of one of these two systems of measurement has been concluded and the use of one or the other system is a question of custom. Table 1 gives values for calculations from wavelengths to wave numbers and vice versa in the region of 1–$30\,\mu$, rounded off to tenths of a micron or to the nearest cm^{-1}. In the text either wavelengths and wave numbers are used, but in the figures both units are always given. Where abbreviations are used instead of the full name, the Greek letter λ denotes the wavelength and the Greek letter ν the wave number. The relationship between them is therefore:

$$\nu\,[\text{cm}^{-1}] = \frac{1}{\lambda\,[\text{cm}]} = \frac{10000}{\lambda\,[\mu]} \qquad [0.1]$$

The wave number ν should not be confused with the *frequency* $\tilde{\nu}$.‡ The latter has the dimensions of a reciprocal time and is usually measured in cycles per

---

† Following a proposal of Meggers the Joint Commission for Spectroscopy recommended, at its meeting of September 1952 in Rome [*J. opt. Soc. Amer.* **43**, 410 (1953)], the use of the description 'kayser' (K) for the unit cm⁻¹. Candler (*109*) has proposed the name 'rydberg' (R); and other names have also been suggested. The decision of the Joint Commission gave rise to some heated discussion, which can be found in detail in *J. opt. Soc. Amer.* **46**, 145 (1956) and *Spectr. Mol.* **5** (1956). In view of this, the recommendation was more or less withdrawn. Because of the uncertain situation the short, unmistakable description 'cm⁻¹' has been retained.

‡ The Joint Commission for Spectroscopy recommended, according to a decision at the Rome meeting of September 1952 [*J. opt. Soc. Amer.* **43**, 410 (1953)], the use of the Greek letter σ for wave number in order to reserve ν for the frequency. In this case confusion is also possible, e.g. with the internationally recognized representation of the constant of the Stefan-Boltzmann law by the letter σ.

TABLE 1

λ	0	1	2	3	4	5	6	7	8	9
1	10000	9091	8333	7692	7143	6667	6250	5882	5556	5263
2	5000	4762	4545	4348	4167	4000	3846	3704	3571	3448
3	3333	3226	3125	3030	2941	2857	2778	2703	2632	2564
4	2500	2439	2381	2326	2273	2222	2174	2128	2083	2041
5	2000	1961	1923	1887	1852	1818	1786	1754	1724	1695
6	1667	1639	1613	1587	1563	1539	1515	1493	1471	1449
7	1429	1408	1389	1370	1351	1333	1316	1299	1282	1266
8	1250	1235	1220	1205	1190	1176	1163	1149	1136	1124
9	1111	1099	1087	1075	1064	1053	1042	1031	1020	1010
10	1000	990	980	971	962	952	943	935	926	917
11	909	901	893	885	877	870	862	855	847	840
12	833	826	820	813	806	800	794	787	781	775
13	769	763	758	752	746	741	735	730	725	719
14	714	709	704	699	694	690	685	680	676	671
15	667	662	658	654	649	645	641	637	633	629
16	625	621	617	613	610	606	602	599	595	592
17	588	585	581	578	575	571	568	565	562	559
18	556	552	549	546	543	541	538	535	532	529
19	526	524	521	518	515	513	510	508	505	503
20	500	498	495	493	490	488	485	483	481	478
21	476	474	472	469	467	465	463	461	459	457
22	455	452	450	448	446	444	442	441	439	437
23	435	433	431	429	427	426	424	422	420	418
24	417	415	413	412	410	408	407	405	403	402
25	400	398	397	395	394	392	391	389	388	386
26	385	383	382	380	379	377	376	375	373	372
27	370	369	368	366	365	364	362	361	360	358
28	357	356	355	353	352	351	350	348	347	346
29	345	344	342	341	340	339	338	337	336	334

second or Hertz units (1 Hz = 1 sec^{-1} or 1 vibration each second); the relation to the wavelength is:

$$\lambda . \tilde{\nu} = c \qquad [0.2]$$

where c is the velocity of light (in a vacuum $c_o = 2 \cdot 99776 \times 10^{10}$ or, rounded off, 3×10^{10} cm sec^{-1}). It follows that the relationship between the wave number and the frequency $\tilde{\nu}$ is:

$$\nu = \frac{\tilde{\nu}}{c} \qquad [0.3]$$

i.e. the wave number of a vibration is proportional to the frequency. The proportionality factor is the reciprocal of the velocity of light. It must be taken into account that the frequency $\tilde{\nu}$ is the fundamental constant for the process, and that the velocity of light varies according to the medium in

3

which the vibration takes place. Therefore, a different wave number and wavelength will be found for the same vibration according to the medium in which the experimental measurement is conducted. In order to make the usual correction from another medium to a vacuum, the refractive index n at the particular wavelength in that medium is required. In any medium the velocity of light is $c_n = c_o/n$; therefore the wave number in a vacuum, indicated by the subscript o, is calculated from that in a medium (subscript n) by dividing the latter by the refractive index: $\nu_o = \nu_n/n$. For practical purposes, the only important medium of measurement is the air. For air the refractive index n, which depends on the wavelength, is given by the equation:

$$(n-1).10^6 = 272 \cdot 599 + 1 \cdot 5358 . \lambda^{-2} + 0 \cdot 01318 . \lambda^{-4} \qquad [0.4]$$

where λ is in μ units† [Edlen (*184*)]. It is easily seen that the difference between ν_o and ν_n is extremely small; but it is still large enough to be sometimes detected by the high-resolution diffraction gratings which are now frequently used in infrared spectroscopy. However, from the purely practical and technical point of view, the difference is negligible. Calculated values of the vacuum correction according to Downie *et al.* (*176*) are given in Table 2.

TABLE 2. *Vacuum Correction for Wave Numbers Measured in Air in* cm^{-1} (*176*)

ν_n	$\nu_n - \nu_0$
500	0·13
1000	0·26
1500	0·39
2000	0·53
2500	0·65
3000	0·78
3500	0·91
4000	1·04
4500	1·17
5000	1·30

Following this account of the characterization of waves, the regions of the spectrum with which infrared spectroscopy is concerned will be more closely defined (Table 3). It is now known that the human eye can detect waves up to a wavelength of 1μ in favourable circumstances. The true infrared region

† According to Rank and Shearer (*572*) this formula gives values of $(n-1)$ that are about $1 \cdot 5$ per cent too small for the infrared region. Values of $n-1$ have been calculated from Edlen's formula for various temperatures by Penndorf (*783*).

thus begins at this value, but wavelength $0 \cdot 76$ μ is usually taken as the division between the visible and the infrared region. The upper boundary with the short electric waves is not so easily defined, because it depends on the state of the experimental techniques for the production and detection of infrared radiation. Values between 400 and 1000 μ, i.e. 25 and 10 cm^{-1}, are given in the literature. This very long-wavelength region, little investigated as yet, is of small importance to the present treatise. Another division of the infrared region, more important for the present account, is according

TABLE 3. *Division of the Infrared Spectral Region*

According to detector:	Photographic IR to $1 \cdot 4$ μ Photocell IR to about 6 μ Thermally detectable: whole region
According to method of dispersion:	Prism IR to about 50 μ (LiF to 6, CaF$_2$ to 9, NaCl to 15, KBr to 25, CsBr to 40, CsI to 50 μ) Grating IR whole region, principally above 50 μ
From the viewpoint of molecular spectroscopy:	Region of overtone vibrational frequencies (particularly of CH bonds) up to 2 μ Region of fundamental vibrational frequencies $2 \cdot 5$ to about 30 μ Region of rotational frequencies above about 50 μ
Usual nomenclature:	Near IR up to about 5 μ Middle IR about 5 to 25 μ Far IR over about 25 μ

to the nature of the spectra found therein. The following can be differentiated: the region of pure rotation spectra found in the wavelength range above *ca.* 50 μ (below *ca.* 200 cm^{-1}), the region of fundamental vibrational spectra from *ca.* $2 \cdot 5$–25 μ (4000 to 400 cm^{-1}), and the region of overtone frequencies (especially of the CH vibration) below about $2 \cdot 5$ μ (above 4000 cm^{-1}). These figures are naturally to be regarded as approximate indications, and not as sharp boundaries. Later, other divisions will be discussed which depend on the experimental techniques involved in the measurements.

The vast majority of the spectra of interest result from *absorption* experiments. Classical electrodynamics shows that interaction between waves and

molecules in the form of emission or absorption is only possible if the interaction is connected with an *alteration in the electrical dipole moment* of
the interacting molecules. The same result is obtained by application of
quantum theory. Only those rotations and vibrations which cause an alteration in the electric dipole moment are excited by radiation, i.e. can be
measured by absorption methods. These are generally called *infrared active*.
Nothing can be learned from the infrared absorption spectrum about rotations and vibrations which do not cause alterations in the dipole moment.
In order to investigate these, it is necessary to use another phenomenon, the
molecular scattering of electromagnetic waves known as the Raman effect.
Although Raman spectroscopy is the natural complement to infrared
spectroscopy, it cannot be considered here for reasons of space; for it,
reference should be made to appropriate accounts.† It must be remembered
that infrared spectroscopy is usually only one of many ways of investigating
the structure of a molecule. It is clear that all the available methods must be
used for a final appraisal.

The appearance of the infrared spectra of molecules (apart from the
differences introduced by using different techniques) is quite different from
that of the emission spectra of atoms in the visible and ultraviolet regions.
The latter consist of series of lines which are first widely spaced but then
fairly quickly converge to a limit. Infrared spectra, especially the important
rotation-vibration spectra of gases, consist of a group of many more or less
equidistant lines which often lie in quite a narrow spectral region, and which
frequently cannot individually be resolved experimentally. Each such
group is called a *band*. This name is used in the context of molecular spectroscopy for each localized emission or absorption in the spectrum without its
being necessarily such a conglomeration of near-lying lines. Without considering the matter in detail, this definition takes into account the fact that
such spectral effects can include a large variety of energy variations. In
most cases, such a band occupies a region of the spectrum that is indeed
narrow, but the width of which certainly cannot be neglected. The position
and the appearance (the so-called contour) of the bands are one main subject
of this book. Where possible, their fine structure and their intensity are also
discussed. The use of the information thus obtained on certain problems of
fundamental and applied chemical research is the other main subject.

Infrared spectroscopy initially included all knowledge of the infrared

---

† Literature: K. W. F. Kohlrausch, Ramanspektren. *Hd. u. Jahrb. d. Chem. Phys.*
(Leipzig, 1943); J. H. Hibben and E. Teller, *The Raman Effect and its Chemical
Applications* (New York, 1939); H. Pajenkamp, Fortschritte in der wissenschaftlichen
und praktischen Anwendung des Raman-Effektes. *Fortschr. Chem. Forsch.* (Berlin,
1950); W. Otting, *Der Raman-Effect* (Berlin, 1952); J. Brandmüller and H. Moser,
Einführung in die Raman-Spektroskopie (Darmstadt, 1961, in preparation).

region of the spectrum. This definition applied from the discovery of infra-red waves by W. Herschel in 1800 until about 1930. Until about then, infrared spectroscopy was a specialized branch of the section of physics dealing with waves, and the general aim was the inclusion of this spectral region in the overall pattern of natural science, and the discovery of its own special laws. As such, infrared spectroscopy was left to relatively few special-ists. During the last thirty years, the meaning of the term infrared spectro-scopy has changed; it has become both narrower and broader. Today, the term infrared spectroscopy has come to mean a special branch of spectro-scopy which is used principally to investigate chemical problems by physical methods. Infrared spectroscopy in its older, broader meaning is now more or less confined to the region of long and very long waves, where the applica-tions are not yet developed and the intensive physical investigation is about to proceed.

The first infrared experiments in the present narrow meaning of the term were made by Abney and Festings who in 1881 measured by photographic means the first infrared absorption spectra of organic molecules in the region of $0 \cdot 7$ to $1 \cdot 2 \mu$. Julius may be considered to be the founder of the infrared spectroscopy of today, for in 1892 he made the first assignment of an absorption band (near 3000 cm^{-1}) to a characteristic chemical structural feature (the methyl group). The first catalogue of infrared spectra of a large number of compounds in the region of up to $ca.$ 15 μ was published by W. W. Coblentz in 1905–8. Although his experiments are now only of historical interest, at that time they were a fundamental advance in experimentation, at the beginning of what has since become a massive development. The discovery and development of the quantum theory in the second decade and the quantum mechanics in the third decade of the present century drew sudden attention to the importance of all types of infrared spectra for the investigation of molecular structure. In these investigations, constantly improving experimental techniques and theoretical interpretation allowed the recognition of numerous effects.

The industrial and technical use of infrared spectroscopy in analysis prob-ably first began in about 1920 in the Oppau Laboratories of the Badische Anilin- und Soda-Fabrik A.G., Ludwigshafen, Germany. From about 1928, the spectra of many substances important in technical processes were recorded and used analytically, at first naturally on a small scale and with-out self-recording apparatus. A great step forward was the introduction to infrared spectroscopy by Lehrer of the use of modulated radiation and continuous recording on paper with ink, in contrast to the earlier use of photographic recording. The first modern infrared spectrograph to be used industrially and come up to technical requirements was made by Lehrer at

Introduction

the end of the third decade of this century at Ludwigshafen-Oppau. It incorporated the recording of percentage absorption, a variable slit-width, and a linear wavelength scale. About the same time, research workers of the American Cyanamid Co. and the Dow Chemical Corp., especially N. Wright, were making use of the same developments. The Second World War brought infrared spectroscopy to its present importance as a reliable and indispensable technique of the experimental chemist.

PART I

Fundamentals of the Theory of Infrared Spectra in the Gaseous State

1. Quantum-Theoretical Basis

The principles required for a survey of the theory of infrared molecular spectra belong in part to classical physics and in part to the quantum theory. It is thus necessary to give a brief survey of the quantum theory; however, what follows is merely a cursory treatment and reference must be made to textbooks of atomic theory for further details.

1. The Bohr Atom Model

The atom can be visualized as a sort of planetary system: a central nucleus with a positive electrical charge, surrounded by the number of negatively charged electrons required to render the whole system electrically neutral. The electrons encircle the nucleus in ellipses of varying eccentricity. The number of electrons is the same as the atomic number of the atom in the periodic classification of the elements. According to the laws of classical electrodynamics such a system is not stable. Electrons moving in the coulombic field of the central nucleus are subject to continuous acceleration because they are not moving in straight lines. Hence they should be radiating continually and this would lead to loss of energy and finally to 'falling into' the nucleus. However, experience presents us with a stable atom and the emission of radiation can only be observed under very special circumstances. To resolve this discrepancy, Bohr proposed the existence of radiation-less stable so-called '*stationary*' states of electrons in their movements around the central atomic nucleus. Each of these stationary states is characterized by a definite energy level and is distinguished from the large number of possible states by satisfying certain *quantum conditions*. These quantum conditions can be expressed by a series of (usually integral) numbers, the so-called *quantum numbers*. For an atom, four quantum numbers are necessary to characterize completely one stationary state or one energy level. In one and the same atom, there cannot be more than one electron for which all four quantum numbers are identical (Pauli's exclusion principle).

In addition to the stationary states, Bohr also postulated the so-called *frequency condition*. This is that interaction of an atom with radiation is only possible with a simultaneous energy exchange and that the frequency

11

Fundamentals of the Theory of Infrared Spectra

of the radiation is connected with the energy levels in question by the equation

$$h \cdot \tilde{\nu} = E_1 - E_2 \qquad [\text{I, 1.1}]$$

where $h = 6 \cdot 626 \times 10^{-27}$ erg sec is the Planck constant. It follows that the frequency of the radiation is determined directly by the energy difference between the two stationary states in question. Emission occurs when the atom loses energy $(E_2 > E_1)$ and absorption occurs when it gains energy $(E_2 < E_1)$. The alteration in energy is connected with an alteration in the quantum numbers. Because this alteration in energy is non-continuous (for the quantum numbers are usually integers) the term 'quantum transition' is also used. Not all possible quantum transitions within the possible values of the quantum numbers are allowed; certain *selection rules* hold, by which, starting from any one value of the quantum numbers (the initial state of the atom), only special neighbouring values are possible end states.

Further, the Bohr postulate of the *correspondence principle* should be mentioned. This requires that for sufficiently large values of the quantum numbers the quantum theoretical laws become identical with the classical laws. In other words, the classical foundation is contained in the quantum laws as a limiting case for sufficiently large values of the quantum numbers. The explanation is that for sufficiently large values of the quantum numbers their alteration by one unit is no longer important. Hence, the change of energy of an atomic system (which according to the quantum theory is fundamentally discontinuous) becomes quasi-continuous, in the region of large quantum numbers. The value of the correspondence principle lies in the fact that many classical results can be used in the quantum theory, in particular in the domain of intensity calculations. These can be taken *en-bloc* from the classical theory and this had to be done in the older quantum theory because no other treatment was possible.

If the relation given in the introduction between the frequency $\tilde{\nu}$ and the wave number ν of a vibration is used, then the Bohr frequency condition can be expressed in wave numbers:

$$\nu = \frac{E_1}{hc} - \frac{E_2}{hc} \qquad [\text{I, 1.2}]$$

The right-hand side is the difference of two quantities which appear as energies divided by the product of the Planck constant and the velocity of light. Each such quantity is called a '*term*' or a term value. It follows that the wave number of the radiation emitted or absorbed is given as a term difference and that each term is a definite function of the quantum numbers.

Therefore, it is possible to express the wave number directly as a function of at least two quantum numbers of similar type. However, the selection rules already mentioned apply to the alterations of the quantum numbers and thus the final quantum number can be deduced from the initial quantum number. Hence, ν can be written as a function of a single quantum number.

2. Quantum-mechanical Treatment

The Bohr theory is somewhat unsatisfactory. It is based on principles which in part are completely new and in part are classical. Further, for systems which are unlike hydrogen, it gives results which are only in qualitative agreement with experiment. These deficiencies are removed by quantum mechanics and wave mechanics. This is not the place for a discussion of this comprehensive theory or its formal mathematical treatment. It will be sufficient to indicate the salient points.

According to Schrödinger, following the introduction of the so-called wave-function (to be interpreted later),

$$\Psi = \psi(x, y, z) \cdot e^{2\pi i \tilde{\nu} t} \qquad [\text{I}, 1.3]$$

the condition of a system of one mass m can be described by the equation

$$\frac{\partial^2 \psi}{\partial x^2} + \frac{\partial^2 \psi}{\partial y^2} + \frac{\partial^2 \psi}{\partial z^2} + \frac{8\pi^2 m}{h^2}(E - V)\psi = 0 \qquad [\text{I}, 1.4]$$

In this equation, E is the total energy and V is the potential energy of the given point-mass. The theory of differential equations shows that such an equation gives solutions of Ψ, which are finite, unique, and different from zero, only for definite values of the total energy E. Such energy values are called the *eigenvalues* of the given problem; they correspond exactly to the stationary energy states which were suggested by Bohr to explain his hypothesis. However, they are not now merely postulated, but appear as an intrinsic result of the treatment. The equation [I, 1.4] can easily be extended to problems involving many particles.

The wave-function in [I, 1.3] is split up there into a position-dependent amplitude ψ and a time-dependent factor $e^{2\pi i \tilde{\nu} t}$. The amplitude ψ is called the *eigenfunction* of the problem. Its square is interpreted as the probability that the given particle occupies a definite place characterized by the co-ordinates x, y, z. The definiteness of the Bohr orbits is therefore lost, because the *eigenfunction* ψ generally possesses values which, although small, differ from zero outside these orbits. This is frequently described by saying that the electron which was previously localized is now smeared out over all

13

space—the so-called *electron cloud*. However, the probability is especially large in the vicinity of the Bohr orbits.

The quantum numbers and the selection rules which were introduced into the Bohr theory merely to give agreement with experiment now follow logically from the mathematical treatment. The Bohr frequency relation [I, 1.2] remains valid in quantum mechanics, but now the energy values appear as eigenvalues of the mathematical equation of the problem in question. The probability of the transition from one energy state to another is defined by the eigenfunction. Here a fundamental advance is made with respect to the older quantum theory in that the intensities of the lines and bands emitted or absorbed can be calculated, at least in principle.

The interaction between electromagnetic radiation, characterized by an electric vector E, with an atomic system in a definite energy state is essentially described by the interaction of the wave with the *electric dipole moment* M of the system. The electric dipole moment is the product of the electric charge e of the system with the distance r between positive and negative charge-centres (this distance is a vector quantity).

$$M = e \cdot r \qquad [\text{I}, 1.5]$$

This equation can be written in the following components

$$M_x = \sum e_i \cdot x_i, \qquad M_y = \sum e_i \cdot y_i, \qquad M_z = \sum e_i \cdot z_i \qquad [\text{I}, 1.6]$$

where e_i is the charge of the i-th particle with the co-ordinates x_i, y_i, z_i. The interaction energy is equal to the product $M \cdot E$. This can be inserted in the Schrödinger equation and the probability that the system goes from state n to state m by this effect be calculated. It can thus be shown (the calculation is outside the scope of this text) that this probability is proportional to certain quantities R^{nm}. These quantities are called the *matrix elements* of the electric dipole moment or *transition moments* and can be calculated from the dipole moment components and the eigenfunctions valid for the states n and m of the system in question. A transition between the two states, i.e. emission or absorption of radiation, is only possible when $R^{nm} \neq 0$. Otherwise the transition is said to be forbidden. The requirement that R^{nm} (or rather at least one of its three components) should not become zero can be shown mathematically to lead to certain conditions for the quantum numbers which characterize the states, and to the possibilities of alterations in these quantum numbers. These conditions are none other than the selection rules mentioned above.

The electric dipole moment is the most important but not the only possible means of enabling interaction between radiation and an atomic system

14

to take place. This can also be brought about by the magnetic dipole moment, the quadrupole moment, and an induced dipole moment (caused for instance by electronic, ionic, or atomic interaction). However, the transition probabilities for these are all much less (by many powers of ten) than those for the electric dipole moment. They can thus usually be neglected in the present treatise.

2. The Infrared Spectrum of Diatomic Molecules

1. Preliminary Remarks

It has become customary, and it is very useful, to treat the diatomic molecules separately from the polyatomic ones in a discussion of the theory of molecular spectra. This has the advantage that easily visualized models can be given for diatomic molecules which reproduce many of the features of the observed spectra. In addition, it is possible to test and improve the theoretical framework on relatively simple empirical findings. Then this can be extended to polyatomic molecules, the treatment of which is more complicated and less easy to visualize.

In accordance with the aim of this book, the *infrared spectra* only of diatomic molecules are discussed. This limitation with respect to the visible and ultraviolet region at first appears quite arbitrary. However, it must be remembered that the appearance of a spectrum is connected with energy alterations in the atomic or molecular species which produces the spectrum, as mentioned in Chapter 1. The fundamental possibilities of energy uptake by a molecule are: (a) the energy of electron promotion E_e; (b) the energy of vibration E_v; (c) the energy of rotation E_r. Each of these energy states is dependent on certain quantum numbers, an alteration of which leads to a specific spectrum. As these spectra are found in quite different regions, the energy changes connected with the emission or absorption of radiation must be of quite different orders of magnitude. The largest values are found for the energies of electron promotion, the characteristic spectra of which are found in the visible or ultraviolet regions. Next follow the term differences of the vibrations; their spectra are found in the infrared region up to about 25 or 30 μ wavelength. Finally, the energy differences of rotations cause spectra in the far infrared region.

However, energy alterations of all types can occur simultaneously. The quantum transitions of electron promotion cause such a spectrum to lie in the ultraviolet region while the vibration and rotation contributions define respectively the coarse and the fine structure of the bands. For the purposes of this book it is taken that $\Delta E_e = 0$ always; now the vibrational contribution causes the spectrum to lie in the middle infrared region, while the rotational contribution, in the main, influences the band contours. Finally, if a vibrationless molecule in a fixed electronic state (usually the ground state) is considered, then pure rotational spectra in the far infrared appear. The spectra become more complicated in the inverse order: the pure rotation spectra are the simplest, the pure vibrational spectra are already more complicated, and the simultaneous occurrence of both, inclusive of second order interaction, is the most complicated subject discussed in this book. The following discussion will start with the more simple cases and gradually lead to the more difficult topics.

As already mentioned in the introduction, infrared spectra only occur for such movements (vibrations and rotations) as are connected with an alteration of the dipole moment of the molecule in question. For a diatomic molecule this necessitates the existence of a *permanent dipole*, i.e. the non-superimposition of the centres of positive and negative electric charges, because only in this case can rotation and vibration cause dipole moment alterations. Therefore, all homonuclear diatomic molecules, such as O_2, H_2, N_2, are excluded, as these have obviously no permanent electric dipole and therefore no infrared spectrum. Their spectra can only be obtained in connection with electron promotions†; the same applies to the inert gases. The molecules possessing a permanent dipole are referred to as polar. In order to neglect intermolecular interaction, the gas state is principally considered in this, the more theoretical, part of the account. The spectra that will be of further interest here have been investigated almost entirely by absorption methods; emission measurements are available for only very few compounds.

2. Rotational Spectra

The rigid rotator. The simplest model of a diatomic molecule is a dumbbell: two atoms (considered to be point masses) of masses m_1 and m_2 at a fixed distance r (Fig. 2). Such a structure can only rotate (i.e. vibration is not permitted). In free space the rotation axis passes through the centre of gravity S of the two atoms. S lies on the straight line connecting the atoms; its position on this line is defined by the equations

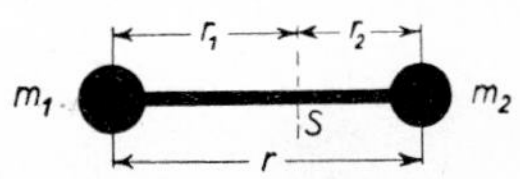

FIG. 2. *Dumbbell model of a diatomic molecule.*

$$r_1 = \frac{m_2}{m_1+m_2}.r, \qquad r_2 = \frac{m_1}{m_1+m_2}.r, \qquad r_1+r_2 = r \qquad [I, 2.1]$$

† This applies only at sufficiently low pressures (which is almost always the case). In condensed phases homonuclear diatomic molecules also show a definite vibration spectrum in the infrared (see Chapter 4 of Part IV).

C

Fundamentals of the Theory of Infrared Spectra

The rotational energy is

$$E_r = \tfrac{1}{2}Iw^2 \qquad\qquad \text{[I, 2.2]}$$

where the angular velocity of the rotation is w and

$$I = \sum m_i r_i^2 = m_1 r_1^2 + m_2 r_2^2 = \frac{m_1 m_2}{m_1 + m_2} r^2 \qquad\qquad \text{[I, 2.3]}$$

gives the moment of inertia about the axis passing through S and perpendicular to the line connecting the two atoms. If the so-called reduced mass

$$\mu = \frac{m_1 m_2}{m_1 + m_2} \qquad\qquad \text{[I, 2.4]}$$

is substituted in the expression for the moment of inertia, it is found that

$$I = \mu r^2 \qquad\qquad \text{[I, 2.5]}$$

That means that instead of the rotation of a dumbbell, the rotation of a point mass μ about a definite axis at distance r can be considered. Such a model is described as a *simple rigid rotator*.

The energy states of such a simple rigid rotator are given by

$$E_r = \frac{h^2}{8\pi^2 \mu r^2} J(J+1) = \frac{h^2}{8\pi^2 I} J(J+1) \qquad\qquad \text{[I, 2.6]}$$

Here J is a quantum number, the so-called 'rotational quantum number' which can take all positive integral values 0, 1, 2, etc. The energy states therefore form a series of definite levels with energies that increase approximately as J^2 (cf. Fig. 3a).

According to the general definition of a term [I, 1.2] it follows that for rotation

$$F(J) = \frac{E_r}{hc} = \frac{h}{8\pi^2 cI} J(J+1) = BJ(J+1) \qquad\qquad \text{[I, 2.7]}$$

where

$$B = \frac{h}{8\pi^2 cI} = \frac{27 \cdot 986}{I} . 10^{-40} [\text{cm}^{-1}] \qquad\qquad \text{[I, 2.8]}$$

B is called the rotational constant and is inversely proportional to the moment of inertia. The wave number of the radiation emitted or absorbed by an alteration in the rotational quantum number is therefore (according to [I, 1.2]):

$$\nu = F(J') - F(J'')$$

$$= BJ'(J'+1) - BJ''(J''+1) \qquad\qquad \text{[I, 2.9]}$$

18

In this equation, the quantum number pertaining to the higher energy state is characterized by ′, and that pertaining to the lower energy state by ″. Only when the selection rule for the rotational quantum number J is taken into account does equation [I, 2.9] give the wave numbers of the rotational lines which actually appear. This selection rule states that only those transitions are allowed that involve an alteration of the rotational quantum number of:

$$\Delta J = \pm 1 \qquad [\text{I, 2.10}]$$

or, with the convention used in this book, that

$$J' = J'' \pm 1 \qquad [\text{I, 2.11}]$$

The rotational quantum number can therefore alter by one unit only. Because J' has been taken as the rotational quantum number of the higher energy state $(J' > J'')$, only positive values can occur in [I, 2.11], i.e. $J' = J'' + 1$. It follows that the wave numbers of the rotational spectrum are given by

$$
\begin{aligned}
\nu &= F(J''+1) - F(J'') \\
&= B(J''+1)(J''+2) - BJ''(J''+1) \\
&= 2B(J''+1) \qquad [\text{I, 2.12}]
\end{aligned}
$$

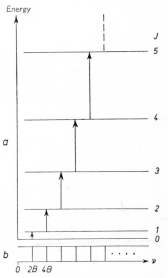

FIG. 3. *Energy levels* (a) *and spectrum* (b) *of the rigid rotator.*

where $J'' = 0, 1, 2, \ldots$ i.e. all positive integral values. As long as only the quantum number of the lower state is considered, the indication ″ can be omitted without causing confusion:

$$\nu = 2B(J+1), \qquad J = 0, 1, 2, \ldots \qquad [\text{I, 2.13}]$$

This indicates that the simple rigid rotator possesses a spectrum consisting of a series of equidistant lines; the first line lies at wave number $2B$ and the distance between each line is $2B$ (Fig. 3b).

The non-rigid rotator. The dumbbell model is retained, but the condition that the distance between the two masses is constant, i.e. the condition of rigidity, is relinquished. It must now be expected that, with increasing rotational energy, the distance r between the two masses increases under the action of centrifugal forces. Therefore, the moment of inertia I increases according to [I, 2.5]. It follows that the rotational constant B, as defined by [I, 2.8], is no longer a constant but decreases with increasing rotational

19

energy, i.e. with an increase in the rotational quantum number J. To a sufficient approximation, the rotational term is given by

$$F(J) = \frac{E_r}{hc} = B[1 - uJ(J+1)]J(J+1) \qquad [I, 2.14]$$

where B is as in [I, 2.8] and $u \ll 1$. It is customary to substitute

$$uB = D \qquad [I, 2.15]$$

and to write the rotational term of a non-rigid rotator as

$$F(J) = BJ(J+1) - DJ^2(J+1)^2 \qquad [I, 2.16]$$

The influence of D, or the centrifugal factor, means that as J increases the term values decrease. However, as $D \ll B$ this effect is negligible for small values of J.

Thus the equation for the radiation of the non-rigid rotator is

$$\nu = F(J+1) - F(J) = 2B(J+1) - 4D(J+1)^3 \qquad [I, 2.17]$$

It follows that the rotational lines are no longer equidistant. The distance between them becomes smaller with increase in the rotational quantum number, as a consequence of the influence of the centrifugal term.

The spectra found in the long-wavelength infrared region have been assigned to the rotational spectrum expected from this model, and expectation and findings are in agreement. Experiment shows that the long-wave infrared spectra of diatomic polar molecules, e.g. HCl, consist of a series of almost equidistant lines, and conform with the expectations indicated above [Czerny (*156*), McCubbin (*454*)]. With the treatment justified in this way, it is possible to obtain a series of important molecular constants directly from these spectra. The position of the first rotational line (when it can be observed), or the distance between neighbouring lines, gives the rotational constant B. From B and equation [I, 2.8], the moment of inertia I about an axis passing through the centre of gravity and perpendicular to the molecular axis can be obtained. Finally, the distance between the two atoms in the molecule in question can be obtained from the moment of inertia, because the reduced mass can be calculated from the masses of the atoms of the molecule. The value of the model used can be tested by a comparison of the molecular distances obtained in this way with those found by other methods. The deviation of the rotational lines from equidistant values leads to the rotational constant D. However, because of its very small size, D cannot be obtained very accurately. D is a measure of the influence of centrifugal forces during rotation, and gives at least an approximate value of the vibrational frequency when used together with B according to equation [I, 2.42] (see page 26).

The Infrared Spectrum of Diatomic Molecules

3. Vibrational Spectra

The harmonic oscillator. A dumbbell is again taken as the simplest model, but this is considered to vibrate and not rotate. The simplest possible form of vibration is the movement of the two atoms along the line joining them in the form of a periodic motion. The distance of each atom from the equilibrium position is then given by a sine-function of the time:

$$x = x_0 \sin (2\pi\tilde{\nu}_s t) \qquad\qquad [\text{I, 2.18}]$$

where x is the distance, x_0 is the amplitude of the vibration, t is the time and $\tilde{\nu}_s$ is the vibrational frequency. Just as the model of the rotating rigid dumbbell was shown to be equivalent to that of a simple rigid rotator, it can now be shown that the vibrating dumbbell is equivalent to the vibration of a simple point-mass μ according to [I, 2.14]. This model is called a harmonic oscillator. Its vibration is sustained by the action of a restoring force proportional to the displacement

$$\boldsymbol{F} = -kx = \mu \frac{d^2 x}{dt^2} \qquad\qquad [\text{I, 2.19}]$$

The equation [I, 2.18] given above is a direct solution of this differential equation from which the classically calculated vibrational frequency follows as

$$\tilde{\nu}_s = \frac{1}{2\pi} \sqrt{\left(\frac{k}{\mu}\right)} \qquad\qquad [\text{I, 2.20}]$$

The constant k is called the *force constant*; the calculation of force constants from spectral data is one of the main aims of infrared spectroscopy, as they determine the behaviour of the binding force between the atoms. According to the laws of classical mechanics such a harmonic oscillator possesses the potential energy

$$V = \tfrac{1}{2}kx^2 = 2\pi^2 \mu\tilde{\nu}_s^2 x^2 \qquad\qquad [\text{I, 2.21}]$$

i.e. a function which is quadratic in the displacement from the equilibrium position, and which can be visualized as an open parabola which lies in the direction of the energy axis with its apex at the equilibrium position of the atoms.

The Schrödinger equation, one-dimensional in this case, gives as the eigenvalues or the possible energy states of the harmonic oscillator the following:

$$E(v) = \frac{h}{2\pi} \sqrt{\left(\frac{k}{\mu}\right)} (v+\tfrac{1}{2}) = h\tilde{\nu}_s(v+\tfrac{1}{2}) \qquad\qquad [\text{I, 2.22}]$$

where v is a quantum number, the so-called vibrational quantum number which can have all integral values from 0 on. The energy levels of the har-

21

monic oscillator are therefore equidistant (Fig. 4). It is of especial importance that, even at the lowest possible level when $v = 0$, a finite vibrational energy $E(0) = \frac{1}{2}h\tilde{\nu}_s$ exists, which is described as the zero-point energy. Nevertheless, the state $v = 0$ is referred to as vibrationless (zero-point vibration). The terms of the harmonic oscillator are therefore

$$G(v) = \frac{E(v)}{hc} = \frac{\tilde{\nu}_s}{c}(v + \tfrac{1}{2}) \qquad [\text{I, 2.23}]$$

It is customary to substitute

$$\frac{\tilde{\nu}_s}{c} = \omega \qquad [\text{I, 2.24}]$$

from which

$$G(v) = \omega(v + \tfrac{1}{2}) \qquad [\text{I, 2.25}]$$

here ω is identical with the wave number of the classically calculated vibrational frequency [I, 2.2].

The wave number of the radiation emitted during the vibration (provided that the vibration is connected with the alteration of a dipole moment) is therefore

$$\nu = G(v') - G(v'') \qquad [\text{I, 2.26}]$$

The selection rule applicable to the harmonic oscillator is

$$\Delta v = \pm 1 \qquad [\text{I, 2.27}]$$

and therefore the wave number of the radiation is

$$G(v+1) - G(v) = \omega \qquad [\text{I, 2.28}]$$

i.e. identical with the result obtained classically. The initial state of the system is obviously immaterial; for each transition, radiation of the same frequency is emitted or absorbed, to whatever degree the vibration is excited. This conclusion could already have been deduced from the equi-distance of the energy levels (Fig. 4).

The anharmonic oscillator. Actual diatomic molecules behave only approximately as harmonic oscillators. The most important discrepancy between expectation and experimental result is that instead of one vibrational frequency several are often observed in the spectrum, the wave numbers of

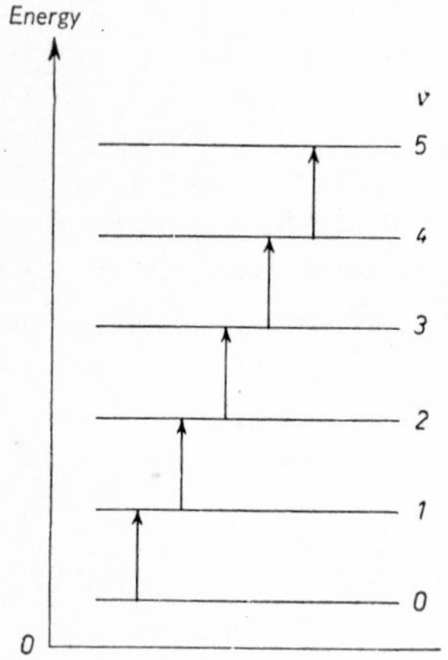

FIG. 4. *Energy levels of the harmonic oscillator.*

which stand in an integral relationship. According to the expressions for the restoring force [I, 2.19] or the potential energy [I, 2.21], these quantities should increase without limit as the two atoms move further apart. This is contrary to experience, the binding force between the two atoms is known to disappear for sufficiently large distances between them and the potential energy becomes constant. Instead of taking the form of a parabola as indicated by [I, 2.21] (shown as a dashed line), the actual potential curve is more like that shown as a full line in Fig. 5. Near the equilibrium position

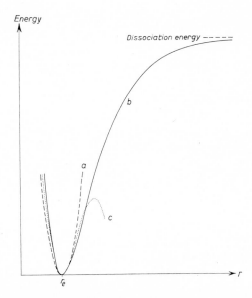

FIG. 5. *Potential curve of the harmonic oscillator* (a), *and the anharmonic oscillator* (b), *the latter approximated by a parabola of higher order* (c) [*taken from Herzberg* (*I*)].

the two curves lie very close together, but they deviate more and more with increasing distance between the two atoms. No compact mathematical expression can be found for the real potential curve. Empirically, the quadratic potential function [I, 2.21] can be extended by an additional term of the third order, and by further terms of still higher orders:

$$V = fx^2 - gx^3 + \ldots \qquad \text{[I, 2.29]}$$

Here x is still the displacement from the equilibrium position and $g \ll f$. The dotted curve in Fig. 5 depicts this new potential function which takes into account the *anharmonicity* of the vibration in question. Even so, complete

23

agreement with the true potential curve is not reached. However, the agreement is much better than when using the quadratic function. An oscillator with a potential curve of this form is described as *anharmonic*. A better agreement with the observed spectrum can be expected using this model than the harmonic oscillator.

The classical description of the movement of such an anharmonic oscillator consists of the superposition of a fundamental vibration of amplitude and frequency given by [I, 2.20] and overtones in the form of a Fourier series:

$$x = x_1 \sin 2\pi \tilde{\nu}_s t + x_2(3 + \cos 2\pi . 2\tilde{\nu}_s t) + x_3 \sin 2\pi . 3\tilde{\nu}_s t + \ldots \quad \text{[I, 2.30]}$$

Here x_1, x_2, x_3, ... are the amplitudes of the fundamental and the first, second ... overtones. As long as $g \ll f$ they decrease very quickly in the order given. However, the influence of the overtones is correspondingly larger as the amplitude x_1 of the fundamental vibration increases. The vibrational frequency $\tilde{\nu}_s$ according to [I, 2.20] is valid only for very small amplitudes and it decreases as these increase. The vibrational frequency of the overtones is always exactly double, treble, etc., that of the fundamental vibrational frequency.

The quantum mechanical treatment of the anharmonic oscillator gives as eigenvalues

$$E(v) = hc\omega_e(v + \tfrac{1}{2}) - hc\omega_e x_e(v + \tfrac{1}{2})^2 + hc\omega_e y_e(v + \tfrac{1}{2})^3 + \ldots \quad \text{[I, 2.31]}$$

according to how many and which higher powers of x are taken into account in [I, 2.29]. The vibrational terms are therefore

$$G(v) = \omega_e(v + \tfrac{1}{2}) - \omega_e x_e(v + \tfrac{1}{2})^2 + \omega_e y_e(v + \tfrac{1}{2})^3 + \ldots \quad \text{[I, 2.32]}$$

where $\omega_e \gg \omega_e x_e \gg \omega_e y_e$. The subscript e on these quantities indicates that they apply to the equilibrium position, i.e. to infinitely small vibrational amplitudes; this is an international convention. According to the expression [I, 2.29] (following Herzberg) with g positive, $\omega_e x_e$ is always positive, whereas $\omega_e y_e$ can be either positive or negative. It follows that the energy levels of the anharmonic oscillator, instead of being equidistant, slowly become closer to each other as v increases. Here also a zero-point energy occurs which is obtained by putting $v = 0$.

$$G(0) = \tfrac{1}{2}\omega_e - \tfrac{1}{4}\omega_e x_e + \tfrac{1}{8}\omega_e y_e \quad \text{[I, 2.33]}$$

Instead of calculating the energy from the real zero this 'vibrationless' level can be used as the energy zero and thus

$$G_0(v) = \omega_0 v - \omega_0 x_0 v^2 + \omega_0 y_0 v^3 + \ldots \quad \text{[I, 2.34]}$$

if the internationally agreed abbreviations

$$\omega_0 = \omega_e - \omega_e x_e + \tfrac{3}{4}\omega_e y_e + \ldots,$$

$$\omega_0 x_0 = \omega_e x_e - \tfrac{3}{2}\omega_e y_e + \ldots, \qquad \omega_0 y_0 = \omega_e y_e \qquad [\text{I, 2.35}]$$

are used.

The classical description of the anharmonic oscillator corresponds to a spectrum in which, in addition to the fundamental, overtones occur with double, treble, . . . the frequency of the fundamental and with intensities which rapidly decrease. The quantum theoretical treatment, in contrast, alters the selection rules of the vibrational quantum number; in addition to $\Delta v = \pm 1$ quantum transitions of $\Delta v = \pm 2, \pm 3, \ldots$ are allowed, but with rapidly decreasing probability. If all the molecules in question are in the ground state $(v = 0)$, then the quantum transitions $\Delta v = 1, 2, \ldots$ are all possible in absorption. A series of vibrational bands therefore arises which can be characterized by specifying the vibrational quantum number of the final and initial states; 1–0, 2–0, ... The frequency of the 2–0 band is very nearly double that of the fundamental frequency. Corresponding relationships hold for the bands of still higher transitions, so that here it has become conventional to use the term 'overtones'. The wave numbers of the absorption bands can be obtained from the expression for the term

$$\nu_{\text{abs}} = G(v') - G(0) = G_0(v') = \omega_0 v' - \omega_0 x_0 v'^2 + \omega_0 y_0 v'^3 + \ldots$$
$$[\text{I, 2.36}]$$

Thus, the absorption frequencies give directly the height of the energy levels above the lowest level $(v = 0)$. The distance between neighbouring bands (or energy levels), neglecting terms of the third order, is given by

$$\Delta G_{v+1/2} = G(v+1) - G(v) = G_0(v+1) - G_0(v)$$
$$= \omega_e - 2\omega_e x_e - 2\omega_e x_e v = \omega_0 - \omega_0 x_0 - 2\omega_0 x_0 v \quad [\text{I, 2.37}]$$

The differences of these values for neighbouring bands can now be calculated:

$$\Delta^2 G_{v+1/2} = \Delta G_{v+3/2} - \Delta G_{v+1/2} = -2\omega_e x_e = -2\omega_0 x_0 \quad [\text{I, 2.38}]$$

The vibrational constants $\omega_e x_e = \omega_0 x_0$ can thus be obtained, giving a measure of the anharmonicity of the oscillator. The values of ω_e and ω_0 are themselves calculable from [I, 2.37] if the wave number of one absorption band is measured, e.g. the 1–0 band:

$$\nu_{\text{abs}}^{1-0} = \Delta G_{1/2} = \omega_e - 2\omega_e x_e = \omega_0 - \omega_0 x_0 \qquad [\text{I, 2.39}]$$

Further, it can be shown that, for infinitely small amplitudes, the vibrational constant ω agrees with the wave number of the vibration that the oscillator possesses on the classical treatment

$$c\omega_e = \tilde{\nu}_s \qquad [\text{I, 2.40}]$$

25

The value of the force constant for infinitely small displacements can be calculated

$$k_e = 4\pi^2 \mu c^2 \omega_e^2 \qquad \text{[I, 2.41]}$$

The value thus obtained is somewhat larger than that calculated from the vibration frequency of a harmonic oscillator.

As has already been mentioned, the vibrational spectra of diatomic molecules are found in the so-called near to middle infrared region, from approximately $2\cdot5$ to $25\,\mu$ wavelength; this assignment is confirmed by experience. Polar molecules show in this region, in general, one intense absorption band, that can be assigned to the fundamental vibration. In addition, less intensive absorption bands are found at approximately double, treble, etc., the fundamental frequency; these correspond to the overtones as already discussed. If, by coincidence, a molecule behaves as an almost perfect harmonic oscillator, then the sole absorption band observed gives directly the vibrational frequency, and thus the force constants. The spectrum of an anharmonic oscillator gives, as already indicated, the vibrational frequency for small amplitudes ω_e, the anharmonicity $\omega_e x_e$ and the force constant k_e. The rotational spectrum alone also gives a value of the vibrational frequency, although not a very exact one. The relationship between the vibrational frequency, measured in wave numbers, and the two rotational constants B and D is:

$$D = \frac{4B^3}{\omega^2} \qquad \text{[I, 2.42]}$$

The vibrational frequency can therefore be calculated from the deviation of the rotational lines from equidistance, but the accuracy is poor because of the small size of the effect.

4. Rotational-vibrational Spectra

The rotating oscillator. Having treated rotation and vibration separately, the next step is to consider simultaneous alterations in the rotational and vibrational states. For this, it is necessary to allow the anharmonic oscillator model to rotate as a non-rigid rotator, while still vibrating. This is the *rotating oscillator*, or the *vibrating rotator*, as both forms of movement are equally important.

As long as the interaction between rotation and vibration is neglected, the energy of the rotating oscillator is given simply by the sum of the energies of the anharmonic oscillator, from [I, 2.31], and the non-rigid rotator, from [I, 2.14]. According to this treatment, each vibrational level of Fig. 4 possesses a whole series of rotational levels, as in Fig. 3. In fact, the interaction between rotation and vibration cannot be neglected. This interaction may be expressed as an alteration of the internuclear distance, and hence the moment of inertia and the rotational constant B, during the vibration. The period of vibration (the reciprocal of the vibrational frequency) is much smaller than the period of the rotation (the reciprocal of the rotational frequency). Therefore, a mean value of the distance r can be taken for the rotation; this value is calculated by averaging in a particular way. A rotational constant

$$B_v = \frac{h}{8\pi^2 c\mu} \overline{\left[\frac{1}{r^2}\right]} \qquad \text{[I, 2.43]}$$

has to be used in which $\overline{\left[\dfrac{1}{r^2}\right]}$ is an average value of $\dfrac{1}{r^2}$ during the vibration. The value of B_v is smaller than B_e, the value of the rotational constant for a constant equilibrium distance of the atoms, because the internuclear distance increases during the vibration and therefore so does the moment of inertia. However, the difference between B_v and B_e is not important. B_e is found by substituting the values for the equilibrium position of the atoms in the definitive equation [I, 2.8]:

$$B_e = \frac{h}{8\pi^2 c\mu r_e^2} = \frac{h}{8\pi^2 cI_e} = \frac{27\cdot986}{I_e}.10^{-40}\,[\text{cm}^{-1}] \qquad \text{[I, 2.44]}$$

The rotational constant B_v, for a vibrational state characterized by the quantum number v, is given to a sufficient approximation by

$$B_v = B_e - \alpha_e(v+\tfrac{1}{2}) + \ldots \qquad \text{[I, 2.45]}$$

In this α_e is a constant, which is small compared with B_e, just as the alteration in the internuclear distance during the vibration is small compared to the equilibrium nuclear distance itself. As a rule of thumb

$$\frac{\alpha_e}{B_e} \approx \frac{\omega_e x_e}{\omega_e} \qquad \text{[I, 2.46]}$$

can be used. The rotational constant D is dealt with in a similar manner:

$$D_v = D_e + \beta_e(v+\tfrac{1}{2}) + \ldots \qquad \text{[I, 2.47]}$$

Here again $\beta_e \ll D_e$, where D_e is that value of D which is obtained by substitution of B_e and ω_e in [I, 2.42].

The rotational term for a given vibrational state can now be written as:

$$F_e(J) = B_v J(J+1) - D_v J^2(J+1)^2 + \ldots \qquad \text{[I, 2.48]}$$

This, together with the vibrational term of the anharmonic oscillator, gives the term of the rotating oscillator:

$$T = G(v) + F_v(J) = \omega_e(v+\tfrac{1}{2}) - \omega_e x_e(v+\tfrac{1}{2}) + \ldots$$
$$+ B_v J(J+1) - D_v J^2(J+1)^2 + \ldots \qquad [\text{I}, 2.49]$$

The form of this term is illustrated in Fig. 6: above each vibrational level v a series of rotational levels J is built up. The distance between successive vibrational levels gradually decreases; that between successive rotational levels gradually increases. The lowest 'vibrationless' state ($v = 0$), i.e. a simple rotation, gives a value $B_0 < B_e$ for the rotational constant which

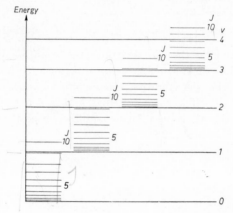

FIG. 6. *Diagram of the terms of a rotating oscillator.*

leads to a value of r which is not the equilibrium distance r_e between the two atoms but the slightly larger average value r_0 of the 'vibrationless' state.

In order to understand the spectrum of the rotating oscillator, the selection rules for the quantum transitions of J and v are still needed. It can be shown that the selection rule already obtained for J, i.e. $\Delta J = \pm 1$, still holds, just as does the selection rule for the anharmonic oscillator $\Delta v = \pm 1, \pm 2, \pm 3, \ldots$. What is new is that $\Delta v = 0$ is also allowed, but this transition gives only a pure rotational spectrum. The vibrational transition from v' to v'' will now be considered. Neglecting the very small rotational constant D_v, it is found that:

$$\nu = G(v') - G(v'') + B_v' J'(J'+1) - B_v'' J''(J''+1) \qquad [\text{I}, 2.50]$$

gives the wave numbers of the lines which occur in such a rotational-vibrational band. In [I, 2.50], a substitution can be made for the pure vibration without taking into account the rotation ($J' = J'' = 0$):

$$G(v') - G(v'') = \nu_0 \qquad [\text{I}, 2.51]$$

If the two allowed quantum transitions of J, i.e. $\Delta J = +1$ and $\Delta J = -1$, which are now both significant, are considered, and if J is again written in place of J'', the following results are obtained:

$$\nu_R = \nu_0 + 2B'_v + (3B'_v - B''_v)J + (B'_v - B''_v)J^2,$$
$$J = 0, 1, 2, \ldots \qquad\qquad [\text{I, 2.52a}]$$

and

$$\nu_P = \nu_0 - (B'_v + B''_v)J + (B'_v - B''_v)J^2, \qquad J = 1, 2, \ldots \quad [\text{I, 2.52b}]$$

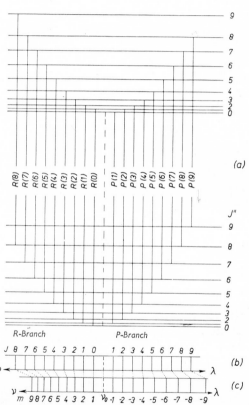

FIG. 7. (a) *Diagram of the term of a rotating oscillator for a given vibrational transition.* (b) *The corresponding rotational-vibrational band neglecting the interaction of rotation and vibration.* (c) *Ditto, taking this interaction into consideration* [*from Herzberg (I)*].

It follows that two series of lines are found, which are described as the R-branch and the P-branch. Fig. 7 gives a graphical representation. The lowest J-value, in the higher as well as in the lower state, must be $J = 0$. It

29

follows that $J'' = 0$ is the lowest in the R-branch, but that $J'' = 1$ is the lowest for the P-branch.

In order to clarify the picture, interaction between rotation and vibration will be neglected temporarily; i.e. it will be assumed that $B'_v = B''_v = B$, when it follows that:

$$\nu_R = \nu_0 + 2B + 2BJ, \qquad \nu_P = \nu_0 - 2BJ \qquad [\text{I, 2.53}]$$

These are two series of equidistant lines, of which ν_R leads towards wavelengths shorter than ν_0 (larger wave numbers), and ν_P leads to longer wavelengths (smaller wave numbers). Because the transition $\Delta J = 0$ is forbidden, the line ν_0 does not itself occur in the spectrum. Therefore a gap is left between the two branches (Fig. 7b). When the interaction between vibration and rotation is taken into account, the equidistance of these lines disappears. Because this interaction is somewhat different in the two vibrational levels in question ($B'_v \neq B''_v$), the term in [I, 2.52] which is quadratic in J comes into action, with the consequence that for $B'_v B''_v$ the lines in the R-branch lie successively nearer together and those in the P-branch further apart (Fig. 7c).

Instead of representing the two branches by two separate equations, as has been done up till now, it is possible to use a single equation. To do this, an integral number m is introduced, which can take on positive values for the R-branch beginning from 1 ($m = J + 1$), and for the P-branch negative values beginning from -1 ($m = -J$). The rotational-vibrational band can now be represented by

$$\nu = \nu_0 + (B'_v + B''_v)m + (B'_v - B''_v)m^2 \qquad [\text{I, 2.54}]$$

This is a single series of lines with a gap at $m = 0$. This gap $\nu = \nu_0$ is called the null line of the band or the *band centre*.

The rotational-vibrational bands of dipolar diatomic molecules observed experimentally fully confirm the treatment indicated above. Especially near the band centre, agreement between prediction and observation is excellent. From this region, the distance between neighbouring lines gives the best values of $2B$, while the distance between the first R-line and the first P-line yields $4B$. The rotational constants in the vibrational states B'_v and B''_v are most simply determined by taking the wave number differences between neighbouring lines and obtaining the second-order differences between these quantities. Using equation [I, 2.54] it is found that

$$\Delta\nu(m) = \nu(m+1) - \nu(m) = 2B'_v + 2(B'_v - B''_v)m \qquad [\text{I, 2.55}]$$

and that

$$\Delta^2\nu(m) = \Delta\nu(m+1) - \Delta\nu(m) = 2(B'_v - B''_v) \qquad [\text{I, 2.56}]$$

30

The latter difference gives directly the difference between the rotational constants in the higher and lower rotational states. From the definitive equation [I, 2.45] for B_v, it follows that this difference $B_v' - B_v''$ increases as the quantum transition Δv increases. In agreement, the convergence of the rotational lines caused by the influence of this term is more pronounced in the overtones with increasing Δv. The direction of the lines in the R-branch can even become reversed, and form a so-called *band-head*. When $B_v' - B_v''$ has been calculated, B_v' can itself be obtained, e.g. graphically by plotting $\Delta v(m)$ according to [I, 2.55] as a function of m; the ordinate intersected by this straight line gives $2B_v'$ directly.

5. Intensity Considerations

The final section of this discussion of diatomic molecules consists of a note about intensities. This will be limited to absorption, and, to avoid certain complications, to sufficiently thin films of the absorbing medium. In this treatment, Δx is the thickness of the irradiated layer, ρ is the intensity of the beam of parallel radiation of wave number ν. The *absorption intensity* is defined as the radiation energy absorbed, i.e. converted into heat, per square centimetre. It is given by

$$J_{\text{abs}} = \rho . N_m . B_{nm} . \Delta x . hc\nu \qquad [\text{I, 2.57}]$$

where N_m is the number of molecules in the initial state, i.e. in the lower state, and B_{nm} is Einstein's *transition probability*. The derivation of this equation cannot be considered here. The transition probability is related to the transition moment by the equation

$$B_{nm} = \frac{8\pi^3}{3h^2 c} . |\boldsymbol{R}^{nm}|^2 \qquad [\text{I, 2.58}]$$

It follows that the absorption intensity is proportional to the *square of the transition moment*, to the frequency of the radiation $\tilde{\nu} = c\nu$, and to the number of the molecules in the initial state.

In order to calculate the absorption intensity, knowledge of the transition moment is therefore necessary, this is obtained from quantum mechanical calculations. The number of molecules in the initial state is also needed; this can be obtained (based on the correspondence principle) from the well-known Maxwell-Boltzmann *energy distribution law*, because the spectra observed are always those of systems in thermodynamical equilibrium. According to this distribution law the number N of molecules of energy between E and $E + \Delta E$ is proportional to $e^{-E/kT} . \Delta E$, where k is the Boltzmann constant (i.e. the gas constant per molecule) and T is the absolute temperature of the system. If a vibrational state is in question, then the value from [I, 2.31] must be substituted for the energy. It can be shown that at a given temperature T the number of molecules in the individual energy states decreases very rapidly with increasing vibrational quantum number v. At normal temperatures almost all the molecules are in the lowest state $v = 0$. Simply substituting the energies from [I, 2.6] in the Boltzmann expression is insufficient to calculate the population of the rotational levels; for reasons which cannot be discussed here, a factor $2J + 1$ has to be intro-

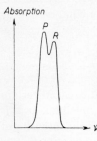

FIG. 8. *Bjerrum-double band.*

duced. It is then found that the lowest rotational level possesses a population differing from zero; with increasing rotational quantum number the population first increases, at a certain small J-value (which depends on the temperature and the rotational constant) it reaches a maximum, and finally decreases with increasing J.

If the available spectral resolution does not suffice for the separation of the individual rotational lines, then the so-called Bjerrum-band is found for polar diatomic molecules. This band (Fig. 8) possesses two maxima which correspond to the P- and R-branches, and which are separated by a certain spectral distance. In agreement with the above discussion, this distance depends on the population of the rotational states and thus on the temperature of the gas being investigated and on its rotational constant:

$$\varDelta \nu_{PR} = \sqrt{\left(\frac{8BkT}{hc}\right)} = 2 \cdot 3583 \sqrt{(BT)} \qquad \text{[I, 2.59]}$$

The measurement of this degree of splitting gives, at a known temperature, an approximate value of the rotational constant B, and hence the moment of inertia I. The first moments of inertia to be measured spectroscopically were found in this way.

3. Symmetry Properties of Molecules and Vibrations

This chapter deals with several properties of polyatomic molecules which are determined by the geometry of the positions of the atoms (or their nuclei) and do not depend on the special values of the internuclear distances, force constants, etc. A discussion of these properties, called quite generally *symmetry properties*, is needed as a preliminary to a treatment of the infrared rotational and vibrational spectra of polyatomic molecules. It will be shown that these purely geometrical symmetry properties of a molecule's nuclear framework are a key to the understanding of the molecule's spectrum. This approach requires no knowledge of the rotational or vibrational frequencies; it gives general information which applies to all molecules of the same symmetry.

1. Symmetry Elements and Symmetry Operations

If the positions of the atomic nuclei of a molecule, e.g. the non-linear molecule XY_2 of Fig. 9, are considered, certain elements of symmetry can be seen.

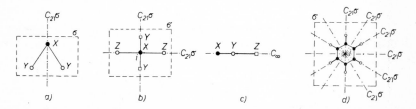

FIG. 9. *Symmetry properties of molecules.*

For more complicated molecules the symmetry is apparent only after more careful inspection. In the example mentioned, the molecule is unaltered by reflection in a plane perpendicular to the plane of the molecule, and which bisects the molecular angle YXY. This operation simply exchanges the two atoms Y, which are indistinguishable individuals. The existence of such a property is quite generally called a *symmetry element*. An operation, which if it is carried out leads from the initial system to a new system

D

indistinguishable from the first, is called a *symmetry operation*. The possible symmetry elements and the corresponding symmetry operations are now listed together with their internationally agreed symbols.

The *identity* is designated by I or E. This symmetry element is not a genuine one, for the corresponding symmetry operation leaves the co-ordinates of the atoms completely unaltered. It is obvious that every molecule possesses this symmetry.

The *plane of symmetry* is designated by σ; a subscript to this symbol describes the position in relation to an axis of symmetry. The corresponding symmetry operation is the reflection of the molecule in this plane of symmetry. Example: XY_2 (Fig. 9a).

The *centre of symmetry* is designated by i. The corresponding symmetry operation is the reflection at this centre, also called inversion. By this, each atom is transposed to the diametrically opposite position relative to the centre; any atom which lies at the centre retains its place. It follows that, if a centre of symmetry is present, all the atoms occur in pairs except any atom lying at the centre of symmetry. Example: XY_2Z_2 (Fig. 9b).

The *axis of symmetry* is designated by C_p, where p indicates the multiplicity. The corresponding symmetry operation is the rigid rotation of the molecule about this axis through an angle of $360°/p$. Obviously, repetition of this operation gives a molecular position indistinguishable from the initial position. Only 2, 3, 4, 5, 6, and ∞ fold axes of symmetry are of importance; the last, also known as an infinite-axis, means that an equivalent molecular position is reached by rotation through an angle as small as is desired. It is obviously only found in linear molecules. Example: linear XYZ (Fig. 9c).

The *rotation-reflection axis* is designated by S_p, where p again gives the multiplicity. The corresponding symmetry operation is the rotation about the axis through the angle $360°/p$, followed by reflection at a plane of symmetry perpendicular to this axis. In actual molecules only the values $p = 2$, 4, and 6 are found. It is obvious that $S_2 \equiv i$. If a molecule possesses one C_p and one σ_h which is perpendicular to it, then the molecule also possesses one S_p; the converse holds only for uneven p. Example: planar X_6Y_6 (Fig. 9d). If in this example the Y-atoms are alternately above and below the plane of the X-atoms (C_3 instead of C_6 perpendicular to the molecular plane), then the S_6 remains but not the C_6.

Further, there are four higher definitions of symmetry:

Dihedral symmetry is designated by D_p. The corresponding symmetry operation contains one p-fold axis of symmetry, and two 2-fold axes $C_2 \perp C_p$.

Tetrahedral symmetry is designated by T. The symmetry operation contains three mutually perpendicular 2-fold axes C_2 (corresponding to D_2), and four 3-fold axes C_3.

Octahedral symmetry is designated by O. The corresponding symmetry operation contains three mutually perpendicular 4-fold axes C_4, and four 3-fold axes C_3.

Icosahedral symmetry is designated by I (not to be confused with the identity). The symmetry operation is composed of six 5-fold axes C_5, ten 3-fold axes C_3, and fifteen 2-fold axes C_2.

2. Point Groups

A molecule usually possesses several of the symmetry elements described above. The geometrical symmetry of the molecule is greater the larger the number of symmetry elements present. The presence of certain symmetry elements follows as a consequence of the presence of two other symmetry elements; e.g. two planes of symmetry perpendicular to each other cause a C_2, the line of their intersection. In the same way, not all symmetry elements can occur together, e.g. a C_3 excludes a C_4 in the same direction, and conversely. Any possible combination of symmetry elements, or symmetry operations, which leaves at least one point of the molecule unaltered, is called a *point group*. The construction and description of such point groups is an essential part of crystallography, the results of which can be largely transferred to this subject, although not all the point groups found in crystallography occur in molecular structures and conversely there are point groups which are found in molecules but not in crystals. The point groups of importance to the following discussion are classified in Table 4 according to the system of Schönfliess. In this table, in order to differentiate the symmetry operations and the symmetry elements from the point groups (which have identical abbreviations), the latter are given in heavy type. Further discussion of the table is omitted because it is self-explanatory, and because molecular examples are given throughout. Some of the data in the table will only become meaningful at a later stage of this treatise.

3. Symmetry Properties of Vibrations

The above classification of the symmetry properties of molecules (which have been considered only as a definite geometrical arrangement of atoms) will be of full value only when its relationship to the vibrations of the molecules has been considered. Based on the results for diatomic molecules, it is here assumed that every molecule possesses definite vibrations which are designated as its *normal vibrations*. Their number is $3N - 6$, or $3N - 5$ for linear molecules, where N is the number of atoms in the molecule. Chapter 5 considers their determination and their special properties. Each such normal

TABLE 4. *Symmetry Properties of the Point Groups According to Schönfliess*

Number	Symbol	Characteristics	Symmetry elements				Symmetry number	Vibrational species	Examples
			Rotational axes C_p	Rotation-reflection axes S_p	Planes of symmetry σ	Symmetry centre i			
1	C_1	Asymmetrical	(C_1)	—	—	—	1	A	asymmetrical C-atom
2	C_2	One symmetry axis (C_{pz}) in one direction (z-axis) (cycl. symmetry)	C_{2z}	—	—	—	2	$A,\ B$	non-planar H_2O_2
3	C_3		C_{3z}	—	—	—	3	A, E	staggered $H_3C\cdot CCl_3$
4	C_4		C_{4z}	—	—	—	4	A, B, E	—
5	C_6		C_{6z}	—	—	—	6	$A, B, 2E$	—
6	D_2 (V)	p 2-fold axes perpendicular to a p-fold C_{pz}-axis (dihedral symmetry)	$2C_2 \perp C_{2z}$	—	—	—	4	$A, 3B$	non-planar $H_2C{=}CH_2$
7	D_3		$3C_2 \perp C_{3z}$	—	—	—	6	$2A, E$	partly staggered $H_3C\cdot CH_3$
8	D_4		$4C_2 \perp C_{4z}$	—	—	—	8	$2A, 2B, E$	—
9	D_6		$6C_2 \perp C_{6z}$	—	—	—	12	$2A, 2B, 2E$	—
10	T	Four 3-fold axes at 109° 28' 16", several C_2-axes perpendicular to each other (cubic symmetry)	$4C_3$, $3C_2$	—	—	—	12	A, E, F	unsymmetrical $C(CH_3)_4$
11	O		$4C_3$, $3C_4$, $6C_2$	—	—	—	12	$2A, E, 2F$	$X(XZ_4)_6$

			C_{pz}	S	σ	i			
12	S_2 (C_i)	One rotation-reflection axis (S_{pz}) in one direction (S-symmetry)	—	S_{2z}	—	i	1	2A	trans-BrClHC.CHClBr
13	S_4		(C_{2z})	S_{4z}	—	—	2	A, B, E	—
14	S_6 (C_{3i})		(C_{3z})	S_{6z}	—	i	3	A, B, 2E	undulating C_6H_6
15	C_s (S_1)	One plane of symmetry in one direction	(C_{1z})	—	σ	—	1	2A	$H_2C=CHCl$
16	C_{2h}	One plane of symmetry σ_h perpendicular to a p-fold axis C_{pz}	C_{2z}	(S_{2z})	σ_h	i	2	2A, 2B	trans-ClHC=CHCl
17	C_{3h}		C_{3z}	(S_{3z})	σ_h	—	3	2A, 2E	1,3,5-$C_6H_3(CH_3)_3$
18	C_{4h}		C_{4z}	—	σ_h	i	4	2A, 2B, 2E	—
19	C_{6h}		C_{6z}	—	σ_h	i	6	2A, 2B, 4E	—
20	C_{2v}	p Planes of symmetry through a p-fold axis (pyramidal symmetry)	C_{2z}	—	$2\sigma_v$	—	2	2A, 2B	H_2O, H_2CO, H_2CCl_2
21	C_{3v}		C_{3z}	—	$3\sigma_v$	—	3	2A, E	NH_3, $HCCl_3$
22	C_{4v}		C_{4z}	$4\sigma_v$	$4\sigma_v$	—	4	2A, 2B, 2E	$PtCl_4$
23	C_{6v}		C_{6z}	—	$6\sigma_v$	—	6	2A, 2B, 2E	C_6H_6 with C and H atoms in different planes

TABLE 4—continued. Symmetry Properties of the Point Groups According to Schönfliess

Number N	Symbol	Characteristics	Symmetry elements Rotational axes C_p	Rotation-reflection axes S_p	Planes of symmetry σ	Symmetry centre i	Symmetry number	Vibrational species	Examples
24	$D_{2h}(V_h)$	p 2-fold axes perpendicular to one p-fold axis, one symmetry plane in the plane of the molecule, p symmetry planes perpendicular to this (dipyram. dihedral symmetry)	$3C_2 \perp †$	—	$3\sigma_v \perp$	i	4	2A, 6B	planar $H_2C{=}CH_2$
25	D_{3h}		$C_{3z},$ $3C_2 \perp C_{3z}$	—	$\sigma_h,$ $3\sigma_v$	—	6	4A, 2E	1,3,5-$C_6H_3Cl_6$
26	D_{4h}		$C_{4z},$ $4C_2 \perp C_{4z}$	(S_{4z})	$\sigma_h,$ $4\sigma_v$	i	8	4A, 4B, 2E	cyclobutane
27	D_{6h}		$C_{6z},$ $6C_2 \perp C_{6z}$	(S_{6z})	$\sigma_h,$ $6\sigma_v$	i	12	4A, 4B, 4E	planar C_6H_6
28	$D_{2d}(V_d)(S_{4v})$	p 2-fold axes perpendicular to one p-fold rotation-reflection axis, p planes of symmetry (dihedral symmetry)	$2C_2 \perp S_4,$ (C_{2z})	S_{4z}	$2\sigma_v$	—	4	2A, 2B, E	$H_2C{=}C{=}CH_2$
29	$D_{2d}(S_{6v})$		$3C_2 \perp S_6,$ (C_{3z})	S_{6z}	$3\sigma_v$	i	6	4A, 2E	C_6H_{12}

No.	Symbol	Description	$3C_2\perp$ / $(4C_3)$	$4S_6$	3σ	i		Representations	Examples
30	T_h	Four 3-fold axes at 109° 28′ 16″, several planes of symmetry (hexagonal symmetry)	$3C_2\perp$, $(4C_3)$	$4S_6$	$3\sigma\perp$	i	12	2A, 2E, 2F	—
31	T_d		$4C_3$	$3S_4$	$6\sigma_v$ through S_4	—	12	2A, E, 2F	CH$_4$, C(CH$_3$)$_4$, P$_4$
32	O_h	Several 2-, 3-, and 4-fold axes, several 4- and 6-fold rotation-reflection axes	$3C_4\perp$, $C_2\equiv C_4$, $6C_2$	$4S_6$ $(\equiv C_3)$, $3S_4$	$3\sigma_h\perp C_4$, $6\sigma_v\perp C_2$	i	24	4A, 2E, 4F	SF$_6$, (PtCl$_6$)$^-$
33	$C_{\infty v}$	One ∞-fold axis C_∞, ∞ symmetry planes σ_v through C_∞						Σ, Π	CN, HCN, HC≡CCl
34	$D_{\infty h}$	One ∞-fold axis C_∞, ∞ 2-fold axes $\perp C_\infty$, ∞ symmetry planes σ_v through C_∞, one plane of symmetry $\sigma_h\perp C_\infty$, centre of symmetry i (C_p and S_v coincide with C_∞)						2Σ, 2Π	O$_2$, CO$_2$, C$_2$H$_2$, C$_3$O$_2$
35	D_{5h}	One 5-fold axis C_5, five 2-fold axes $C_2\perp C_5$, five planes of symmetry σ_v through C_5, one plane of symmetry $\sigma_h\perp C_5$.						4A, 4E	cyclopentane
36	D_{4d} (S_{8v})	One 4-fold axis C_4, four 2-fold axes $C_2\perp C_4$, four planes of symmetry σ_d through C_4, and bisecting the angle between the C_2 (S_8 C_2 coincide with C_4)						2A, 2B, 3E	S$_8$ (undulating Ring)
37	I	Six 5-fold C_5, ten 3-fold C_3, fifteen 2-fold C_2						A, 2F, G, H	—
38	I_h	In addition to the symmetry elements of I, fifteen planes of symmetry σ_v through C_p and $\sigma_h\perp C_2$						2A, 4F, 2G, 2H	—

† $\perp$ by itself means perpendicular to each other.

vibration stands in a special relationship to every symmetry operation of the symmetry group in question. A symmetry operation carried out on a non-vibrating, i.e. rigid, molecule leads to a geometrical position identical with the initial position. It follows that the potential energy of the molecule, and the force field inside the molecule, is completely unchanged by this process. However, it does not follow that the displacements of the atoms from their equilibrium positions during a vibration are identical before and after carrying out a symmetry operation. The nature of the normal vibration of the transformed molecule can be different to that of the initial one. For the alteration caused by a definite symmetry operation allowed by the molecular symmetry, there are only three possibilities. Firstly, the normal vibration may be unaltered; secondly, the normal vibration can be altered only in sign, and thirdly, the normal vibration can be altered in other ways. For the first two possibilities the following internationally agreed terminology is used: a normal vibration that is unaltered with respect to a symmetry operation, is called *symmetrical* with respect to this symmetry operation; if only the sign is altered, then it is called *antisymmetrical* with respect to the symmetry operation in question. It is found that every non-degenerate vibration† can only be symmetrical or antisymmetrical with respect to every allowed symmetry operation. The behaviour of degenerate vibrations is considerably more complicated: for a discussion of this behaviour, and its derivation, reference must be made to more detailed works of molecular spectroscopy.

A series of results of practical importance can already be drawn from this general treatment. For example, if an atom in its equilibrium position lies on a plane of symmetry, then it can vibrate only in this plane, if the vibration is symmetrical with respect to this plane, or perpendicular to this plane, if the vibration is antisymmetrical. Correspondingly, an atom with its equilibrium position on an axis of symmetry can only move along this axis, if the vibration is symmetrical with respect to this axis, or only perpendicular to it, if it is antisymmetrical. A non-degenerate vibration must be symmetrical with respect to a C_p with uneven p, while for even p the relationship can be either symmetrical or antisymmetrical; this follows from the requirement that the atom must be transformed into itself after p rotations through the angle $360°/p$, i.e. after one full rotation.

As a further explanatory example, the planar molecule XYZ_2 will be considered. Its normal vibrations, before and after reflection at the plane of symmetry perpendicular to the molecular plane and through XY, are depicted in Fig. 10. It can be seen that the vibrations designated as ν_1, ν_2, ν_3, and ν_6, are symmetrical with respect to this symmetry operation.

† For the concept of degeneracy, see p. 63.

Vibrations ν_4 and ν_5 are antisymmetrical with respect to it because the direction of the vibrational vector is reversed by the symmetry operation. For another symmetry operation, the behaviour can be completely different. For example, ν_4, ν_5, and ν_6 are antisymmetrical with respect to rotation through $180°$ about the 2-fold axis passing through XY.

Only a few quite definite types of vibrations can occur. These are classified according to their general behaviour towards the symmetry properties of the molecule. These types of vibrations depend on the point group to which the molecule belongs and are designated as *vibrational species*. The convention of Placzek for the notation of these species is used internationally. With respect to a definite axis C_p, usually that with the largest multiplicity, symmetrical vibrations are designated by the letter A, all antisymmetrical ones by B, doubly degenerate ones by E, trebly degenerate ones by F, and more than trebly degenerate ones by G, H, etc., according to the degree of degeneration. The subscripts g ($\equiv$ gerade, i.e. German for even) for symmetrical vibrations and u ($\equiv$ ungerade, i.e. uneven) for antisymmetrical vibrations describe the relationship to a centre of symmetry i. Symmetry towards a plane of symmetry is designated by $'$, antisymmetry by $''$. For degenerate vibrations, $+$ or $-$ signify that they are symmetrical or antisymmetrical with respect to a 2-fold symmetry axis C_2. If several vibrational species of the same type are found in the same molecule they are differentiated in their abbreviations by a numerical subscript. The abbreviations just described are used except for the two point groups $\boldsymbol{C_{\infty v}}$ and $\boldsymbol{D_{\infty h}}$ (linear molecules). For reasons which will not be entered into here, large Greek letters are used: Σ for non-degenerate vibrations, in

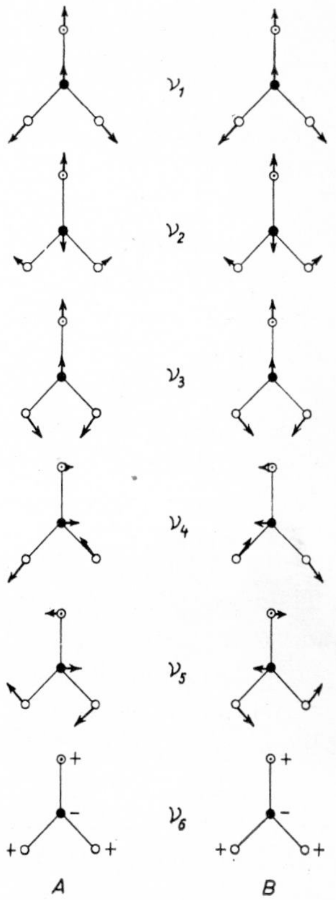

FIG. 10. *Normal vibrations of an XYZ_2 molecule* (A) *before and* (B) *after reflection at the plane of symmetry through* XY *and perpendicular to the plane of the molecule.* $+$ *and* $-$ *signify movement perpendicular to the plane of the paper,* $+$ *from above to below,* $-$ *in the opposite direction* [*from Herzberg* (I)].

$\bullet$ = X-atom, $\odot$ = Y-atom, $\bigcirc$ = Z-atom.

41

particular Σ^+ for symmetrical, and Σ^- for antisymmetrical vibrations with respect to a plane of symmetry through the molecular axis; Π, Δ, Φ, ... for degenerate vibrations with a degree of degeneracy increasing in this order.

The vibrational species for the individual point groups are given in Table 4 according to this convention. It can be seen that the point group depends on the symmetry of the molecule, and thus so does the nature of the molecule's normal vibrations. The symmetry properties can give still more information, e.g. the number of vibrations of each species for a molecule can be derived. In order to understand this calculation, it must be noted that the atoms in a polyatomic molecule which can be transformed into one another by means of the symmetry operations form a *set*. Of course, a set can consist of only one atom. As an example H_3C—CCl_3 can be taken, belonging to the point group C_3. In this, the three H-atoms and the three Cl-atoms each form a set, because they can be transformed into one another by successive rotations of 120° about the molecular axis C–C; each of the C-atoms forms an independent set because there is no symmetry operation that transforms the one into the other.

Instead of considering the set of atoms that belong together as a whole, one *representative* or *generating* member of the set can be considered, because the whole set is then unambiguously defined. The representative member of the set can have a quite general position in the molecule in that it lies on no element of symmetry, or it can have a special position in that it lies on at least one element of symmetry, or on two, or on still more. Obviously, the number of members in a set is completely determined when the position of a representative member is known. The number is largest when the representative member lies on no element of symmetry because then each symmetry operation generates a new member. The larger the number of symmetry elements on which the representative member of a set lies, the smaller the membership of the set is. It is smallest, i.e. 1, when the representative atom lies on all the symmetry elements of the point group in question.

These positional considerations are important for the calculation of the number of possible vibrations (degrees of freedom) that a set of atoms can contribute to every non-degenerate species of vibrations. When the representative atom lies on no element of symmetry this number is three, if it lies on certain symmetry elements the number is 2, 1, or 0 according to the symmetry type. By ascertaining the positional symmetry of the atoms with respect to all the possible symmetry elements, the total number of vibrations of each species can be found, and, by summation for all species, the total number of vibrations for the molecule in question. However, the so-called non-genuine vibrations with a vibrational frequency of zero (certain translations and rotations that will be discussed further later) are included in this number, and they must be subtracted in order to obtain the total number $3N - 6$ (for linear molecules $3N - 5$) already mentioned.

The results of these calculations for every point group and for every vibrational species are given in Table 5. The table also contains data about the spectral behaviour of the different vibrational species, i.e. whether infrared active (a) or not (ia), and about their properties in the Raman effect (p = polarized, p^* = partly polarized, dp = depolarized, v = forbidden). These data will be of importance later; the data on the Raman effect are given as an example of the mutual supplementation of infrared and Raman spectra. Finally, some of the abbreviations for the vibrational species of Placzek as supplemented by Mecke are given; these classify the observed vibrations from a somewhat different standpoint. So-called stretching vibrations (ν) and so-called deformation vibrations (δ) are differentiated; if a plane of symmetry is present δ indicates a deformation parallel to this plane and γ one perpendicular to it. Additional characters π or σ indicate that the vibration in question is parallel or perpendicular to special symmetry elements (rotational axes or planes of symmetry). The letters s and a mean symmetrical and antisymmetrical with respect to a centre of symmetry. These symbols, originating from Mecke, have not been accepted internationally.

The planar molecule X_2Y_4 (Fig. 11) will be considered in detail as a practical example. It belongs to the point group $V_h = D_{2h}$ and therefore possesses a centre and three planes of symmetry, which are the plane of the molecule itself, the plane perpendicular to this through the XX-line and bisecting the YXY-angle, and the plane perpendicular to the molecular plane and bisecting the XX-line and perpendicular to it. Further, there are three 2-fold axes, i.e. the lines of mutual intersection of the symmetry planes previously mentioned; one of these axes coincides with the XX-line, and both the others bisect this line, one in the molecular plane and the other perpendicular to it. For convenience, the co-ordinates are orientated so that the x-axis coincides with the XX-line, the y-axis falls in the molecular plane, and the z-axis is perpendicular to it. The quantities needed for calculating the vibrations are (using the notation of Table 4): $n = 0$, $n_0 = 0$, $n_{xy} = 1$, $n_{xz} = n_{yz} = 0$, $n_{2x} = 1$, $n_{2y} = n_{2z} = 0$. The molecule possesses two A- and six B-vibrational species, i.e. A_g, A_u, B_{1g}, B_{1u}, B_{2g}, B_{2u}, B_{3g}, B_{3u}. The number of vibrations is found to be $3A_g$, $1A_u$, $2B_{1g}$, $1B_{1u}$, $1B_{2g}$, $2B_{2u}$, $0B_{3g}$, $2B_{3u}$, a total of twelve normal vibrations. These are indicated in Fig. 11 by the arrows; with the help of Fig. 11, the symmetry of the normal vibrations can be studied with respect to the symmetry elements which are present. Only

Fundamentals of the Theory of Infrared Spectra

TABLE 5. *Number of Properties of the Normal Vibrations*

Explanation: In the column 'activity', IR indicates infrared spectrum, R = RAMAN effect, a = active, ia = inactive, p = polarised, p^* = partly polarised, dp = depolarised, v = forbidden.

In the column 'number of vibrations', n = number of representative atoms of sets which lie on no symmetry elements; n_0 = number of representative atoms of a set which lie on all elements of symmetry; n_p = number of representative atoms of a set which lie only on one p-fold symmetry axis, a further subscript indicates the orientation of this axis where necessary; n_v, n_d, n_h = number of representative atoms of a set which lie on only the symmetry planes, σ_v, σ_d, σ_h; if the symmetry plane is the plane of the coordinates, then this is indicated by xy, xz, or yz, instead of v, d, and h. The non-genuine vibrations of each species, which are given for completeness, are characterised by the indication in brackets $T_{x,y,z}$ or $R_{x,y,z}$ which mean transation or rotation about the indicated axes; the degree of degeneracy of the species in question should be taken into consideration for the number of these non-genuine vibrations.

Point group N = total number of atoms	Vibrational species	Activity IR	Activity R	Number of vibrations
C_1 $(N = n)$	A	a	p^*	$3n - 6\ (T_{x,y,z},\ R_{x,y,z})$
$C_i\ (S_2)$ $(N = 2n + n_0)$	$A_g(s)$	ia	p^*	$3n - 3\ (R_{x,y,z})$
	$A_u(a)$	a	v	$3n + 3n_0 - 3\ (T_{x,y,z})$
$C_s\ (C_{1v})$ $(N = 2n + n_0)$	$A'(v, \delta)$	a	p^*	$3n + 2n_0 - 3\ (T_{x,y},\ R_z)$
	$A''(\gamma)$	a	dp	$3n + n_0 - 3\ (T_z,\ R_{x,y})$
C_2 $(N = 2n + n_0)$	$A(s)$	a	p^*	$3n + n_0 - 2\ (T_z,\ R_z)$
	$B(a)$	a	dp	$3n + 2n_0 - 4\ (T_{x,y},\ R_{x,y})$
C_{2h} $(N = 4n + 2n_h$ $+ 2n_2 + n_0)$	$A_g(s)$	ia	p^*	$3n + 2n_h + n_2 - 1\ (R_z)$
	$A_u(\gamma, a)$	a	v	$3n + n_h + n_2 + n_0 - 1\ (T_z)$
	$B_g(\gamma, s)$	ia	dp	$3n + n_h + 2n_2 - 2\ (R_{x,y})$
	$B_u(a)$	a	v	$3n + 2n_h + 2n_2 + 2n_0 - 2\ (T_{x,y})$
C_{2v} $(N = 4n + 2n_{xz}$ $+ 2n_{yz} + n_0)$	$A_1(\pi)$	a	p	$3n + 2n_{xz} + 2n_{yz} + n_0 - 1\ (T_z)$
	$A_2(\gamma, s)$	ia	dp	$3n + n_{xz} + n_{yz} - 1\ (R_z)$
	$B_1(\sigma)$	a	dp	$3n + 2n_{xz} + n_{yz} + n_0 - 2\ (T_x,\ R_y)$
	$B_2(\gamma, a)$	a	dp	$3n + n_{xz} + 2n_{yz} + n_0 - 2\ (T_y,\ R_x)$

TABLE 5 (continued)

Point group N = total number of atoms	Vibrational species	Activity		Number of vibrations
		IR	R	
$D_2(V)$ $(N = 4n + 2n_{2x}$ $+ 2n_{2y} + 2n_{2z}$ $+ n_0)$	$A\ (s)$	ia	p	$3n + n_{2x} + n_{2y} + n_{2z}$
	$B_1(a)$	a	dp	$3n + 2n_{2x} + 2n_{2y} + n_{2z} + n_0 - 2$ (T_z, R_z)
	$B_2(a)$	a	dp	$3n + 2n_{2x} + n_{2y} + 2n_{2z} + n_0 - 2$ (T_y, R_y)
	$B_3(a)$	a	dp	$3n + n_{2x} + 2n_{2y} + 2n_{2z} + n_0 - 2$ (T_x, R_x)
$D_{2h}\ (V_h)$ $(N = 8n + 4n_{xy}$ $+ 4n_{xz} + 4n_{yz}$ $+ 2n_{2x} + 2n_{2y}$ $+ 2n_{2z} + n_0)$	$A_g(\pi, s)$	ia	p	$3n + 2n_{xy} + 2n_{xz} + 2n_{yz} + n_{2x}$ $+ n_{2y} + n_{2z}$
	$A_u(\gamma, s)$	ia	v	$3n + n_{xy} + n_{xz} + n_{yz}$
	$B_{1g}(\sigma, s)$	ia	dp	$3n + 2n_{xy} + n_{xz} + n_{yz} + n_{2x} + n_{2y}$ $- 1\ (R_z)$
	$B_{1u}(\gamma, a)$	a	v	$3n + n_{xy} + 2n_{xz} + 2n_{yz} + n_{2x} + n_{2y}$ $+ n_{2z} + n_0 - 1\ (T_z)$
	$B_{2g}(\gamma, s)$	ia	dp	$3n + n_{xy} + 2n_{xz} + n_{yz} + n_{2x} + n_{2z}$ $- 1\ (R_y)$
	$B_{2u}(\sigma, a)$	a	v	$3n + 2n_{xy} + n_{xz} + 2n_{yz} + n_{2x} + n_{2y}$ $+ n_{2z} + n_0 - 1\ (T_y)$
	$B_{3g}(\pi, a)$	ia	dp	$3n + n_{xy} + n_{xz} + 2n_{yz} + n_{2y} + n_{2z}$ $- 1\ (R_x)$
	$B_{3u}(\gamma, a)$	a	v	$3n + 2n_{xy} + 2n_{xz} + n_{yz} + n_{2x} + n_{2y}$ $+ n_{2z} + n_0 - 1\ (T_x)$
C_3 $(N = 3n + n_0)$	$A(\pi)$	a	p	$3n + n_0 - 2\ (T_z, R_z)$
	$E(\sigma)$	a	p^*	$3n + n_0 - 2\ (T_{x,y}, R_{x,y})$
$C_{3i}(S_6)$ $(N = 6n + 2n_3$ $+ n_0)$	$A_g(\pi, s)$	ia	p	$3n + n_3 - 1\ (R_z)$
	$B_u(\pi, a)$	a	v	$3n + n_3 + n_0 - 1\ (T_z)$
	$E_{1u}(\sigma, a)$	a	v	$3n + n_3 + n_0 - 1\ (T_{x,y})$
	$E_{2g}(\sigma, s)$	ia	p^*	$3n + n_3 - 1\ (R_{x,y})$
C_{3v} $(N = 6n + 3n_v$ $+ n_0)$	$A_1(\pi)$	a	p	$3n + 2n_v + n_0 - 1\ (T_z)$
	$A_2(\pi)$	ia	v	$3n + n_v - 1\ (R_z)$
	$E(\sigma)$	a	p^*	$6n + 3n_v + n_0 - 2\ (T_{x,y}, R_{x,y})$
C_{3h} $(N = 6n + 3n_h$ $+ 2n_3 + n_0)$	$A'(\pi)$	ia	p	$3n + 2n_h + n_3 - 1\ (R_z)$
	$A''(\pi)$	a	v	$3n + n_h + n_3 + n_0 - 1(T_z)$
	$E'(\sigma)$	a	p^*	$3n + 2n_h + n_3 + n_0 - 1\ (T_{x,y})$
	$E''(\sigma)$	ia	dp	$3n + n_h + n_3 - 1\ (R_{x,y})$

TABLE 5 (continued)

Point group N = total number of atoms	Vibrational species	Activity		Number of vibrations
		IR	R	
$\boldsymbol{D_3}$ $(N = 6n + 3n_2 + 2n_3 + n_0)$	$A_1(\pi)$	ia	p	$3n + n_2 + n_3$
	$A_2(\pi)$	a	v	$3n + 2n_2 + n_3 + n_0 - 2\,(T_z, R_z)$
	$E(\sigma)$	a	p^*	$6n + 3n_2 + 2n_3 + n_0 - 2\,(T_{x,y}, R_{x,y})$
$\boldsymbol{D_{3h}}$ $(N = 12n + 6n_v + 6n_h + 3n_2 + 2n_3 + n_0)$	$A'_1(\pi)$	ia	p	$3n + 2n_v + 2n_h + n_2 + n_3$
	$A''_1(\pi)$	ia	v	$3n + n_v + n_h$
	$A'_2(\pi)$	ia	v	$3n + n_v + 2n_h + n_2 - 1\,(R_z)$
	$A''_2(\pi)$	a	v	$3n + 2n_v + n_h + n_2 + n_3 + n_0 - 1(T_z)$
	$E'(\sigma)$	a	p^*	$6n + 3n_v + 4n_h + 2n_2 + n_3 + n_0 - 1$ $(T_{x,y})$
	$E''(\sigma)$	ia	dp	$6n + 3n_v + 2n_h + n_2 + n_3 - 1\,(R_{x,y})$
$\boldsymbol{D_{3d}(S_{6v})}$ $(N = 12n + 6n_d + 6n_2 + 2n_6 + n_0)$	$A_{1g}(\pi, s)$	ia	p	$3n + 2n_d + n_2 + n_6$
	$A_{1u}(\pi, a)$	ia	v	$3n + n_d + n_2$
	$A_{2g}(\pi, s)$	ia	v	$3n + n_d + 2n_2 - 1\,(R_z)$
	$A_{2u}(\pi, a)$	a	v	$3n + 2n_d + 2n_2 + n_6 + n_0 - 1\,(T_z)$
	$E_g(\sigma, s)$	ia	p^*	$6n + 3n_d + 3n_2 + n_6 - 1\,(R_{x,y})$
	$E_u(\sigma, a)$	a	v	$6n + 3n_d + 3n_2 + n_6 + n_0 - 1\,(T_{x,y})$
$\boldsymbol{C_4}$ $(N = 4n + n_0)$	A	a	p	$3n + n_0 - 2\,(T_z, R_z)$
	B	ia	p^*	$3n$
	E	a	dp	$3n + n_0 - 2\,(T_{x,y}, R_{x,y})$
$\boldsymbol{S_4}$ $(N = 4n + 2n_2 + n_0)$	A	a	p	$3n + n_2 - 1\,(R_z)$
	B	ia	p^*	$3n + n_2 + n_0 - 1\,(T_z)$
	E	a	dp	$3n + 2n_2 + n_0 - 2\,(T_{x,y}, R_{x,y})$
$\boldsymbol{C_{4v}}$ $(N = 8n + 4n_v + 4n_d + n_0)$	A_1	a	p	$3n + 2n_v + 2n_d + n_0 - 1\,(T_z)$
	A_2	ia	v	$3n + n_v + n_d - 1\,(R_z)$
	B_1	ia	p^*	$3n + 2n_v + n_d$
	B_2	ia	dp	$3n + n_v + 2n_d$
	E	a	dp	$6n + 3n_v + 3n_d + n_0 - 2\,(T_{x,y}, R_{x,y})$
$\boldsymbol{C_{4h}}$ $(N = 8n + 4n_h + 2n_4 + n_0)$	A_g	ia	p	$3n + 2n_h + n_4 - 1\,(R_z)$
	A_u	a	v	$3n + n_h + n_4 + n_0 - 1\,(T_z)$
	B_g	ia	p^*	$3n + 2n_h$
	B_u	ia	v	$3n + n_h$
	E_g	ia	dp	$3n + n_h + n_4 - 1\,(R_{x,y})$
	E_u	a	v	$3n + 2n_h + n_4 + n_0 - 1\,(T_{x,y})$

TABLE 5 (continued)

Point group N = total number of atoms	Vibrational species	Activity		Number of vibrations
		IR	R	
$D_{2d} \equiv V_d \equiv S_{4v}$ $(N = 8n + 4n_d$ $+ 4n_2 + 2n_4$ $+ n_0)$	A_1	ia	p	$3n + 2n_d + n_2 + n_4$
	A_2	ia	v	$3n + n_d + 2n_2 - 1\,(R_z)$
	B_1	ia	p^*	$3n + n_d + n_2$
	B_2	a	dp	$3n + 2n_d + 2n_2 + n_4 + n_0 - 1\,(T_z)$
	E	a	dp	$6n + 3n_d + 3n_2 + 2n_4 + n_0 - 2$ $(T_{x,y}, R_{x,y})$
D_4 $(N = 8n + 4n_2$ $+ 4n'_2 + 2n_4$ $+ n_0)$	A_1	ia	p	$3n + n_2 + n'_2 + n_4$
	A_2	a	v	$3n + 2n_2 + 2n'_2 + n_4 + n_0 - 2$ (T_z, R_z)
	B_1	ia	p^*	$3n + n_2 + 2n'_2$
	B_2	ia	dp	$3n + 2n_2 + n'_2$
	E	a	dp	$6n + 3n_2 + 3n'_2 + 2n_4 + n_0 - 2$ $(T_{x,y}, R_{x,y})$
D_{4h} $(N = 16n + 8n_v$ $+ 8n_d + 8n_h$ $+ 4n_2 + 4n'_2$ $+ 2n_4 + n_0)$	A_{1g}	ia	p	$3n + 2n_v + 2n_d + 2n_h + n_2 + n'_2 + n_4$
	A_{1u}	ia	v	$3n + n_v + n_d + n_h$
	A_{2g}	ia	v	$3n + n_v + n_d + 2n_h + n_2 + n'_2 - 1$ (R_z)
	A_{2u}	a	v	$3n + 2n_v + 2n_d + n_h + n_2 + n'_2$ $+ n_4 + n_0 - 1\,(T_z)$
	B_{1g}	ia	p^*	$3n + 2n_v + n_d + 2n_h + n_2 + n'_2$
	B_{1u}	ia	v	$3n + n_v + 2n_d + n_h + n'_2$
	B_{2g}	ia	dp	$3n + n_v + 2n_d + 2n_h + n_2 + n'_2$
	B_{2u}	ia	v	$3n + 2n_v + n_d + n_h + n_2$
	E_g	ia	dp	$6n + 3n_v + 3n_d + 2n_h + n_2 + n'_2$ $+ n_4 - 1\,(R_{x,y})$
	E_u	a	v	$6n + 3n_v + 3n_d + 4n_h + 2n_2$ $+ 2n'_2 + n_4 + n_0 - 1\,(T_{x,y})$
C_6 $(N = 6n + n_0)$	A	a	p	$3n + n_0 - 2\,(T_z, R_z)$
	B	ia	v	$3n$
	E_1	a	dp	$3n + n_0 - 2\,(T_{x,y}, R_{x,y})$
	E_2	ia	p	$3n$
C_{6v} $(N = 12n + 6n_v$ $+ 6n_d + n_0)$	A_1	a	p	$3n + 2n_v + 2n_d + n_0 - 1\,(T_z)$
	A_2	ia	v	$3n + n_v + n_d - 1\,(R_z)$
	B_1	ia	v	$3n + 2n_v + n_d$
	B_2	ia	v	$3n + n_v + 2n_d$
	E_1	a	dp	$6n + 3n_v + 3n_d + n_0 - 2\,(T_{x,y}, R_{x,y})$
	E_2	ia	p	$6n + 3n_v + 3n_d$

TABLE 5 (continued)

Point group N = total number of atoms	Vibrational species	Activity		Number of vibrations
		IR	R	
C_{6h} ($N = 12n + 6n_h$ $+ 2n_6 + n_0$)	A_g	ia	p	$3n + 2n_h + n_6 - 1\,(R_z)$
	A_u	a	v	$3n + n_h + n_6 + n_0 - 1\,(T_z)$
	B_g	ia	v	$3n + n_h$
	B_u	ia	v	$3n + 2n_h$
	E_{1g}	ia	dp	$3n + n_h + n_6 - 1\,(R_{x,y})$
	E_{1u}	a	v	$3n + 2n_h + n_6 + n_0 - 1\,(T_{x,y})$
	E_{2g}	ia	p*	$3n + 2n_h$
	E_{2u}	ia	v	$3n + n_h$
D_6 ($N = 12n + 6n_2$ $+ 6n'_2 + 2n_6$ $+ n_0$)	A_1	ia	p	$3n + n_2 + n'_2 + n_6$
	A_2	a	v	$3n + 2n_2 + 2n'_2 + n_6 + n_0 - 2$ (T_z, R_z)
	B_1	ia	v	$3n + n_2 + 2n'_2$
	B_2	ia	v	$3n + 2n_2 + n'_2$
	E_1	a	dp	$6n + 3n_2 + 3n'_2 + 2n_6 + n_0 - 2$ $(T_{x,y}, R_{x,y})$
	E_2	ia	p*	$6n + 3n_2 + 3n'_2$
D_{6h} ($N = 24n + 12n_v$ $+ 12n_d + 12n_h$ $+ 6n_2 + 6n'_2$ $+ 2n_6 + n_0$)	A_{1g}	ia	p	$3n + 2n_v + 2n_d + 2n_h + n_2 + n'_2 + n_6$
	A_{1u}	ia	v	$3n + n_v + n_d + n_h$
	A_{2g}	ia	v	$3n + n_v + n_d + 2n_h + n_2 + n'_2 - 1$ (R_z)
	A_{2u}	a	v	$3n + 2n_v + 2n_d + n_h + n_2 + n'_2$ $+ n_6 + n_0 - 1\,(T_z)$
	B_{1g}	ia	v	$3n + n_v + 2n_d + n_h + n'_2$
	B_{1u}	ia	v	$3n + 2n_v + n_d + 2n_h + n_2 + n'_2$
	B_{2g}	ia	v	$3n + 2n_v + n_d + n_h + n_2$
	B_{2u}	ia	v	$3n + n_v + 2n_d + 2n_h + n_2 + n'_2$
	E_{1g}	ia	dp	$6n + 3n_v + 3n_d + 2n_h + n_2 + n'$ $+ n_6 - 1\,(R_{x,y})$
	E_{1u}	a	v	$6n + 3n_v + 3n_d + 4n_h + 2n_2$ $+ 2n'_2 + n_6 + n_0 - 1\,(T_{x,y})$
	E_{2g}	ia	p*	$6n + 3n_v + 3n_d + 4n_h + 2n_2 + 2n'_2$
	E_{2u}	ia	v	$6n + 3n_v + 3n_d + 2n_h + n_2 + n'_2$
T ($N = 12n + 6n_2$ $+ 4n_3 + n_0$)	A	ia	p	$3n + n_2 + n_3$
	E	ia	dp	$3n + n_2 + n_3$
	F	a	dp	$9n + 5n_2 + 3n_3 + n_0 - 2$ $(T_{x,y,z}, R_{x,y,z})$

TABLE 5 (continued)

Point group N = total number of atoms	Vibrational species	IR	R	Number of vibrations
$\boldsymbol{T_h}$ $(N = 24n + 8n_3$ $+ 6n_2 + 12n_d$ $+ n_0)$	A_g	ia	p	$3n + n_3 + n_2 + 2n_d$
	A_u	ia	v	$3n + n_3 + n_d$
	E_g	ia	dp	$3n + n_3 + n_2 + n_d$
	E_u	ia	v	$3n + n_3 + n_d$
	F_g	ia	dp	$9n + 3n_3 + 2n_2 + 4n_d - 1$ $(T_{x,y,z})$
	F_u	a	v	$9n + 3n_3 + 3n_2 + 5n_d + n_0 - 1$ $(R_{x,y,z})$
$\boldsymbol{T_d}$ $(N = 24n + 12n_d$ $+ 6n_2 + 4n_3$ $+ n_0)$	A_1	ia	p	$3n + 2n_d + n_2 + n_3$
	A_2	ia	v	$3n + n_d$
	E	ia	dp	$6n + 3n_d + n_2 + n_3$
	F_1	ia	v	$9n + 4n_d + 2n_2 + n_3 - 1$ $(R_{x,y,z})$
	F_2	a	dp	$9n + 5n_d + 3n_2 + 2n_3 + n_0 - 1$ $(T_{x,y,z})$
$\boldsymbol{O}$ $(N = 24n + 8n_3$ $+ 6n_4 + 12n_2$ $+ n_0)$	A_1	ia	p	$3n + n_3 + n_4 + n_2$
	A_2	ia	v	$3n + n_3 + 2n_2$
	E	ia	dp	$6n + 3n_3 + n_4 + 3n_2$
	F_1	a	v	$9n + 3n_3 + 3n_4 + 5n_2 + n_0 - 1$ $(R_{x,y,z})$
	F_2	ia	dp	$9n + 3n_3 + 2n_4 + 4n_2 - 1$ $(T_{x,y,z})$
$\boldsymbol{O_h}$ $(N = 48n + 24n_h$ $+ 24n_d + 12n_2$ $+ 8n_3 + 6n_4$ $+ n_v)$	A_{1g}	ia	p	$3n + 2n_h + 2n_d + n_2 + n_3 + n_4$
	A_{1u}	ia	v	$3n + n_h + n_d$
	A_{2g}	ia	v	$3n + 2n_h + n_d + n_2$
	A_{2u}	ia	v	$3n + n_h + 2n_d + n_2 + n_3$
	E_g	ia	dp	$6n + 4n_h + 3n_d + 2n_2 + n_3 + n_4$
	E_u	ia	v	$6n + 2n_h + 3n_d + n_2 + n_3$
	F_{1g}	ia	v	$9n + 4n_h + 4n_d + 2n_2 + n_3 + n_4$ $- 1 (R_{x,y,z})$
	F_{1u}	a	v	$9n + 5n_h + 5n_d + 3n_2 + 2n_3 + 2n_4$ $+ n_0 - 1 (T_{x,y,z})$
	F_{2g}	ia	dp	$9n + 4n_h + 5n_d + 2n_2 + 2n_3 + n_4$
	F_{2u}	ia	v	$9n + 5n_h + 4n_d + 2n_2 + n_3 + n_4$
$\boldsymbol{I}$ $(N = 60n + 30n_2$ $+ 20n_3 + 12n_5$ $+ n_0)$	A	ia	p	$3n + n_2 + n_3 + n_5$
	F_1	a	v	$9n + 5n_2 + 3n_3 + 3n_5 + n_0 - 1$ $(T_{x,y,z})$
	F_2	ia	dp	$9n + 5n_2 + 3n_3 + n_5 - 1$ $(R_{x,y,z})$
	G	ia	dp	$12n + 6n_2 + 4n_3 + 2n_5$
	H	ia	dp	$15n + 7n_2 + 5n_3 + 3n_5$

E

TABLE 5 (continued)

Point group N = total number of atoms	Vibrational species	Activity IR	Activity R	Number of vibrations
$\boldsymbol{I_h}$ $(N = 120n$ $+ 60n_h + 30n_2$ $+ 20n_3 + 12n_5$ $+ n_0)$	A_g	ia	p	$3n + 2n_h + n_2 + n_3 + n_5$
	A_u	ia	v	$3n + n_h$
	F_{1g}	ia	v	$9n + 4n_h + 2n_2 + n_3 + n_5 - 1(T_{x,y,z})$
	F_{1u}	a	v	$9n + 5n_h + 3n_2 + 2n_3 + 2n_5 + n_0$ $- 1(R_{x,y,z})$
	F_{2g}	ia	dp	$9n + 4n_h + 2n_2 + n_3$
	F_{2u}	ia	v	$9n + 5n_h + 3n_2 + 2n_3 + n_5$
	G_g	ia	dp	$12n + 6n_h + 3n_2 + 2n_3 + n_5$
	G_u	ia	v	$12n + 6n_h + 3n_2 + 2n_3 + n_5$
	H_g	ia	dp	$15n + 8n_h + 4n_2 + 3n_3 + 2n_5$
	H_u	ia	v	$15n + 7n_h + 3n_2 + 2n_3 + n_5$
$\boldsymbol{C_{\infty v}}$ $(N = n_0)$	$\Sigma^+(v)$	a	p	$n_0 - 1\ (T_z, R_z)$
	$\Pi(\delta)$	a	dp	$n_0 - 2\ (T_{x,y}, R_{x,y})$
	$\Delta(\delta)$	ia	p	0
$\boldsymbol{D_{\infty h}}$ $(N = 2n_{oo} + n_0)$	$\Sigma^+_g(\gamma, s)$	ia	p	n_∞
	$\Sigma^+_u(v, a)$	a	v	$n_\infty + n_0 - 1\ (T_z, R_z)$
	$\Pi_g(\delta, s)$	ia	dp	$n_\infty - 1\ (R_{x,y})$
	$\Pi_u(\delta, a)$	a	v	$n_\infty + n_0 - 1\ (T_{x,y})$
	$\Delta_g(\delta, s)$	ia	p^*	0
$\boldsymbol{C_{5v}}$ $(N = 10n + 5n_v$ $+ n_0)$	A_1	a	p	$3n + 2n_v + n_0 - 1\ (T_z)$
	A_2	ia	v	$3n + n_v - 1\ (R_z)$
	E_1	a	dp	$6n + 3n_v + n_0 - 2\ (T_{x,y}, R_{x,y})$
	E_2	ia	p	$6n + 3n_v$
$\boldsymbol{D_{5h}}$ $(N = 20n + 10n_v$ $+ 10n_h + 5n_2$ $+ 2n_5 + n_0)$	A'_1	ia	p	$3n + 2n_v + 2n_h + n_2 + n_5$
	A''_2	ia	v	$3n + n_v + n_h$
	A'_2	ia	v	$3n + n_v + 2n_h + n_2 - 1\ (R_z)$
	A''_2	a	v	$3n + 2n_v + n_h + n_2 + n_5 + n_0 - 1$ $(T_{x,y})$
	E'_1	a	v	$6n + 3n_v + 4n_h + 2n_2 + n_5 + n_0$ $- 1\ (T_z)$
	E''_1	ia	dp	$6n + 3n_v + 2n_h + n_2 + n_5 - 1\ (R_{x,y})$
	E'_2	ia	p^*	$6n + 3n_v + 4n_h + 2n_2$
	E''_2	ia	p	$6n + 3n_v + 2n_h + n_2$
$\boldsymbol{D_{4d}}$ $(N = 16n + 8n_d$ $+ 8n_2 + 2n_8$ $+ n_0)$	A_1	ia	p	$3n + 2n_d + n_2 + n_8$
	A_2	ia	v	$3n + n_d + 2n_2 - 1\ (R_z)$
	B_1	ia	v	$3n + n_d + n_2$
	B_2	a	v	$3n + 2n_d + 2n_2 + n_8 + n_0 - 1\ (T_z)$
	E_1	a	v	$6n + 3n_d + 3n_2 + n_8 + n_0 - 1\ (T_{x,y})$
	E_2	ia	p^*	$6n + 3n_d + 3n_2$
	E_3	ia	dp	$6n + 3n_d + 3n_2 + n_8 - 1\ (R_{x,y})$

five are infrared active ($1B_{1u}$, $2B_{2u}$, $2B_{3u}$), six are found in the Raman spectrum ($3A_g$, $2B_{1g}$, $1B_{2g}$), and one ($1A_u$) is forbidden both in the infrared and in the Raman spectrum. For further details see later.

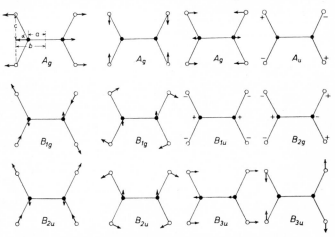

FIG. 11. *Classification of the normal vibrations and the geometrical parameters of a X_2Y_4 molecule [according to Herzberg (I)]. For explanation, see text.*

4. Symmetry Properties of Dipole Moments and Vibrational States

Still using symmetry considerations, a further step can be taken. The fact has already been mentioned, and will later be discussed in detail, that a vibration or rotation of a molecule has to be connected with an alteration of dipole moment for the corresponding band to appear in the infrared spectrum. The dipole moment M is a vector and can be resolved into its three components, M_x, M_y, M_z, along the co-ordinates. For each of these components, the behaviour towards the symmetry properties of the molecule can be given, expressed through the point group, i.e. the combination of symmetry elements, to which the molecule belongs. Exactly the same symmetry classes, with the same symbols, occur as for the vibrations themselves. The symmetry classes of the dipole moment components are given in Table 6 for the most important point groups.

This does not entirely exhaust the information that can be obtained from symmetry considerations: for each rotational or vibrational state, a dominating symmetry type can similarly be determined. By this the following is meant: the Schrödinger equation [I, 1.4], written with the expression for the potential energy V which applies to the problem in question, is, in general (as already mentioned), soluble only for certain discrete values of the total

51

TABLE 6

	C_s	C_i	C_2	C_{2h}	C_{2v}	V D_2	V_h D_{2h}	V_d D_{2d}	C_3	C_{3h}	C_{3v}	C_4	C_{4h}
M_x	A'	A_u	B	B_u	B_1	B_3	B_{3u}	E	E	E'	E	E	E
M_y	A'	A_u	B	B_u	B_2	B_2	B_{2u}	E	E	E'	E	E	E
M_z	A''	A_u	A	A_u	A_1	B_1	B_{1u}	B_2	A	A''	A_1	A	A

	C_{4v}	C_5	C_{5h}	C_{5v}	C_6	C_{6h}	C_{6v}	$C_{\infty v}$	D_3	D_{3h}	D_{3d}	D_4	D_{4h}
M_x	E	E	E'	E_1	E	E_u	E_1	Π	E	E'	E_u	E	E_u
M_y	E	E	E'	E_1	E	E_u	E_1	Π	E	E'	E_u	E	E_u
M_z	A_1	A	A''	A_1	A	A_u	A_1	Σ^+	A_2	A''_2	A_{2u}	A_2	A_{2u}

	D_{4d}	D_5	D_{5h}	D_6	D_{6h}	$D_{\infty h}$	S_4	S_6	T	T_d	O	O_h
M_x	E_1	E_1	E'_1	E_1	E_{1u}	Π_u	E	E_{1u}	F	F_2	F_1	F_{1u}
M_y	E_1	E_1	E'_1	E_1	E_{1u}	Π_u	E	E_{1u}	F	F_2	F_1	F_{1u}
M_z	B_2	A_2	A''_2	A_2	A_{2u}	Σ^+_u	A	B_u	F	F_2	F_1	F_{1u}

energy E, the so-called eigenvalues. In general, a special solution or eigenfunction ψ belongs to every possible eigenvalue. This can be written as a function of special co-ordinates, the so-called normal co-ordinates, which will be discussed later. According to whether or not the normal co-ordinates, and therefore the eigenfunctions, retain their sign by a transformation of the co-ordinates corresponding to a symmetry element allowed by the molecular symmetry, they are called (analogously to the vibrations) symmetrical or antisymmetrical with respect to this element of symmetry. This convention, initially applying only to non-degenerate normal vibrations, can also be extended to degenerate vibrations. Thus definite symmetry types are obtained for eigenfunctions and therefore for the possible states of a molecule. The behaviour of the vibrational levels is given directly by the corresponding values of the vibrational quantum number v_i. For example, the following rules hold. All values are symmetrical for a vibration which is symmetrical with respect to a plane of symmetry, independent of v_i. The levels of a vibration antisymmetrical with respect to a plane of symmetry are alternately symmetrical and antisymmetrical, according to whether the value of the vibrational quantum number is even or uneven. As will be seen later, the vibrational eigenvalues of a polyatomic molecule can, under certain conditions, be given as the sum of $3N$ eigenvalues of simple harmonic oscillators. The total eigenfunction is symmetrical or antisymmetrical for an antisymmetrical vibration in the case in question according to whether Σv_i is even or uneven. For degenerate vibrations the eigenfunction is transformed by a symmetry operation into a linear combination of eigenfunctions of the same level. Similar considerations apply to rotational states. The symbols for the symmetry behaviour of states are the same as those already given.

The value of this consideration of the symmetry of dipole moments and states will become clear later. In combination with the vibrational species, it allows a decision as to which vibrations are active in the infrared spectrum. Further, the levels able to combine with one another, i.e. which term differences are possible, can be determined from it. In addition to the purely numerical selection rules of the quantum numbers, further selection rules which allow a complete survey of the expected spectrum can sometimes be thus obtained; however, this subject cannot be considered in detail in this treatise.

4. Rotational Spectra of Polyatomic Molecules

1. Preliminary Remarks

Just as for diatomic molecules so also for polyatomic molecules, the attempt can be made to divide the total energy into electronic, vibrational, and rotational energy, in addition to that due to interaction, and to treat each separately while keeping the other quantities constant. However, the approximation to reality achieved is less than that with diatomic molecules, because the term differences of the possible energy states can sometimes be of the same order of magnitude for each type, and hence interaction becomes more marked. Nevertheless, this chapter deals with a vibrationless molecule in a fixed electronic state, and treats only the alterations of the rotational energy, from which the pure rotational spectra of polyatomic molecules are derived.

The rotational spectrum of a molecule is mainly determined by the magnitude of its moment of inertia. The moment of inertia of a rigid system of atoms (point masses) of masses m_i, about an axis is defined by

$$I = \sum m_i r_i^2 \qquad \text{[I, 4.1]}$$

where r_i is the perpendicular distance of the i-th mass from the rotational axis. From classical mechanics it is shown that the inertial behaviour of such a body is completely defined by the moments of inertia about three definite, mutually perpendicular axes. These pass through the centre of gravity, and are called the *main inertia axes* or the *principal axes*. The corresponding moments of inertia are called the *principal moments of inertia*. The symmetry of the molecule is important in determining the main inertia axes; axes of symmetry are always main inertia axes, and planes of symmetry are always perpendicular to main inertia axes. Thus the symmetry of a molecule also characterizes its rotational behaviour.

If all three principal moments of inertia are different, the molecule is called an *asymmetrical top*. If two of them are equal, then it is called a *symmetrical top*. If all three of the principal moments of inertia are identical, then a *spherical top* is present. Finally, the special case should be mentioned

54

in which one principal moment of inertia of a symmetrical top is zero or very small in comparison with the other two equally large ones.

2. Linear Molecules

Linear molecules belong to the point group $D_{\infty h}$ if they contain a plane of symmetry perpendicular to the line of the nuclei, and to the point group $C_{\infty v}$ if this plane of symmetry is not present. In almost all cases, the moment of inertia about the molecular axis (the line of the nuclei) can be neglected; thus the fourth of the possibilities just enumerated is present. If this is so, the model of the simple rotator suffices, with a possible correction for non-rigidity. Thus the rotational term is

$$F(J) = \frac{E_r}{hc} = BJ(J+1) - DJ^2(J+1)^2 + \ldots \qquad [\text{I, 4.2}]$$

where E_r is the rotational energy and J is again the rotational quantum number with the possible values $J = 0, 1, 2, \ldots$. The equation

$$B = \frac{h}{8\pi^2 c I_B} \qquad [\text{I, 4.3}]$$

which is analogous to [I, 2.8], defines the rotational constant B. Here I_B is the moment of inertia with respect to an axis perpendicular to the molecular axis and passing through the centre of gravity of the molecule:

$$I_B = \sum m_i r_i^2 \qquad [\text{I, 4.4}]$$

where the summation is carried out over all the atoms present, and r_i is the distance of the i-th atom from the centre of gravity of the molecule. The rotational constant D, which takes into account the non-rigidity, is again very small compared with B. In certain simple cases, D can be calculated from B and the known wave numbers of the molecule's vibrations, in a way similar to that in [I, 2.42]. The presence of a pure rotational spectrum is again dependent on the presence of a permanent dipole moment, as for diatomic molecules. Molecules of point groups $D_{\infty h}$ are non-polar, because of the presence of a centre of symmetry, and therefore possess no pure rotational spectrum. Elsewhere, the selection rule $\Delta J = \pm 1$ holds, and the wave number of the radiation emitted or absorbed is given by

$$\nu = 2B(J+1) - 4D(J+1)^3 \qquad [\text{I, 4.5}]$$

i.e. a series of almost equidistant lines, which for known molecules lie in the far infrared, or in the region of microwaves.

3. Symmetrical-top Molecules

Each molecule with one more than 2-fold rotational axis is a symmetrical top. One of the main inertia axes coincides with this rotational axis and the other two lie in a plane perpendicular to this C_3, C_4, ..., pass through the centre of gravity of the molecule, and are identical by reason of the molecular symmetry. Naturally, molecules not conforming with these symmetry requirements can be symmetrical tops because, by chance, two principal moments of inertia are equal. The two equal principal moments of inertia are denoted by I_B, and the third by I_A; the axis belonging to the latter (for genuine symmetrical-top molecules, the symmetry axis) is also called the body axis of the top. From classical mechanics, it is known that this body axis rotates about the axis of angular momentum which is fixed in space (nutation movement). The rotational term of the symmetrical top is

$$F(J,K) = BJ(J+1) + (A-B)K^2 \qquad [\text{I, 4.6}]$$

with the two rotational constants

$$B = \frac{h}{8\pi^2 c I_B}, \qquad A = \frac{h}{8\pi^2 c I_A} \qquad [\text{I, 4.7}]$$

The new quantum number K, which appears here, is connected with the component of the total angular momentum in the direction of the body axis; K can never be greater than J, i.e. for a given J we have

$$K = J, J-1, J-2, \ldots \qquad [\text{I, 4.8}]$$

Corresponding to the two possible directions of rotation about the axis of angular momentum, all levels with $K > 0$ are doubly degenerate, i.e. split into two levels with little difference between them.

According to whether $A > B$ or $A < B$, a top with elongated poles or with flattened poles is present. For every value of J there is a whole series of levels with increasing K. For $A > B$, the energy increases for a given J with increasing K, but for $A < B$ it decreases with increasing K. Taking non-rigidity into account extends the term by several factors

$$F(J,K) = BJ(J+1) + (A-B)K^2 - D_J J^2(J+1)^2 - \qquad [\text{I, 4.9}]$$
$$- D_{JK} J(J+1)K^2 - D_K K^4$$

As all D-values are very small compared with A and B, the corresponding factors of the term can almost always be neglected.

For the presence of a rotational spectrum in the infrared a permanent dipole moment is again needed. If the body axis is at the same time a symmetry axis, the dipole moment must lie in this axis. The selection rules are:

$$\Delta K = 0, \qquad \Delta J = 0, \pm 1 \qquad [\text{I, 4.10}]$$

However, $\Delta J = 0$ indicates (at constant K) no transition at all, and $\Delta J = -1$ is unimportant for the same reasons as described on p. 19. Thus, the final selection rule is that only transitions between neighbouring J-levels with the same K are allowed. From this we obtain, with $K' = K''$ and $J' = J''+1 \equiv J+1$

$$\nu = F(J',K') - F(J'',K'') = 2B(J+1) \qquad [\text{I}, 4.11]$$

i.e. as in linear molecules, a series of equidistant lines at a distance $2B$ from each other, from which the moment of inertia I_B can be determined. Taking into account centrifugal force factors, this becomes

$$\nu = 2B(J+1) - 2D_{JK}(J+1)K^2 - 4D_J(J+1)^2 \qquad [\text{I}, 4.12]$$

Here, the lines are no longer equidistant, and in addition, each line with $J > 0$ is split into $J+1$ lines, corresponding to all the possible values of K, with increasing distance from each other. However, this effect is so small that it is not usually observed.

Cases of the necessary permanent dipole moment not coinciding with the symmetry axis, which can only happen for molecules which are symmetric tops by chance, are of theoretical interest only.

4. Spherical-top Molecules

The presence of two or more 3- or more-fold rotational axes causes a molecule to be a spherical top with three equal principal moments of inertia; the moment of inertia about each axis passing through the centre of gravity has the same value. The point groups of tetrahedral- and octahedral-symmetry are those in question. Here also, molecules occur which are spherical tops, despite a different symmetry, because their three principal moments of inertia are equal by chance. Formula [I, 4.6] applies to the rotational term of the spherical top if the substitution $I_A = I_B$, i.e. $A = B$, is made:

$$F(J) = BJ(J+1) \qquad [\text{I}, 4.13]$$

The centrifugal correction corresponds exactly to [I, 4.9]. Because the permanent dipole moment necessary for a rotational spectrum in the infrared must lie in one of the symmetry axes, it must be zero for genuine spherical tops. Therefore, the latter possess no pure rotational infrared spectrum. Only molecules which are spherical tops by chance can possess such a spectrum, because their dipole moments need not disappear for reasons of molecular symmetry. The selection rule for them is $\Delta J = 0, \pm 1$, where again only $\Delta J = \pm 1$ is meaningful. Corresponding to this, the spectrum familiar from the symmetrical top is found.

5. Asymmetrical-top Molecules

Molecules, which do not possess rotational axes of symmetry higher than *2-fold*, are asymmetrical tops with three different principal moments of inertia. The great majority of all molecules belongs to this type. The rather complicated motion of such an asymmetrical top, according to the laws of classical mechanics, cannot be treated here. It is found that there is no longer a compact formulation for the energy states, and therefore for the rotational term. Each level is characterized by a rotational quantum number J. $J+1$ sub-levels belong to each level, corresponding to the quantum number $K = 0, 1, \ldots, J$. Each sub-level is doubly degenerate except that for which $K = 0$. Thus for every J-value there are a total of $2J+1$ different energy levels, which are differentiated by an index

$$\tau = -J, -J+1, -J+2, \ldots 0 \ldots, +J \qquad [\text{I, 4.14}]$$

The three principal moments of inertia, in the order of increasing magnitude, are symbolized by I_A, I_B, I_C. Corresponding to these three rotational constants

$$A = \frac{h}{8\pi^2 c I_A}, \qquad B = \frac{h}{8\pi^2 c I_B}, \qquad C = \frac{h}{8\pi^2 c I_C} \qquad [\text{I, 4.15}]$$

are defined. Two limiting cases of the asymmetrical top can now be considered, for which $I_B = I_C$ and $I_B = I_A$. Both of these are symmetrical tops to which formula [I, 4.6] must apply (in the second case after substitution of A by C); together with the conclusions made about the energy alterations in relationship to J and K. The term of the asymmetrical top must lie between the terms thus described. Several authors† have given formulae for their quantitative description:

$$F(J) = \tfrac{1}{2}(B+C)J(J+1) + [A - \tfrac{1}{2}(B+C)]W \qquad [\text{I, 4.16}]$$

$$F(J) = \tfrac{1}{2}(A+C)J(J+1) + \tfrac{1}{2}(A-C)E \qquad [\text{I, 4.17}]$$

The quantities W and E, which are found in these, depend in a complicated way on A, B, C, and J; this cannot be discussed here. In accordance with these formulae, the term diagram of the asymmetric top no longer consists of simple series as in the other cases. Regarding the spectrum which is to be expected, the selection rule $\Delta J = 0, \pm 1$ again holds. In addition, however, there are others, because each level J is split into $2J+1$ sub-levels of which not all can combine together (even when the above selection rule is obeyed); special rules apply according to their symmetry properties. This will not be discussed more closely. As always, a permanent dipole moment must be present if a rotational spectrum of an asymmetric-top molecule is to appear in the infrared spectrum. Molecules of the point group V_h do not possess a permanent dipole moment, but those of the point groups C_{2v}, C_2, or those with still less symmetry, do.

† For literature, see Herzberg (*I*).

5. The Vibrations and Vibrational Spectra of Polyatomic Molecules

1. Calculation of Vibrational Frequencies

For the mathematical description of the motions in space of the N atoms of a polyatomic molecule, by means of a mutually perpendicular co-ordinate system, a total of $3N$ co-ordinates are needed,

$$x_i, y_i, z_i \quad (i = 1, \ldots, N)$$

Correspondingly, such a system possesses $3N$ *degrees of freedom*. In addition to vibrational degrees of freedom, i.e. the possibility of motion of the atoms relative to each other, these include 3 degrees of freedom that describe translational motion of the system as a whole, e.g. as expressed by the co-ordinates of the molecular centre of gravity. If the latter are chosen as the origin of the co-ordinate system, which is the case when the equation

$$\sum m_i x_i = \sum m_i y_i = \sum m_i z_i = 0 \qquad \text{[I, 5.1]}$$

applies, then $3N - 3$ co-ordinates suffice to determine the relative position of the atoms with respect to their centre of gravity. Among these, a further 3 co-ordinates determine the orientation of the system (considered to be rigid) in space, i.e. describe the rotation of the system about an axis passing through the centre of gravity. If these are disregarded, $3N - 6$ co-ordinates remain for the description of purely vibrational motion, corresponding to the same number of vibrational modes. This number is $3N - 5$ for linear molecules, because only 2 co-ordinates are necessary to describe their orientation in space.

Rather than the co-ordinates of the atoms (which are considered as mass points), their alterations, the so-called *displacements*, are of interest to the following discussion. For clarity, these will be symbolized in future by x_i, y_i, z_i, where the subscript i indicates the atom in question. To take a specific physical case, it will be assumed that the atoms are bound together by elastic forces, which can be visualized as helical springs. Naturally, this does not indicate the true nature of the binding forces in the molecule. Under the action of such forces, and without disturbance from outside, the molecule takes up an equilibrium position, in which all the atoms lie in definite positions and at definite distances from each other. It is now assumed that this equilibrium is disturbed from outside, for example one or more atoms are displaced from the equilibrium position, and then the system is again left undisturbed. It tries to return to its equilibrium position under the action of the binding forces (considered to be elastic) and this occurs in the form of a vibration. Because of the system of forces present, *restoring forces* appear, causing the molecule as a whole to perform a more or less complicated periodic movement (vibration). Such motions can be mathematically described as the superposition of simple harmonic vibrations, co-directional with the co-ordinate axes, and with suitably chosen frequencies, amplitudes, and phases. It is of interest to determine under what conditions each atom of a molecule in question can carry out a simple harmonic motion with all the particles having the same vibrational frequency and vibrating all in phase. Such forms of molecular motion are called *normal vibrations*. As a fundamental condition, it will be assumed that the displacements from the equilibrium position are small.

For such a simple harmonic motion of a point mass m_i with frequency $\tilde{\nu}$, the displacement is given, as a function of the time t, by the equation

$$s_i = s_{i0} \cos{(2\pi\tilde{\nu}t + \phi)} \qquad [\text{I}, 5.2]$$

where ϕ is the phase of the vibration. By two successive differentiations with respect to time, it follows that the acceleration is given by

$$b_i = -4\pi^2\,\tilde{\nu}^2 s_i \qquad [\text{I}, 5.3]$$

and therefore the restoring force is

$$F^i = m_i b_i = -4\pi^2\,\tilde{\nu}^2 m_i s_i \qquad [\text{I}, 5.4]$$

i.e. always proportional to $m_i s_i$. This result holds in the same way for the component of the motion in any direction. The restoring forces, which occur on the displacement of one particle (subscript 1) out of a total N, can be described as linear functions of these displacements x_1, y_1, z_1, provided the latter are sufficiently small:

$$
\begin{aligned}
F_x^1 &= -k_{xx}^{11} x_1 - k_{xy}^{11} y_1 - k_{xz}^{11} z_1, \\
F_y^1 &= -k_{yx}^{11} x_1 - k_{yy}^{11} y_1 - k_{yz}^{11} z_1, \\
F_z^1 &= -k_{zx}^{11} x_1 - k_{zy}^{11} y_1 - k_{zz}^{11} z_1
\end{aligned}
\qquad [\text{I}, 5.5]
$$

Here the factors k_{ab}^{il} correspond to the quantities k in [I, 2.19] and are certain *force constants*, the meaning of which will presently become clear.

60

However, disturbing the system from outside generally displaces all N atoms from the equilibrium position instead of only one. In the expression for the forces acting on particle 1, terms are then found that derive from the displacements of the other particles and vice versa. Finally, for the components of the forces acting on all N atoms, the system of equations is obtained

$$F_x^1 = -k_{xx}^{11} x_1 - k_{xy}^{11} y_1 - k_{xz}^{11} z_1 - k_{xx}^{12} x_2 - \ldots - k_{xz}^{1N} z_N,$$

$$F_y^1 = -k_{yx}^{11} x_1 - k_{yy}^{11} y_1 - k_{yz}^{11} z_1 - k_{yx}^{12} x_2 - \ldots - k_{yz}^{1N} z_N,$$

$$F_z^1 = -k_{zx}^{11} x_1 - k_{zy}^{11} y_1 - k_{zz}^{11} z_1 - k_{zx}^{12} x_2 - \ldots - k_{zz}^{1N} z_N,$$

$$F_x^2 = -k_{xx}^{21} x_1 - k_{xy}^{21} y_1 - k_{xz}^{21} z_1 - k_{xx}^{22} x_2 - \ldots - k_{xz}^{2N} z_N,$$

$$\ldots$$
$$\ldots$$
$$\ldots$$

$$F_z^N = -k_{zx}^{N1} x_1 - k_{zy}^{N1} y_1 - k_{zz}^{N1} z_1 - k_{zx}^{N2} x_2 - \ldots - k_{zz}^{NN} z_N \qquad \text{[I, 5.6]}$$

In this the force constants k_{xy}^{il} indicate how strongly the x-component of the binding force on the i-th particle depends on the y-component of the displacement of the first particle. It can be shown that

$$k_{ab}^{il} = k_{ba}^{li} \qquad \text{[I, 5.7]}$$

holds, where a and b are respectively two of the co-ordinate directions, x, y, z. Because of this, the number of coefficients in [I, 5.6] is much reduced.

To obtain conditions under which a disturbed system of atoms in a molecule describes a simple harmonic motion with all the particles at the same frequency $\tilde{\nu}$, the expressions for the restoring force corresponding to [I, 5.4] must be substituted for the left-hand components:

$$F_x^i = -4\pi^2 \tilde{\nu}^2 m_i x_i, \qquad F_y^i = -4\pi^2 \tilde{\nu}^2 m_i y_i, \qquad F_z^i = -4\pi^2 \tilde{\nu}^2 m_i z_i$$
$$\text{[I, 5.8]}$$

When this is substituted in [I, 5.6], a system of $3N$ linear, homogeneous equations is obtained for the $3N$ unknowns

$$x_1, y_1, z_1, x_2, \ldots, z_N$$

Such a system has a unique solution only when the determinant formed from its coefficients is equal to zero

$$\begin{vmatrix} k_{xx}^{11} - 4\pi^2 \tilde{\nu}^2 m_1 & k_{xy}^{11} & \cdots & k_{xz}^{1N} \\ k_{yx}^{11} & k_{yy}^{11} - 4\pi^2 \tilde{\nu}^2 m_1 & \cdots & k_{yz}^{1N} \\ \cdot & \cdot & & \cdot \\ \cdot & \cdot & & \cdot \\ k_{zx}^{1N} & k_{zy}^{1N} & \cdots & k_{zz}^{NN} - 4\pi^2 \tilde{\nu}^2 m_N \end{vmatrix} = 0 \qquad \text{[I, 5.9]}$$

by who?

This is a definitive equation for $\tilde{\nu}^2$, the values of k_{xy}^{il} being given. A simultaneous simple harmonic motion of all the atoms in the molecule is possible only for values of the vibrational frequency $\tilde{\nu}$ which are roots of this equation. The solution of this equation [I, 5.9], called a *secular equation*, therefore gives the frequencies of the normal vibrations of the molecule in question.

The determinant [I, 5.9], written as an equation, is of degree $3N$. Therefore it possesses $3N$ roots for $\tilde{\nu}^2$ and thus gives $3N$ values of the vibrational frequency $\tilde{\nu}$ (only the positive sign of the root from $\tilde{\nu}^2$ is meaningful), for which the motion in question occurs. However, among these, six (for linear molecules five) roots are equal to zero. These are called *non-genuine* vibrations. They correspond to the components of pure translation and rotations mentioned at the beginning of this chapter. There remain, therefore, $3N - 6$ (or $3N - 5$) genuine vibrations, in agreement with the above calculation of the number of degrees of freedom. When the possible $\tilde{\nu}$-values have been determined from [I, 5.9], then the displacements can be obtained by substitution of these values in [I, 5.6] and solution of this equation for x_i, y_i, z_i. Because of the homogeneous character of the equations, instead of the displacements, the corresponding *ratios* $x_1 : y_1 : z_1 : x_2 : \ldots : z_N$ are obtained. For given $\tilde{\nu}$, these ratios are independent of the time, and simultaneously represent the ratios of the vibrational amplitudes and of the maximal vibrational velocities. If the normal vibrations of a molecule are known, every periodic motion of the molecule, however complicated, can be represented as a *super-position* of these normal vibrations. It is a characteristic property of the normal vibrations that this should be possible.

The coefficients of the displacement ratios just mentioned, expressed in Cartesian co-ordinates, $x_1 : \ldots : z_N$, are also known as *normal co-ordinates*. It can be shown that the normal co-ordinates of a vibrating system of $3N$ point masses can be obtained from the $3N$ Cartesian displacement co-ordinates by an unambiguous, reversible, linear transformation. For the following discussion they will rarely be needed; they are of theoretical importance only and will not be discussed here in any detail.

The method just discussed for the determination of the normal vibrations of a polyatomic molecule is of fundamental importance. However, there is another method of classical physics which is usually used in practice. Because this discussion is mainly concerned with fundamental issues, this second method will merely be outlined. Its basis is the *law of conservation of energy*, which in this special case states that the total energy, of the system of point masses in question, is always equal to the sum of its potential and kinetic energy. The first is expressed as a quadratic function of the displacements (assumed to be small); the latter as a quadratic function of the derivative of the displacements with respect to time. By means of a transformation of co-ordinates, it is possible to eliminate the mixed terms of these quadratic functions, i.e. those terms which contain two different co-ordinates. In the new co-ordinate system, both parts of the energy are found as quadratic functions, with terms which contain only one co-ordinate or its derivative with respect to time. The coefficients of this new expression for the potential energy, connected with the original one by the transformation outlined, are none other than the quantities $4\pi^2 \tilde{\nu}_i^2$ which occur in [I, 5.9], i.e. disregarding the factor, the squares of the normal frequencies. The total energy of the system of N point masses can now be written as the sum of $3N$ independent quantities of the form $\frac{1}{2}(4\pi^2 \tilde{\nu}_i^2 . \eta_i^2 + \dot{\eta}_i^2)$, where η_i are the new co-ordinates and $\dot{\eta}_i$ their derivatives with respect to time. Every one of these terms represents the energy of a harmonic oscillator of mass $m = 1$. In other words, the motion of the system of N particles can be written as the superposition of $3N$ independent harmonic vibrations, in agreement with the result previously obtained.

Up to this point it has been assumed that solving the equation [I, 5.9] gives (disregarding the non-genuine vibrations) $3N - 6$ or $3N - 5$ different values of the frequency $\tilde{\nu}$. However, it can happen that two or more of the roots are identical, i.e. two or more vibrations have the same frequency. Such vibrations with the same frequency are called *degenerate*; the degree of degeneracy is the number of vibrations possessing the same frequency. The superposition of two non-degenerate vibrations, i.e. with different frequencies, does not give a simple harmonic motion but a more or less complicated, so-called Lissajous-motion. On the other hand, the linear combination of two degenerate vibrations with the same frequencies always gives another simple harmonic motion. This is also true of the superposition of degenerate vibrations of different phase, only in this case all the atoms no longer vibrate in phase.

It is obvious that with increase in the number N of atoms in the molecule, the equation [I, 5.9] becomes more and more complicated and more and more difficult to solve. All normal methods of solution fail even for $N = 4$. The ways in which this difficulty can be overcome in practice, at least in certain cases, cannot be discussed here. Reference must be made to the more detailed treatises. Symmetry considerations, which have been briefly mentioned already, give some help, especially the possibility of calculating the number of vibrations of a definite species. In general the values of the vibrational frequencies are available from spectral measurements, and the object is to calculate the force constants, because the determination of the force field acting in molecules is one of the most important objects of such investigations. However, the number of normal vibrations of a molecule is usually smaller than the number of the force constants needed. These cannot therefore be directly determined from the normal vibrations, and additional relationships are necessary, for example those often found by the isotope effects to be discussed shortly. Here, two hypotheses will be mentioned, which frequently have been successful for the evaluation of spectral data. They concern the form in which the potential energy of the molecule in question is expressed.

The first hypothesis assumes only the action of *central forces*, i.e. the force acting on one atom of the molecule is considered to result from the attraction and repulsion of all the other atoms; these forces depend on the distance of the other atoms from that in question, and act along the straight lines connecting the nuclei. Mathematically expressed, the potential energy is a pure quadratic function of the alteration L_i of the distance l_i between the atoms

$$V = \tfrac{1}{2} \sum a_{ii} L_i^2 \qquad \text{[I, 5.10]}$$

The restoring forces are then

$$F_x^k = -\frac{\delta V}{\delta x_k} = -\sum a_{ii} L_i \frac{\delta L_i}{\delta x_k} \qquad \text{[I, 5.11]}$$

Because the expression $\delta L_i/\delta x_k$ is only different from zero for such L_i that concern the atom characterized by k, the contribution of F^k to the force constant is given by $a_{kk} L_k$ alone. The purely formal reasons for this hypothesis are that the number of force constants now frequently becomes less than that of the normal vibrations, especially when some of the a_{kk} become identical by reason of certain symmetry properties. Then, one definite set of force constants can be found corresponding to the spectral data.

The second hypothesis frequently used is based on *valence forces*. This means that a strong restoring force is assumed to act only along the chemical bonds and that this is called into play when the distance between atoms bonded to each other is altered. In addition, further restoring forces are assumed which counteract an alteration of the angle between the valency bonds which connect one atom with two others. For the further treatment of this, the central force field, and further more complicated hypotheses, reference must be made to more detailed works.

The quantum mechanical calculation of the possible vibrational frequencies of a given system of atoms again starts with the Schrödinger equation. For a system of N atoms of masses m_i, this takes on the form

$$\sum \frac{1}{m_i}\left(\frac{\delta^2 \psi}{\delta x_i^2}+\frac{\delta^2 \psi}{\delta y_i^2}+\frac{\delta^2 \psi}{\delta z_i^2}\right) + \frac{8\pi^2}{h^2}(E-V)\psi = 0 \qquad \text{[I, 5.12]}$$

It can be shown that, if the normal co-ordinates are substituted, and the potential energy is expressed in the form [I, 5.10] (instead of a_{ii}, λ_i is written here; instead of L_i, ξ_i is written here), this equation is transformed into the sum of $3N$ equations of the form

$$\frac{1}{\psi_i}\frac{\delta^2\psi_i}{\delta\xi_i^2} + \frac{8\pi^2}{h^2}(E_i - \tfrac{1}{2}\lambda_i\xi_i^2) = 0 \qquad [\text{I, 5.13}]$$

Here

$$E_1 + E_2 + \ldots = E \qquad [\text{I, 5.14}]$$

is the total energy expressed as the sum of the individual energies, and ψ_i is the wave-function of a simple harmonic oscillator of potential energy $\tfrac{1}{2}\lambda_i\xi_i^2$ and mass 1. Accordingly, the vibrational motion of the molecule can be considered to be the superposition of $3N$ simple harmonic vibrations in the $3N$ normal co-ordinates. The eigenvalues of [I, 5.13], and therefore the energy states of the i-th harmonic oscillator, are

$$E_i = h\tilde{\nu}_i(v_i + \tfrac{1}{2}) \qquad [\text{I, 5.15}]$$

with the vibrational quantum number $v_i = 0, 1, 2, \ldots$ and the classically calculated vibrational frequency

$$\tilde{\nu}_i = \frac{1}{2\pi}\sqrt{(\lambda_i)} \qquad [\text{I, 5.16}]$$

It follows that the total vibrational energy of the system as a whole is

$$E(v_1, v_2, \ldots) = h\tilde{\nu}_1(v_1 + \tfrac{1}{2}) + h\tilde{\nu}_2(v_2 + \tfrac{1}{2}) + \ldots \qquad [\text{I, 5.17}]$$

or, written as a term,

$$G(v_1, v_2, \ldots) = \omega_1(v_1 + \tfrac{1}{2}) + \omega_2(v_2 + \tfrac{1}{2}) + \ldots \qquad [\text{I, 5.18}]$$

with the abbreviations

$$\omega_1 = \frac{\tilde{\nu}_1}{c}, \qquad \omega_2 = \frac{\tilde{\nu}_2}{c}, \ldots \qquad [\text{I, 5.19}]$$

Here also six, or for linear molecules five, vibrations can be eliminated as non-genuine vibrations. The zero-point energy is

$$G(0, 0, \ldots) = \tfrac{1}{2}\omega_1 + \tfrac{1}{2}\omega_2 + \ldots \qquad [\text{I, 5.20}]$$

Calculated from the zero point, the vibrational term becomes

$$G(v_1, v_2, \ldots) - G(0, 0, \ldots) = G_0(v_1, v_2, \ldots) = \omega_1 v_1 + \omega_2 v_2 + \ldots \qquad [\text{I, 5.21}]$$

In order to include vibrations of any desired degree of degeneracy, [I, 5.18] can be expressed in the form

$$G(v_1, v_2, \ldots) = \sum \omega_i\left(v_i + \frac{d_i}{2}\right) \qquad [\text{I, 5.22}]$$

F

Here d_i gives directly the degree of degeneracy, i.e. is one for non-degenerate vibrations, two for doubly degenerate, etc.

2. *Vibrational Spectra of Polyatomic Molecules*

The normal vibrations of a molecule, discussed in the preceding section, correspond in the classical treatment directly to the frequency of the radiation emitted or absorbed, as long as the vibration is connected with an alteration of the dipole moment of the molecule. Vibrations which fulfil this fundamental condition are called *infrared active*, in contrast to the infrared inactive vibrations, which are not connected with an alteration in the dipole moment, and therefore do not appear in the infrared spectrum. The dipole moment is symbolized by M and its components in the directions of the co-ordinates by M_x, M_y, M_z, and these values for the equilibrium position of the vibrating molecule are characterized by the subscript o. For sufficiently small displacements x_i, y_i, z_i, can be written

$$M_x = M_{xo} + \sum \left[\left(\frac{\delta M_x}{\delta x_i} \right)_o \cdot x_i + \left(\frac{\delta M_x}{\delta y_i} \right)_o \cdot y_i + \left(\frac{\delta M_x}{\delta z_i} \right)_o \cdot z_i \right] + \ldots$$

$$[\text{I}, 5.23]$$

and corresponding equations for the y- and the z-components. Using normal co-ordinates ξ, this becomes

$$M_x = M_{xo} + \sum \left(\frac{\delta M_x}{\delta \xi_i} \right)_o \cdot \xi_i,$$

$$M_y = M_{yo} + \sum \left(\frac{\delta M_y}{\delta \xi_i} \right)_o \cdot \xi_i,$$

$$M_z = M_{zo} + \sum \left(\frac{\delta M_z}{\delta \xi_i} \right)_o \cdot \xi_i \qquad [\text{I}, 5.24]$$

Under the action of a vibration, the dipole moment alters at the same frequency as the vibration itself. It can be seen from [I, 5.24] that the condition for infrared activity of a vibration is solely that at least one of the three derivatives of the dipole moment components, with respect to the normal co-ordinates, is not equal to zero. The existence of a permanent dipole moment M_o is not itself required; this distinction from the condition for pure rotational spectra should be noted.

An alteration of a dipole moment can be in its direction, or in its size, or in both together. Obviously, the symmetry properties of a molecule give information about the behaviour of the vibrations in this respect. In molecules with no symmetry at all, every normal vibration causes an alteration in the dipole moment, although these can be of quite different magnitude according to the special circumstances; thus all the normal vibrations are infrared active. In molecules with certain symmetry properties some vibrations occur with no change in dipole moment, which are therefore not infrared active. However, not all the vibrations of symmetrical molecules belong to this group. Table 5 indicates, for every point group, and every species of vibrations, whether they are to be expected in the infrared spectrum or not. This information is based on a comparison of the symmetry behaviour of the vibration in question and the dipole moment alteration connected with it. The conditions for the infrared activity of a vibration can be stated somewhat more rigidly than before: a vibration is infrared active, as a fundamental, only if it behaves, with respect to all the symmetry properties of the molecule in question, in the same way as at least one component of the dipole moment. The information in Table 6 was tabulated for this reason. For a decision about the possible infrared activity of a vibration, it needs only to be determined whether or not its symmetry type is that of at least one of the dipole moment components. The point groups that contain a centre of symmetry are of special interest: all infrared active vibrations (these must be of antisymmetrical type in this case) are forbidden in the Raman-effect; and all those allowed in the Raman spectrum are infrared inactive. It does not follow from this rule (which is known as the principle of *mutual exclusion*) that all the vibrations which do not occur in the Raman-effect must appear in the infrared spectrum, and vice-versa; vibrations can be forbidden in both types of spectra. In addition to deciding the activity or non-activity of a vibration, the dipole moment components are also important for the consideration of the intensity of infrared bands. The intensity of a band is proportional to the sum of the squares of the derivatives of the dipole moment components.

According to the classical treatment, it follows that the vibrational spectrum of a molecule consists of a certain number of bands, the frequencies of which are given by the roots of [I, 5.9], as long as they belong to infrared active vibrations. By the quantum theory, the expected spectrum is obtained from [I, 5.18] taking into account the selection rules for the vibrational quantum number v_i. These are

$$\Delta v_i = \pm 1 \qquad\qquad [I, 5.25]$$

In the approximation being considered (see equation [I, 5.13] ff.), the individual oscillators are assumed to be independent, i.e. only one of the vibrational quantum numbers can alter at any one time. Accordingly, the wave number of the corresponding infrared band is

$$\nu = \omega_i \qquad\qquad [I, 5.26]$$

67

Fundamentals of the Theory of Infrared Spectra

A more rigorous definition of the selection rules, made with the help of quantum mechanical investigations, leads to the specific effects of the symmetry properties of the dipole moment components which have just been mentioned.

3. Refined Treatment and Special Effects

The treatment hitherto gives only a rough picture of the real relationships. In the pure vibrational spectra of polyatomic molecules (as found by spectral investigations with moderate resolutions), just as for those of diatomic molecules, in addition to the bands corresponding to the normal vibrations, others also occur, of lower intensity and with approximately double, treble, ... the wave number. These correspond to overtones in contrast to the fundamental vibrations. Further bands occur, usually weak compared to the fundamental bands, whose wave numbers are of the form

$$a.v_i \pm b.v_k \pm c.v_1 \ldots$$

i.e. composed from several normal vibrations, where $a, b, c, \ldots$ are, in general, small whole numbers. The corresponding vibrations are called *combination vibrations*.

These two effects are due to the anharmonicity of the oscillators. Their discussion, as for that of the diatomic molecules, can be based on the form of the potential energy curve. However, this cannot be represented in one plane but only in space of a higher dimension, whereby the clarity of the model is lost. Mathematically, in order to take into account the anharmonicity, terms of the third and higher degrees must be included in the expression for the potential energy. The complicated calculation cannot be given here, but the result is that the division of the vibrational movements into simple harmonic motions, which is the fundamental basis of the conception of the normal vibrations, is no longer rigorously possible. The discrepancy is the larger the greater the anharmonicity and the greater the vibrational amplitude.

In the energy states of the quantum theory, the influence of anharmonicity is felt in that there are no longer only terms with a single quantum number, but also so-called mixed terms containing two or more quantum numbers. The equation [I, 5.18] can now be written, e.g. for a triatomic non-linear molecule without degenerate vibrations:

$$
\begin{aligned}
G(v_1, v_2, v_3) = {} & \omega_1(v_1 + \tfrac{1}{2}) + \omega_2(v_2 + \tfrac{1}{2}) + \omega_3(v_3 + \tfrac{1}{2}) + \\
& + x_{11}(v_1 + \tfrac{1}{2})^2 + x_{22}(v_2 + \tfrac{1}{2})^2 + x_{33}(v_3 + \tfrac{1}{2})^2 + \\
& + x_{12}(v_1 + \tfrac{1}{2})(v_2 + \tfrac{1}{2}) + x_{13}(v_1 + \tfrac{1}{2})(v_3 + \tfrac{1}{2}) + \\
& + x_{23}(v_2 + \tfrac{1}{2})(v_3 + \tfrac{1}{2}) + \ldots
\end{aligned}
\qquad [\mathrm{I}, 5.27]
$$

Here $\omega_1, \omega_2, \omega_3$ are the wave numbers of the normal vibrations for infinitely small amplitudes, and x_{ik} are the anharmonicity constants. If the lowest vibrational level ($v_1 = v_2 = v_3 = 0$) is taken as reference instead of the real vibrational energy zero, there is obtained

$$G(0,0,0) = \tfrac{1}{2}\omega_1 + \tfrac{1}{2}\omega_2 + \tfrac{1}{2}\omega_3 + \tfrac{1}{4}x_{11} + \tfrac{1}{4}x_{22} + \tfrac{1}{4}x_{33} +$$
$$+ \tfrac{1}{4}x_{12} + \tfrac{1}{4}x_{13} + \tfrac{1}{4}x_{23} \qquad [\text{I, } 5.28]$$

and correspondingly

$$G_0(v_1, v_2, v_3) = G(v_1, v_2, v_3) - G(0,0,0) = \omega_1^0 v_1 + \omega_2^0 v_2 +$$
$$+ \omega_3^0 v_3 + x_{11}^0 v_1^2 + x_{22}^0 v_2^2 + x_{33}^0 v_3^2 + x_{12}^0 v_1 v_2 +$$
$$+ x_{13}^0 v_1 v_3 + x_{23}^0 v_2 v_3 \qquad [\text{I, } 5.29]$$

Here, $x_{ik}^0 = x_{ik}$, as long as no terms of degree higher than two in the quantum numbers need be considered; in addition

$$\omega_1^0 = \omega_1 + x_{11} + \tfrac{1}{2}x_{12} + \tfrac{1}{2}x_{13},$$
$$\omega_2^0 = \omega_2 + x_{22} + \tfrac{1}{2}x_{12} + \tfrac{1}{2}x_{23},$$
$$\omega_3^0 = \omega_3 + x_{33} + \tfrac{1}{2}x_{13} + \tfrac{1}{2}x_{23} \qquad [\text{I, } 5.30]$$

The fundamental vibrations, corresponding to the transition

$$v_i = 1 \ \rightarrow \ v_i = 0$$

with all other $v_k = 0$, are given in this convention by

$$\nu_1 = \omega_1^0 + x_{11}^0 = \omega_1 + 2x_{11} + \tfrac{1}{2}x_{12} + \tfrac{1}{2}x_{13},$$
$$\nu_2 = \omega_2^0 + x_{22}^0 = \omega_2 + 2x_{22} + \tfrac{1}{2}x_{12} + \tfrac{1}{2}x_{23},$$
$$\nu_3 = \omega_3^0 + x_{33}^0 = \omega_3 + 2x_{33} + \tfrac{1}{2}x_{13} + \tfrac{1}{2}x_{23} \qquad [\text{I, } 5.31]$$

Usually it is found that, because $x_{ik} < 0$, the values ω_i^0 are just a little greater than those for ν_i; also that $\omega_i > \omega_i^0$.

The generalization of the formulae, just given to the case $N > 3$, should be clear. Only the general case will be given, with consideration of doubly degenerate vibrations, and using the convention introduced in [I, 5.22]:

$$G(v_1, v_2, \ldots) = \sum \omega_i\left(v_i + \frac{d_i}{2}\right) + \sum_i \sum_{k \geqslant i} x_{ik}\left(v_i + \frac{d_i}{2}\right)\left(v_k + \frac{d_k}{2}\right) +$$
$$+ \sum_i \sum_{k \geqslant i} g_{ik} l_i l_k + \cdots \qquad [\text{I, } 5.32]$$

Here, the first two terms express the fact that a small d_i-fold degenerate vibration contributes $v_i + d_i/2$; the last term indicates that certain energy

Fundamentals of the Theory of Infrared Spectra

levels, which are superimposed for the harmonic oscillator, are split by the action of anharmonicity. The number l, here introduced, can take on the following values:

$$l_i = v_i, v_i-2, v_i-4, \ldots, 1 \text{ or } 0 \qquad [\text{I}, 5.33]$$

according to whether v_i is uneven or even. For non-degenerate vibrations, $l_i = g_{ik} = 0$. Corresponding to the equations [I, 5.28] to [I, 5.31], it is found that

$$G(0,0,\ldots) = \sum_i \omega_i \frac{d_i}{2} + \sum_i \sum_{k \geqslant i} x_{ik} \frac{d_i d_k}{4};$$

$$G_0(v_1,v_2,\ldots) = \sum_i \omega_i^0 v_i + \sum_i \sum_{k \geqslant i} x_{ik}^0 v_i v_k + \sum_i \sum_{k \geqslant i} g_{ik} l_i l_k;$$

$$\omega_i^0 = \omega_i + x_{ii} d_i + \tfrac{1}{2} \sum_{k \pm i} x_{ik} d_k + \ldots;$$

$$v_i = \omega_i^0 + x_{ii} + g_{ii} = \omega_i + x_{ii}(1+d_i) + \tfrac{1}{2} \sum_{k \pm i} x_{ik} d_k + g_{ii} \qquad [\text{I}, 5.34]$$

Some further details will now be given concerning the overtones and combination vibrations. The name '*binary combination*' has become customary for those bands for which the total alteration in the vibrational quantum number is two units. This can occur by one quantum number v changing by two units, or by two different ones, v_i and v_k, altering by one unit each. The name '*ternary combination*' is used analogously and so on. It is of especial interest that not all the overtones need be infrared active even if the corresponding fundamental is itself infrared active. For example, the fundamentals of species Σ_u^+ of a linear XY_2 molecule are active in the infrared; the corresponding odd numbered overtones of the state Σ_g^+ are not, yet the even numbered ones of the state Σ_u^+ are infrared active; the converse applies for the Raman effect. Thus the appearance or non-appearance of certain overtones in the spectrum helps in the determination of the point group to which the molecule belongs. For a series of bands composed of a fundamental and its overtones the wave numbers can be calculated by [I, 5.32], by equating all the v_i, except one, to zero:

$$\nu = G(0,0,\ldots,0,v_i,0,\ldots,0) - G(0,0,\ldots,0)$$
$$= G_0(0,0,\ldots,0,v_i,0,\ldots,0) = \omega_i^0 v_i + x_{ii}^0 v^2 + g_{ii} l_i^2 + \ldots$$

$$[\text{I}, 5.35]$$

The individual bands of such a series are almost equidistant, because the x_{ii} and g_{ii} are always very small.

70

The term *summation band* is given to a combination band for which the lower state is the vibrational ground state. It is, at least approximately, the sum of two or more fundamentals or overtones:

$$\nu = v_i.\nu_i + v_j.\nu_j + \dots \qquad [\text{I, } 5.36]$$

Infrared active summation bands can be formed from inactive fundamentals. If the initial state is not the vibrational ground state, but is the vibrational state v_i, and the transition leads to a state $v_k > v_i$, then the band is called *the difference band* $v_k - v_i$. The difference band is allowed or not according to whether the corresponding summation band $v_k + v_i$ is allowed or not.

In addition to the true degeneracy already mentioned, at times it can occur for a polyatomic molecule that two vibrational states, corresponding to different vibrations or to different combinations of vibrations, possess by chance the same energy. These cases of accidental degeneracy are called *Fermi-resonance*. The result of such a chance degeneracy is a mutual disturbance of the levels, which shows itself in a mutual 'repulsion', i.e. a divergence of the levels, and in a mixing of their symmetry behaviour. A more exact treatment of this important effect is outside the scope of this book.

Certain effects connected with special forms of the potential energy curves are of especial importance. First considering a diatomic molecule, because this is easier to visualize, it can be seen from Fig. 5 that the potential energy curve possesses only a single minimum. This corresponds to the single possible equilibrium state of the molecule. Certain polyatomic molecules can possess *several equilibrium positions* and, correspondingly, *several potential energy minima*. These can be at quite different heights. This is the case, for example, when the molecule in question can exist in different *isomeric* forms, such as cis-trans-isomerism. Then, not only are the different minima at different heights, but the course of the potential curve is not the same in the neighbourhood of the minimum. When the potential maximum between two such minima is not too small, two such isomeric forms can be considered as distinct molecules, which in principle, although not always in practice, are chemically distinct. Cases also exist with identical potential minima both with respect to the height and to the course of the curve. There are two important possibilities:

(1) All non-planar molecules which possess two potential minima corresponding to equilibrium positions interconvertible by inversion of all the atoms about the molecular centre of gravity. Obviously, these two forms are not in general interconvertible by a simple rotation. Because of the resonance of the energy states, each level is split into two, this is denoted as

Fundamentals of the Theory of Infrared Spectra

inversion splitting.† The best-known example is the NH_3 molecule, which exists as a pyramid in which the N-atom (the apex of the pyramid) can lie on either side of the plane formed by the H-atoms. Fig. 12 shows the potential energy curve and the vibrational levels. The splitting increases rapidly with the quantum number v.

(2) Molecules containing two or more identical atoms in which the positions of these atoms can be reversed neither by a simple rotation, nor by an

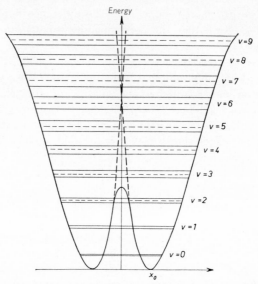

FIG. 12. *Potential energy curve of a XY_3 pyramidial molecule. The inversion splitting of the vibrational levels is indicated; it increases rapidly with the vibrational quantum number v. Abscissa: distance of the atom X from the Y_3-plane [from Herzberg (I)].*

inversion, nor by both successively. To this type belong, for example, molecules of the form

$$\begin{matrix} W \\ \diagdown \\ Z \diagup \end{matrix} X—X \begin{matrix} \diagup Y \\ \diagdown Y \end{matrix}$$

In each case an *inner rotation*, i.e. a rotation of one part of the molecule about another, leads to such an interconversion of positions. Each level is degenerate according to the number of identical potential minima.

The so-called *torsional vibrations* are very important examples of this

---

† Inversion here means a real movement of the molecule and not the mathematical condition of a symmetry operation.

72

class. These consist of a rotational movement to and fro about an axis. Corresponding to this motion, the potential energy is rationally given as a function of the angle of rotation; e.g. for n minima by

$$V = \tfrac{1}{2}V_0(1-\cos n\chi) \qquad \text{[I, 5.37]}$$

where V_0 is the potential barrier between the minima and χ is the relative angle of rotation (Fig. 13). Only for those cases in which the states considered

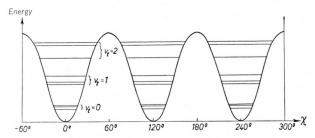

FIG. 13. *Potential energy curve of C_2H_6 or similar molecule with the torsional vibrational levels [from Herzberg (I)].*

lie *below* the potential barrier can the eigenvalues and terms be given in closed form:

$$G(v_t) = \omega_t(v_t+\tfrac{1}{2}) \qquad \text{[I, 5.38]}$$

Here v_t is the quantum number of the torsional vibration and ω_t is the rotational frequency which is given by

$$\omega_t = n\sqrt{\left(\frac{V_0 A_1 A_2}{A}\right)} \qquad \text{[I, 5.39]}$$

where A_1 and A_2 denote rotational constants corresponding to the moments of inertia $I_A^{(1)}$ and $I_A^{(2)}$ of the parts of the molecule which carry out the torsional movement. A is the rotational constant which corresponds to the moment of inertia $I_A = I_A^{(1)} + I_A^{(2)}$ of the whole molecule.

The last effect to be discussed in this connection is the influence of *isotopes* on the vibrational spectrum. This is of interest, firstly because it provides further equations for the calculation of force constants, and secondly because the assignment of the observed bands to definite vibrations of the molecule is frequently possible with the help of this effect. These methods are based on the logical assumption, widely supported by experience, that the substitution of an atom by its isotope alters the mass relationships but not the force relationships of a molecule. In other words, the potential energy is unaltered by the isotope effect, at least to within the experimental precision. However, the spectral position is altered for all the bands for which the corresponding vibration involves the isotopic atoms. In the interest of

73

brevity, the general case will be considered first and then the application to a definite molecule.

With the help of the Teller and Redlich *rule*, the product of the ratios ω_i'/ω_i for all the *normal* vibrations of a given vibrational species can be expressed for two isotopic molecules. This expression is independent of the force constants and depends only on the masses of the atoms and on the geometrical structure of the molecule:

$$\frac{\omega_1'}{\omega_1} \cdot \frac{\omega_2'}{\omega_2} \cdots \frac{\omega_k'}{\omega_k} = \sqrt{\left[\left(\frac{m_1}{m_1'}\right)^\alpha \cdot \left(\frac{m_2}{m_2'}\right)^\beta \cdots \left(\frac{M'}{M}\right)^t \cdot \left(\frac{I_x'}{I_x}\right)^{\delta_x} \cdot \left(\frac{I_y'}{I_y}\right)^{\delta_y} \cdot \left(\frac{I_z'}{I_z}\right)^{\delta_z} \right]}$$

$$[\text{I, } 5.40]$$

All the quantities which relate to one of the isotopic molecules are characterized by $'$, whereas those relating to the other molecule are not specially characterized. ω_1, ω_2, ... are the wave numbers of the specific vibrations of the species in question; m_1, m_2, ... are the masses of the generative atoms of the different sets of atoms. The number of vibrations, including non-genuine vibrations, contributed by each set to the species in question is given by α, β, ... M is the total mass of the molecule; I_x, I_y, I_z are the moments of inertia about the x-, y-, z-axes through the molecular centre of gravity. The number of translations belonging to the species of vibrations in question is indicated by t. The exponents δ_x, δ_y, δ_z are 1 or 0 according to whether or not the rotation about the x-, y-, z-axes is a non-genuine vibration of a species in question. A degenerate vibration is always counted once only.

The application of this complicated formula to a definite molecule will now be explained [from Herzberg (*I*)]. For this, the planar, symmetrical molecule X_2Y_4 and its isotopic analogue X_2Y_4' (Fig. 11) will be considered; both belong to the point group V_h. The possible vibrational species, and the number of vibrations corresponding to each, can be obtained from Table 5 and from the notes to Fig. 11 on p. 51, together with the other information needed for the application of equation [I, 5.40].

Species A_g: no non-genuine vibrations, therefore $t = \delta_x = \delta_y = \delta_z = 0$; two sets of atoms, one on the xy-plane (Y_4) with two vibrations, one on the x-axis (X_2) with one vibration, therefore $\alpha = 1$, $\beta = 2$; because $m_X = m_X'$, it follows that

$$\frac{\omega_1' \, \omega_2' \, \omega_3'}{\omega_1 \, \omega_2 \, \omega_3} = \frac{m_Y}{m_Y'}$$

Species A_u: as before $t = \delta_x = \delta_y = \delta_z = 0$; one set of atoms ($Y_4$) on the xy-plane with one vibration; therefore, it follows that

$$\frac{\omega_4'}{\omega_4} = \sqrt{\left(\frac{m_Y}{m_Y'}\right)}$$

Species B_{1g}: one non-genuine vibration (rotation about the z-axis), therefore $t = \delta_x = \delta_y = 0, \delta_z = 1$; sets of atoms and numbers of vibrations as for the species A_g; using the convention of Fig. 11 it is found that

$$I_z = 2m_X a^2 + 4m_Y(b^2 + c^2)$$

and

$$\frac{\omega'_5\,\omega'_6}{\omega_5\,\omega_6} = \frac{m_Y}{m'_Y}\sqrt{\left(\frac{m_X a^2 + 2m'_Y(b^2 + c)^2}{m_X a^2 + 2m_Y(b^2 + c)^2}\right)}$$

Species B_{1u}: one non-genuine vibration (translation in the z-direction), therefore $t = 1, \delta_x = \delta_y = \delta_z = 0$; each set contributes one vibration and thus

$$\frac{\omega'_7}{\omega_7} = \sqrt{\left(\frac{m_Y}{m'_Y}\cdot\frac{m_X + 2m'_Y}{m_X + 2m_Y}\right)}$$

Correspondingly, it can be shown that for the other vibrational species (B_{2g}, B_{2u}, B_{3u}) the following relationships apply

$$\frac{\omega'_8}{\omega_8} = \sqrt{\left(\frac{m_Y}{m'_Y}\cdot\frac{m_X a^2 + 2m'_Y b^2}{m_X a^2 + 2m_Y b^2}\right)}$$

$$\frac{\omega'_9\,\omega'_{10}}{\omega_9\,\omega_{10}} = \frac{m_Y}{m'_Y}\sqrt{\left(\frac{m_X + 2m'_Y}{m_X + 2m_Y}\right)} = \frac{\omega'_{11}\,\omega'_{12}}{\omega_{11}\,\omega_{12}}$$

Substitution of an atom by its isotope can lower the symmetry of a molecule; in the example above this happens if instead of all four only two of the Y-atoms are changed (point group C_{2v} or C_{2h}). The Teller-Redlich rule then applies only to the vibrational species of the lower symmetry group. The possibilities of changing from one symmetry species to another by symmetry alterations could be tabulated; however, for the details reference must be made to more specialized works.

In addition to the Teller-Redlich rule, there are similar rules called *summation rules* for the normal vibrations of isotopic molecules. These were first formulated for special cases by Crawford (*147*) and by Decius (*165*) and were then generalized by Primas and Günthard (*564*). For the details reference should be made, for example, to (*U*).

6. Rotational-Vibrational Spectra of Polyatomic Molecules

The pure vibrational spectra of polyatomic molecules treated in the last chapter are only observed under certain experimental conditions, especially by using low dispersion and resolution. Usually so-called *rotational-vibrational spectra* are obtained experimentally, i.e. the result of simultaneous excitation of rotation and vibration. A similar situation has already been discussed for diatomic molecules, and the method of treatment which was then applied will also be used here, however the relationships are much more complicated for the polyatomic molecules. Therefore the discussion will generally be limited to what is observed experimentally; the theoretical interpretation can only be mentioned briefly. As in the chapter dealing with pure rotational spectra, it is advantageous to base the treatment of the interaction of rotation and vibration on definite molecular models. The interaction of the two types of excitation results in certain alterations of the rotational terms and in the appearance of special effects.

1. Linear Molecules

As a first approximation, the rotational-vibrational term for polyatomic linear molecules is given by the sum of the vibrational and rotational terms, just as for the diatomic molecules:

$$T = G(v_1, v_2, \ldots) + F(J) \qquad \text{[I, 6.1]}$$

Under the influence of the vibration, the rotational constants B for different vibrational levels now take on somewhat different values in $F(J)$:

$$B_{[v]} = B_e - \alpha_1(v_1 + \tfrac{1}{2}) - \alpha_2(v_2 + \tfrac{1}{2}) + \ldots \qquad \text{[I, 6.2]}$$

where α_i are small compared to B_e and where $[v]$ denotes all the vibrational quantum numbers $v_1, v_2, \ldots$. In order to include degenerate vibrations, this is written using the convention already introduced

$$B_{[v]} = B_e - \sum \alpha_i \left(v_i + \frac{d_i}{2} \right) \qquad \text{[I, 6.3]}$$

The summation is over all vibrations, degenerate vibrations are counted once only.

The action of centrifugal forces again demands the introduction of a further rotational constant D; like B, this is dependent on all the vibrational quantum numbers in a way analogous to that in [I, 6.3]. Because of its small size, D can frequently be neglected. The rotation term of a linear polyatomic molecule for non-degenerate vibrational states is accordingly

$$F_{[v]}(J) = B_{[v]}J(J+1) - D_{[v]}J^2(J+1)^2 \qquad [\text{I, 6.4}]$$

In degenerate levels, the angular momentum is quantized with a quantum number l and this introduces additional terms:

$$F_{[v]}(J) = B_{[v]}[J(J+1) - l^2] - D_{[v]}[J(J+1) - l^2]^2 \qquad [\text{I, 6.5}]$$

It is customary to include the terms in l in the corresponding term of the vibrational term of [I, 5.32]. The rule that J may not be smaller than l applies here also, i.e.

$$J = l, l+1, l+2, \ldots \qquad [\text{I, 6.6}]$$

or, in other words, the states with $J = 0, 1, \ldots, l-1$ do not appear. The complete rotational-vibrational term is therefore

$$T = G(v_1, v_2, \ldots) + F_{[v]}(J)$$
$$= \sum \omega_i \left(v_i + \frac{d_i}{2}\right) + \sum \sum x_{ik}\left(v_i + \frac{d_i}{2}\right)\left(v_k + \frac{d_k}{2}\right) + \sum g_{ii} l_i^2 +$$
$$+ B_{[v]}J(J+1) - D_{[v]}J^2(J+1)^2 \qquad [\text{I, 6.7}]$$

Accordingly, a whole series of rotational levels appears for every vibrational level. The interval between the rotational levels is somewhat different for the individual vibrational states, and the initial numbers up to the $(l-1)$-th are suppressed.

In the discussion of diatomic molecules, a constant α was introduced which described the dependence of the rotational constant B on the vibrational quantum number. This took into consideration the fact that, even for a harmonic oscillator, the mean value of the reciprocal of the moment of inertia $1/I$ differs from $1/I_e$ (although $r_{\text{mean}} = r_e$) and further that, because of the anharmonicity of the vibration, the mean value of r during a vibration is no longer the same as r_e. The constants α_i in equation [I, 6.3] play the same part, and in addition take into consideration a further effect, the so-called *Coriolis interaction* of rotation and vibration.†

† The Coriolis effect applies of course to diatomic molecules, but it was not mentioned in the discussion of these molecules, because of its small magnitude.

Fundamentals of the Theory of Infrared Spectra

As is well known, a steadily rotating body is subject to other accelerations in addition to those stemming from forces acting from without. These are caused by forces initiated by the rotation of the system of co-ordinates: the centrifugal forces and the Coriolis forces. The former is given by

$$F_{centr} = mrw^2 \qquad [\text{I, 6.8}]$$

where m is the mass of the body in question, r is its distance from the axis of rotation and w is the angular velocity of the rotating system of co-ordinates with respect to one fixed in space. The Coriolis force is expressed by

$$F_{Cor} = 2mv_a w \sin \phi \qquad [\text{I, 6.9}]$$

where v_a is the velocity with respect to the rotation co-ordinate system and ϕ is the angle between the direction of v_a and the rotational axis. The essential difference between these two additional forces is the fact that the Coriolis force applies only to a *moving* particle ($v_a \neq 0$), and acts in a direction perpendicular both to the axis of rotation and to the direction of motion. A detailed discussion cannot be given here of the rather complicated motions of the particles subject to the action of these forces. The result of the interaction between rotation and vibration in a molecule will be discussed. Taking into consideration the centrifugal forces leads to the introduction of the rotational constant D, which has already been mentioned, and which is usually very small. The Coriolis force causes a new type of interaction which only appears in vibrating molecules and which can be much stronger in its action than the centrifugal force. It is therefore useful to divide the constants α_i, introduced above, into three parts:

$$\alpha_i = \alpha_i^{harm} + \alpha_i^{anharm} + \alpha_i^{Cor} \qquad [\text{I, 6.10}]$$

Reference must be made to detailed theoretical works for the calculation of these quantities, which is possible under certain conditions. In addition to the Coriolis interaction one further effect can be observed; this corresponds to the Fermi resonance already mentioned and appears when two vibrational levels have nearly the same energy.

To a sufficient approximation, the selection rules for vibration and rotation can be applied in order to deduce the rotational-vibrational spectrum to be expected. Accordingly, only those vibrational transitions are allowed corresponding to an alteration for which:

$$\Delta l = 0, \pm 1 \qquad [\text{I, 6.11}]$$

and only those rotational transitions that involve an alteration

$$\Delta J = 0, \pm 1 \qquad [\text{I, 6.12}]$$

However, transitions $J = 0 \rightarrow J = 0$ are forbidden, and both signs are meaningful. Corresponding to these rules, quite definite types of bands are to be expected in the infrared spectrum:

1. Transitions with $l = 0$ in both the initial and final states (*parallel bands, Σ-Σ-transitions*); for these only $\Delta J = \pm 1$ is allowed, therefore a P- and an R-branch are found but no Q-branch.
2. Transitions with $\Delta l = \pm 1$ (*perpendicular bands, Π-Σ-, Δ-Π-transitions*); now both $\Delta J = 0$ and $\Delta J = \pm 1$ are allowed and therefore all three branches appear with the Q-branch stronger than the other two.
3. Transitions with $\Delta l = 0$, but $l \neq 0$ (*parallel bands, Π-Π-, Δ-Δ-transitions*); by reason of $\Delta J = 0, \pm 1$ all three branches appear, however the Q-branch is weak.

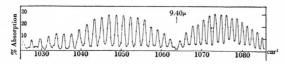

FIG. 14. *The rotational-vibrational band of the CO_2 molecule near $9 \cdot 40 \ \mu$ as an example of a parallel band of a linear molecule* [*from Barker and Adel*, Phys. Rev. **44**, *185 (1933)*].

Of course, all three types of bands can appear in the spectrum of a given molecule. For those of type 1 (Σ-Σ) the expressions for the two branches, analogous to [I, 2.52], are

$$R(J) = \nu_0 + 2B' + (3B' - B'')J + (B' - B'')J^2 \qquad \text{[I, 6.13]}$$

$$P(J) = \nu_0 \qquad -(B' + B'')J + (B' - B'')J^2 \qquad \text{[I, 6.14]}$$

where J is the rotational quantum number of the lower state. Alternatively, using a running number m analogous to [I, 2.54], it follows that

$$\nu = \nu_0 + (B' + B'')m + (B' + B'')m^2 \qquad \text{[I, 6.15]}$$

with $m = J + 1$ for the R-branch and $m = -J$ for the P-branch. It follows that a series of almost equidistant lines, with a zero gap is found here again (Fig. 14). This zero gap is characteristic for linear molecules. The lines are alternately strong and weak for molecules belonging to the point group $\boldsymbol{D}_{\infty h}$; if the symmetry depending on the existence of a centre of symmetry is destroyed (e.g. by introduction of an isotopic atom), then this intensity alternation is no longer observed. Therefore an experimental criterion has been found both for the linearity and for the symmetry of linear molecules.

For bands of type 2 (Π-Σ) a strong Q-branch is found in addition, which is given by

$$Q(J) = \nu_0 + (B' - B'')J + (B' - B'')J^2 \qquad \text{[I, 6.16]}$$

79

In general all the lines of this branch almost coincide so that the band centre appears as a very strong line because of the overlap (Fig. 15). For this type of band the line $P(0)$, and frequently $P(1)$, do not appear, but their positions are usually masked by the Q-branch so that an unambiguous observation of this effect is not possible. For Σ-Π-transitions, the line $R(0)$ is excluded instead of $P(0)$. Here also, an intensity alternation is found for molecules of the point group $\boldsymbol{D}_{\infty h}$ in each branch of a Π-Σ- or Σ-Π-transition, but not for Δ-Π- and Π-Δ-transitions.

The differentiation between parallel and perpendicular bands (type 1 and type 2) is experimentally possible even with moderate resolution, because in the first case the contour of the band has a minimum at the centre and in the second case a pronounced maximum. While the determination of a

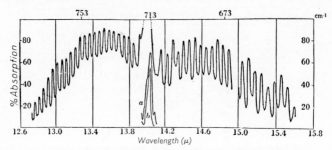

FIG. 15. *Rotational-vibrational bands of the HCN molecule at 14·04 μ as an example of a perpendicular band of a linear molecule. The curves* a *and* b *were measured at a lower gas pressure than the main curves* [*from Choi and Barker*, Phys. Rev. **41**, 777 (1932)].

minimum suffices for a decision about the linearity of the molecule (as already mentioned), this does not apply when a maximum appears, because other types of molecule also possess such bands with three branches, as will be seen.

Bands of type 3 (Π-Π) are not observed as fundamentals, but only as difference bands. Even if the Q-branch is not resolved, the corresponding 'line' attains, at the most, the intensity of the P- or R-branch. Intensity alternation is not found even for $\boldsymbol{D}_{\infty h}$-molecules.

The exact determination of the structure of the P- and R-branches of a rotational-vibrational band gives an accurate method for the determination of the rotational constant B. It follows from equations [I, 6.13] and [I, 6.14] that

$$R(J)-P(J) = F'(J+1)-F'(J-1) = \Delta^2 F'(J) \qquad [\text{I}, 6.17]$$

$$R(J-1)-P(J+1) = F''(J+1)-F''(J-1) = \Delta^2 F''(J) \qquad [\text{I}, 6.18]$$

Taking into consideration the expression [I, 6.4] for F, and assuming that the centrifugal factor can be neglected, it follows that

$$\Delta^2 F(J) = 4B(J + \tfrac{1}{2}) \qquad\qquad [\text{I}, 6.19]$$

Accordingly, $\Delta^2 F(J)$ appears as a linear function of J, the slope of which gives $4B$ directly; this applies both to the lower and to the higher state. The difference $B' - B''$ itself, together with the band centre ν_0, is determined from $R(J-1)+P(J)$, or from a resolved Q-branch, for

$$R(J-1)+P(J) = 2\nu_0 + 2(B' - B'')J^2 \qquad\qquad [\text{I}, 6.20]$$

This means that $R(J-1)+P(J)$ is a linear function of J^2, the slope of which gives the difference $B' - B''$ and the intersect of which on the ordinate gives ν_0.

2. Symmetrical-top Molecules

The rotational-vibrational term for symmetrical-top molecules is also given approximately by the sum of the separate rotational and vibrational terms. However, it must be taken into consideration that both the rotational constants A and B now show a similar dependence on the vibrational quantum number v_i as does B in the previous section:

$$B_{[v]} = B_e - \sum \alpha_i^{(B)}\left(v_i + \frac{d_i}{2}\right) + \ldots; \quad A_{[v]} = A_e - \sum \alpha_i^{(A)}\left(v_i + \frac{d_i}{2}\right) + \ldots$$

$$[\text{I}, 6.21]$$

At first not considering degenerate vibrations, where the relationships are somewhat more complicated, the rotational term is

$$F_{[v]}(J, K) = B_{[v]}J(J+1) + (A_{[v]} - B_{[v]})K^2 \qquad\qquad [\text{I}, 6.22]$$

and the rotational-vibrational term is

$$T = G(v_1, v_2, \ldots) + F_{[v]}(J, K) \qquad\qquad [\text{I}, 6.23]$$

The limitation on the possible values of J and K is the same as before (see page 56). Thus here again a series of rotational states is found for each vibrational level; the interval between the rotational states is almost the same in each individual vibrational state. For a rigorous treatment the influence of the centrifugal forces should be considered here in a way similar to that done before, but the effect is small and may be neglected.

If now the (genuine) *degenerate* vibrations are included in the treatment, it is found that, compared with the cases previously considered, much stronger perturbations are experienced as a result of Coriolis *interactions*. This is because, e.g., the two components of a doubly degenerate vibration

G

can be perturbed in this way. The result is a splitting of the degenerate states into two levels; this splitting, which is zero for $K = 0$, increases with K, i.e. with increasing rotation about the axis of the top. This influence is characterized by a number known as the Coriolis *factor* $-1 \leqslant \zeta_i \leqslant +1$, which is a measure of the magnitude of the angular momentum occurring during the vibration. The rotational term can now be written as

$$F_{[v]}(J, K) = B_{[v]}J(J+1) + (A_{[v]} - B_{[v]})K^2 \mp 2A_{[v]}\zeta_i K \qquad [\text{I}, 6.24]$$

The calculation of the individual ζ_i values is difficult and frequently not possible; however, a statement about the sum of the ζ_i values for all the vibrations of a species is possible, but this will not be given explicitly here. The relationship holds rigorously only as long as neither anharmonicity nor resonance effects have to be taken into consideration.

With respect to the selection rules for the quantum numbers applying to symmetrical-top molecules, several cases have to be differentiated. In these the selection rule for the alteration of the vibrational quantum number is always the same, namely [I, 5.25]. The essential role in the determination of the possibilities of alteration of the rotational quantum number is now played by the direction of the dipole moment alteration or, in other words, by the transition moment.

1. If it is parallel to the axis of the top (*parallel band*) then the following apply

$$\Delta K = 0, \Delta J = 0, \pm 1, \quad \text{if} \quad K \neq 0 \qquad [\text{I}, 6.25\text{a}]$$

$$\Delta K = 0, \Delta J = \pm 1, \quad \text{if} \quad K = 0 \qquad [\text{I}, 6.25\text{b}]$$

2. If the transition moment is perpendicular to the axis of the top (*perpendicular band*) then

$$\Delta K = \pm 1, \qquad \Delta J = 0, \pm 1 \qquad [\text{I}, 6.26]$$

3. If the transition moment has both a component parallel and one perpendicular to the axis of the top, which can occur for molecules which are symmetrical tops by chance (see page 57), then the rules [I, 6.25] and [I, 6.26] apply simultaneously; the bands which so originate are called *mixed*- or *hybrid-bands*.

Inversion splitting must be mentioned as a further effect which complicates the spectrum. This applies to all non-planar symmetrical-top molecules and (if it cannot be neglected) changes each line of a band into a doublet (with a few exceptions).

Initially, only non-degenerate vibrations will be considered. A whole series of transitions of the rotational quantum numbers K and J is possible, according to the conditions of excitation which apply as given in [I, 6.25] and [I, 6.26], for every transition of the vibrational quantum number. For every value of K, all values $J \geqslant K$ are allowed which accord with the selection rules applying to ΔJ. Therefore, for each value of K there is a complete band which can possess 2 or 3 branches. Each such band, determined by a K-value, is called a *sub-band*. The rotational-vibrational band is composed of the superposition of such sub-bands, the number of which depends on how many K-levels are excited (Fig. 16). The types of bands originating

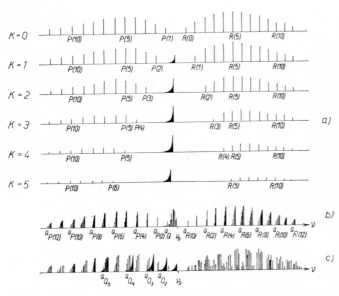

FIG. 16. *The formation of a parallel band of a symmetrical top molecule. (a) Sub-bands, (b) superimposition of the sub-bands where there is little alteration in the rotational constant differences $A - B$ in the two states, (c) superimposition of the sub-bands for greater differences [from Herzberg (I)].*

according to the possibilities discussed above are symbolized in the literature in definite ways. P, Q, or R indicate the branch of a band that is in question; a superscript placed in front, ^{P}, ^{Q}, or ^{R}, refers to the transitions $\Delta K = -1$, $\Delta K = 0$, or $\Delta K = +1$. The K-value of the lower state is given as a subscript, and the rotational quantum number of the particular line is indicated in brackets. For example, $^P R_4(5)$ indicates the line of the R-branch corresponding to $J = 5$ of a sub-band $K'' = 4 \rightarrow K' = 3$ with $\Delta K = -1$.

The relationships are simplest for the parallel bands, because for these K remains constant. Therefore, for each K-value there is only one sub-band.

83

If $K = 0$, it possesses only a P- and an R-branch ($\Delta J = -1$ and $\Delta J = +1$); for $K \neq 0$ there is also a Q-branch ($\Delta J = 0$). Because of the limitation $J \geqslant K$, fewer and fewer lines occur in the P- and R-branches with increase in K. To help visualize the overall appearance of the whole band, the interaction between the rotation and vibration will be momentarily neglected. This means that the rotational constants are the same in the higher and lower vibrational states ($A' = A''$, $B' = B''$), and therefore all the sub-bands exactly coincide, i.e. the centres of all the sub-bands are identical. In addition, the Q-branches agree in all the sub-bands. The superposition of all the sub-bands gives a strong Q-branch and a P- and R-branch; this is just the same as has been found for the perpendicular bands of the linear molecule (see page 79). The interval between the lines in the P- and R-branches is $2B$, i.e. it depends only on the moment of inertia about an axis perpendicular to the axis of the top. If the divergence between A' and A'' and also B' and B'' is now taken into consideration, the lines in the P- and R-branches are no longer equidistant and converge, as has already been found. The lines of the Q-branch are no longer exactly superimposed and the positions of the individual sub-bands are somewhat different corresponding to the relationship

$$\nu_0^{\text{sub}} = \nu_0 + [(A'_{[v]} - A''_{[v]}) - (B'_{[v]} - B''_{[v]})] K^2 \qquad [\text{I, } 6.27]$$

With not too great a resolution, the band as a whole possesses, in addition to a normal P- and R-branch, a narrow, line-like Q-branch. Such experimental findings suffice for a decision whether a molecule is a symmetrical top or not, as long as a linear arrangement can be excluded on other grounds.

For perpendicular bands, the relationships are more complicated. As a result of the two possibilities of alteration of K, i.e. $\Delta K = +1$ and $\Delta K = -1$, two systems of sub-bands occur. The centres of the sub-bands are always different, even if interaction between rotation and vibration be neglected:

$$\nu_0^{\text{sub}} = \nu_0 + (A'_{[v]} - B'_{[v]}) \pm 2(A'_{[v]} - B'_{[v]}) K + [(A'_{[v]} - B'_{[v]}) - (A''_{[v]} - B''_{[v]})] K$$
$$[\text{I, } 6.28]$$

According to whether the third term on the right-hand side takes the positive or negative sign, the R-branch ($\Delta K = +1$ with $K = 0, 1, 2, \ldots$) or the P-branch ($\Delta K = -1$ with $K = 1, 2, \ldots$) is obtained. In the interest of clarity, here also $A' = A''$ and $B' = B''$ will be taken. Then the lines which fall in the Q-branch of ν_0^{sub} form an equidistant series with the interval $2(A' - B')$. The intensity of one of the Q-branches is in general as large as that of the P- and R-branches taken together. When the total band built up by the overlapping sub-bands is considered, the Q-branch therefore dominates the picture. This is so because of the preponderance of its series of lines and because of the weakness of the P- and R-branches which count only as a background. The appearance is similar to that of a perpendicular band of linear molecules without the zero gap; the band centre lies between the two strongest lines. Here also, with increasing K more and more lines of the P- and R-branches are eliminated, because $J \geqslant K$. If the divergence of A' and A'' and of B' and B'' is taken into consideration, the lines of each Q-branch no longer coincide and the series of Q-branches converges, usually towards shorter wavelengths. Perpendicular bands of this type can occur for non-degenerate vibrations only in accidental symmetrical-top molecules. However, in this case the transition moment usually does not lie exactly perpendicular to the axis of the top, so that hybrids occur. These are bands which show the characteristics both of parallel and of perpendicular bands. It is quite obvious that the spectra then become extraordinarily complicated.

It is now necessary to consider degenerate vibrations. Based on symmetry considerations, it can be shown that perpendicular bands (i.e. bands without a dipole moment component lying along the axis of the top) can only occur for symmetrical-top molecules if at least one of the vibrational states taking part is degenerate (except for accidentally symmetrical-top molecules). It will be taken for granted that this degenerate state necessary is the higher one. The relationships are then similar to those discussed above for the perpendicular bands, because symmetry properties of the energy state, which will not be specified here, prevent the Coriolis interaction causing all the possible line splittings. The mathematical expression for the lines of each sub-band with given values of K' and K'' remains, but the zero lines of the sub-bands, or in other words their Q-branches, are altered:

$$\nu_0^{\mathrm{sub}} = \nu_0 + [A'_{[v]}(1-2\zeta_i) - B'_{[v]}] \pm 2[A'_{[v]}(1-\zeta_i) - B'_{[\iota]}]K$$
$$+ [(A'_{[v]} - B'_{[v]}) - (A''_{[v]} - B''_{[v]})]K^2 \qquad [\text{I}, 6.29]$$

Compared with [I, 6.28], terms including the Coriolis factor now occur. Therefore the positional relationships in the band as a whole are somewhat altered, especially because the ζ_i values can differ very much for different degenerate vibrations of one and the same molecule. The general appearance of the band is unaltered. The measurement of rotational vibrational bands is an excellent method for the determination of rotational constants for the symmetrical top molecules, just as it is for linear molecules. Formulae [I, 6.17] and [I, 6.18] can be used on parallel bands with coinciding sub-bands for the direct determination of the rotational constant B, but not A. When B is known from the parallel bands, A can be determined using perpendicular bands by means of the term of ν_0^{sub} which is linear in K from [I, 6.28], or from the quadratic term of this equation, or from

$$^{R}Q_{K-1} - {}^{P}Q_{K+1} = 4(A'' - A'\zeta_i - B'')\,K \qquad [I, 6.30]$$

$$^{R}Q_{K} - {}^{P}Q_{K} = 4(A' - A'\zeta_i - B')\,K \qquad [I, 6.31]$$

If molecules with degenerate vibrations are in question, i.e. those with a more than twofold rotational axis, these methods obviously do not give $A' - B'$, but, on the contrary,

$$A'(1 - \zeta_i) - B' \quad \text{or} \quad A'' - A'\,\zeta_i - B''$$

3. Spherical-top Molecules

The expression [I, 4.13] for the rotational term of a spherical top remains, but by simultaneous consideration of rotation and vibration B is now dependent on $[v]$:

$$B_{[v]} = B_e - \sum \alpha_i\left(v_i + \frac{d_i}{2}\right) \qquad [I, 6.32]$$

Centrifugal influences, if not negligible, extend the rotational term in the usual way. The rotational-vibrational term is again the sum of the pure vibrational term and pure rotational term. For degenerate vibrations, splitting is again found as a result of Coriolis interaction. If the discussion is limited to purely tetrahedral molecules, it can be shown that for the doubly degenerate vibrations of species E there is no Coriolis splitting, but this does occur for the treble degenerate F_1 and F_2. Three sub-levels result, the terms of which, symbolized by F^+, F^0, and F^-, are:

$$F^+(J) = B_{[v]}J(J+1) + 2B_{[v]}\zeta_i(J+1) \qquad [I, 6.33a]$$

$$F^0(J) = B_{[v]}J(J+1), \qquad [I, 6.33b]$$

$$F^-(J) = B_{[v]}J(J+1) - 2B_{[v]}\zeta_i J \qquad [I, 6.33c]$$

Complicated considerations of symmetry (the so-called Jahn rule) lead to the result that only certain combinations of the terms are possible because of the Coriolis interaction. The selection rule for the rotational quantum number J is again

$$\varDelta J = 0, \pm 1 \qquad [I, 6.34]$$

In the infrared spectrum, in contrast to the Raman effect, only the transitions $F_2 - F_1$ are allowed. Disregarding the Coriolis splitting [I, 6.33] of F_2, a band results with P-, Q-, and R-branches. The formulae for these are:

$$R(J) = \nu_0 + 2B'_{[v]}(1 - \zeta_i) + (3B'_{[v]} - B''_{[v]} - 2B'_{[v]}\zeta_i)J +$$
$$+ (B'_{[v]} - B''_{[v]})J^2 \qquad [I, 6.35a]$$

$$Q(J) = \nu_0 + (B'_{[v]} - B''_{[v]})J + (B'_{[v]} - B''_{[v]})J^2 \qquad \text{[I, 6.35b]}$$

$$P(J) = \nu_0 + (B'_{[v]} + B''_{[v]} - 2B'_{[v]}\zeta_i)J + (B'_{[v]} - B''_{[v]})J^2 \qquad \text{[I, 6.35c]}$$

or, written as one expression,

$$\nu = \nu_0 + (B'_{[v]} + B''_{[v]} - 2B'_{[v]}\zeta_i)\,m + (B'_{[v]} - B''_{[v]})\,m^2 \qquad \text{[I, 6.36]}$$

with $m = J + 1$ for the R-branch, $m = -J$ for the P-branch, and a zero gap for $m = 0$. If the vibration has little influence on B, i.e. when $B' - B'' = 0$, the interval between the lines is $2B(1 - \zeta_i)$. This, because of the difference of ζ_i for the individual vibrations, can be quite different in the individual bands. If the corresponding moment of inertia is known, then the different ζ_i values can be determined in this way. *Vice versa* B and I_B can be calculated from experimental values of the line intervals of all the active fundamental vibrations by using the ζ_i summation law. With sufficient resolution, the higher rotational lines of spherical-top molecules are found to have a complex structure, as a result of the Coriolis perturbation which is more marked in the higher rotational levels.

4. Asymmetrical-top Molecules

Following the preceding discussion, the theoretical treatment of the rotational-vibrational spectra of asymmetrical-top molecules can be brief. The vibrational influence must be taken into consideration for all the rotational constants:

$$A_{[v]} = A_e - \sum \alpha_i^{(A)}\left(v_i + \frac{d_i}{2}\right), \qquad B_{[v]} = B_e - \sum \alpha_i^{(B)}\left(v_i + \frac{d_i}{2}\right),$$

$$C_{[v]} = C_e - \sum \alpha_i^{(C)}\left(v_i + \frac{d_i}{2}\right) \qquad \text{[I, 6.37]}$$

The respective α_i are small compared to the rotational constants for the equilibrium position. Corresponding to the latter are the three moments of inertia

$$A_e = \frac{h}{8\pi^2 cI_A^{(e)}}, \qquad B_e = \frac{h}{8\pi^2 cI_B^{(e)}}, \qquad C_e = \frac{h}{8\pi^2 cI_C^{(e)}} \qquad \text{[I, 6.38]}$$

which are so denoted that $I_A^{(e)} < I_B^{(e)} < I_C^{(e)}$. Moments of inertia can be derived from $A_{[v]}$, $B_{[v]}$, $C_{[v]}$ instead of from A_e, B_e, C_e. These effective moments of inertia differ from the mean value of the moment of inertia in the vibrational state $[v]$ in question. The rotational-vibrational term can now be written, with reference to [I, 4.16], as:

$$T = G(v_1, v_2, \ldots) + \tfrac{1}{2}(B_{[v]} + C_{[v]})J(J + 1) + [A_{[v]} - \tfrac{1}{2}(B_{[v]} + C_{[v]})]\,W_{\tau[v]}$$
$$\text{[I, 6.39]}$$

Fundamentals of the Theory of Infrared Spectra

For each value of J there are $2J+1$ different energy levels, however their position varies somewhat for every vibrational state.

The selection rules are again:

$$\Delta J = 0, \pm 1 \qquad\qquad [\text{I, 6.40}]$$

Different types of bands result according to whether the direction of the dipole moment change lies parallel to the axis of the smallest, the intermediate, or the largest moment of inertia. For molecules of the point group C_{2v}, V, or V_h, these are called *type A*, *type B*, or *type C*; for still lower symmetry, mixed types (hybrids) are found. For the description of the individual types, it is advantageous to introduce the ratios of the moments of inertia; thus

$$\rho = \frac{I_A}{I_B} = \frac{B}{A} \qquad\qquad [\text{I, 6.41}]$$

Type A-bands: At not too great a resolution, the band resembles the parallel bands of symmetrical-top molecules. When ρ is small it possesses a

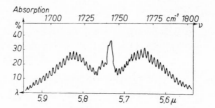

FIG. 17. *Rotational-vibrational band of the H_2CO molecule at $5\cdot73\ \mu$ as an example of a type A-band [from Nielsen,* Phys. Rev. **46**, *117 (1934)*].

non-resolved central maximum with a branch of equidistant lines on each side (Fig. 17); for large ρ the lines of the Q-branch are resolved; they move towards the band centre as ρ decreases, with simultaneous decrease in intensity.

Type B-bands (Fig. 18): These do not possess a strong central branch, rather the lines of the Q-branch overlap the P- and R-branches. For $\rho = 1$, type B is identical with type A; in the limiting case $\rho = 0$ its structure corresponds to the perpendicular band of a symmetrical-top molecule.

Type C-band: the changing dipole moment lies in the axis of the largest moment of inertia. For small ρ-values, the band's structure is almost identical with that of type B. For $\rho \to 1$, they merge into the parallel bands of a symmetrical-top molecule with a strong Q-branch and a line interval of $2A = 2B = 4C$ in the P- and R-branches.

If the available dispersion is not sufficient for the resolution of the rotational structure of the bands, then the contour also suffices to decide the band type: type A shows for every value $\rho \neq 0$ a pronounced central maxi-

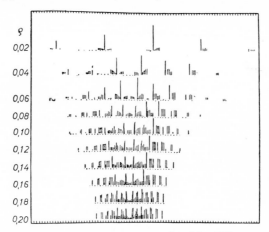

FIG. 18. *The appearance of a type B-band for different values of the ratio of the rotational constants* $\rho = B/A$ [*from Nielsen,* Phys. Rev. **38**, *1432 (1931)*].

mum with an accompanying maximum on each side. Type C has a similar contour, but for small ρ-values the central branch is less pronounced so that the whole band has more the appearance of a broad maximum which gradu-

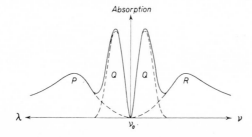

FIG. 19. *Type B-band for a non-planar asymmetrical top molecule with the ratios of the rotational constants* $B/A = 0 \cdot 73$ *and* $C/A = 0 \cdot 64$ *at low spectral resolution* [*from Badger and Zumwalt,* J. Chem. Phys. **6**, *711 (1938)*].

ally falls away. Type B has no central maximum, but the Q-branch forms an equally high maximum on each side of the band centre which, after a more or less flat minimum, is followed by the broad maximum of the P- or R-branch (Fig. 19). With falling ρ, the two middle maxima gradually

Fundamentals of the Theory of Infrared Spectra

recede, in addition, insufficient resolution frequently shows deceptively only one central maximum, i.e. an appearance like type C. The experimental measurement of the fine structure of rotational-vibrational bands of asymmetrical-top molecules can lead, just as for other types of molecule, to the determination of certain molecular constants. However, the method is much more difficult, and therefore the details will not be given here.

5. Molecules with Internal Rotation

It is necessary to give a special treatment to molecules with free, or only somewhat hindered, internal rotation, i.e. rotation of one part of the molecule about another, because this effect leads to new complications. The result of internal rotation on the pure vibrational spectrum has already been considered on page 72. For the special case of a symmetrical-top molecule with the moment of inertia I_A with respect to the axis of the top, and completely free rotation of two parts of the molecule (moments of inertia $I_A^{(1)}$ and $I_A^{(2)}$), a new part

$$F_t(k_1, k) = \frac{A_1 A_2}{A}\left(k_1 - k\frac{A}{A_1}\right)^2 \qquad [\text{I}, 6.42]$$

is added to the normal rotational term. In this the rotational constants corresponding to the moments of inertia are

$$A_1 = \frac{h}{8\pi^2 c I_A^{(1)}}, \qquad A_2 = \frac{h}{8\pi^2 c I_A^{(2)}}, \qquad A = \frac{h}{8\pi^2 c I_A} \qquad [\text{I}, 6.43]$$

and $k = \pm K$, where k_1 denotes the K-value of the first part of the molecule for which the following applies:

$$k_1 = 0, \pm 1, \pm 2, \ldots \qquad [\text{I}, 6.44]$$

Instead of [I, 6.42], the following can be written

$$F_t(k_1, k) = A_1 k_1^2 + A_2 k_2^2 - A k^2 \qquad [\text{I}, 6.45]$$

if $k_2 = k - k_1$ is introduced as the quantum number of the second part of the molecule. Accordingly, the whole rotational term is

$$F(J, K, k_1, k_2) = BJ(J+1) - BK^2 + A_1 k_1^2 + A_2 k_2^2 \qquad [\text{I}, 6.46]$$

By comparison with [I, 4.6], it can be seen that, for the case of free internal rotation, AK^2 is superseded by $A_1 k_1^2 + A_2 k_2^2$. If the two parts of the molecule are identical, then $A_1 = 2A$ and instead of [I, 6.42] we have

$$F_t(k_1, k) = A(2k_1 - k)^2 = A(k_1 - k_2)^2 = A K_i^2 \qquad [\text{I}, 6.47]$$

if $K_i = k_1 - k_2$ is introduced as the quantum number of the inner rotation.

If the internal rotation is not completely free, but is also not completely hindered (in other words, if the potential hill between the two identical potential minima is only moderately high), then the energy states must obviously be intermediate between the two limiting cases of completely free rotation (just discussed) and torsional vibration (see page 72). They prove to be split into two components whose interval increases with decrease in the height of the potential barrier. In certain circumstances, these sub-levels can themselves be degenerate, as, for example, for CH_3OH.

In symmetrical molecules, such as C_2H_6 or C_2H_4, the torsional vibration is inactive, and this applies also to the free internal rotation. Therefore, a pure rotational spectrum does not occur for the free rotation. For the rotational-vibrational spectrum, the following selection rules apply:

$$\Delta K_i = 0 \quad \text{for} \quad \Delta K = 0 \qquad [\text{I, 6.48a}]$$

$$\Delta K_i = \pm 1 \quad \text{for} \quad \Delta K = \pm 1 \qquad [\text{I, 6.48b}]$$

Thus, K_i can alter for perpendicular bands but not for parallel bands. The result is that the structure of the parallel bands of a symmetrical-top molecule is independent of whether free internal rotation occurs or not, as can be seen from the behaviour of the sub-bands which are coincident because $\Delta K_i = 0$. However, for the perpendicular bands a somewhat different structure is to be expected if free internal rotation occurs; each of the line-like Q-branches should split into a series of almost equidistant lines at intervals $2B$. Up to now such a structure has never been found for a band; from this it is concluded that no free internal rotation is present in C_2H_6.

For unsymmetrical molecules, the torsional vibration is infrared active and therefore so is any possible free internal rotation. The example of CH_3OH is of especial interest. Here, only the OH group, with a non-zero transitional moment about the axis of the top, can rotate, therefore the selection rules for the spectrum corresponding to the internal rotation are:

$$\Delta J = 0, \pm 1, \quad \Delta K = \pm 1, \quad \Delta k_1 = \pm 1, \quad \Delta k_2 = 0$$
$$[\text{I, 6.49}]$$

In this, the index 1 corresponds to the OH group, and the index 2 to the CH_3 group. For the lines of the Q-branch of the free internal rotation the following double series applies:

$$\nu = A_1 - B \mp 2BK \pm 2A_1 k_1 \qquad [\text{I, 6.50}]$$

(higher sign for positive, lower for negative ΔK and Δk_1). The selection rules for the rotational-vibrational spectrum are

$$\Delta k_1 = 0, \quad \Delta k_2 = 0 \qquad [\text{I, 6.51a}]$$

for parallel bands ($\Delta K = 0$),

$$\Delta k_1 = \pm 1, \quad \Delta k_2 = 0 \quad \text{or} \quad \Delta k_1 = 0, \quad \Delta k_2 = \pm 1 \quad [\text{I, 6.51b}]$$

for perpendicular bands ($\Delta K = \pm 1$).

The two alternatives of the last equation correspond to whether the dipole moment of the first or second part of the molecule is in question. Accordingly, the familiar structure of the parallel band remains. On the other hand, the perpendicular band shows a more complicated structure; in addition to the usual sub-bands, a further sub-division of each sub-band occurs. Details will not be given here, especially as the necessary theoretical and experimental data are still meagre.

7. Infrared Spectra of Liquids and Solids

1. Alterations in the Rotational Structure of the Bands

Up to this point, the discussion and the calculations have been for substances in the *gaseous* state. The essential characteristic of this state, compared with the liquid and solid states, is the much lower density. This permits the assumption to be made that each molecule exists free and uninfluenced by neighbouring molecules. Even where this is not so, e.g. for gases which deviate greatly from an ideal gas, then a lowering of the pressure generally suffices to obtain the ideal state. This results in the appearance in the spectrum of rotational fine structure in agreement with theory and this is influenced at most by *intra-* and not by *inter-*molecular interaction. The investigation of the effect of gradually increasing the molecular density, e.g. by means of a progressive increase of pressure, allows the study of the intermolecular interactions which appear, by means of the alterations in the spectrum connected with them. With increasing pressure, the number of molecular collisions increases, and the broadening of the individual rotational lines, due to damping by these collisions, is the first observed spectral alteration. For this, it is generally immaterial whether the higher pressure results from the absorbing gas itself or from the addition of a non-absorbing gas (which is usually a non-polar diatomic gas or an inert gas), as long as the number of molecules absorbing in the beam remains the same. By this collisional damping, and by other intermolecular interactions which appear with decreasing intermolecular distance, the rotational structure of the vibrational bands becomes more and more smeared out. Already before reaching the liquid state, only an indication of the former structure remains in the form of two or three maxima and this disappears completely in the liquid state. Fig. 20 shows this gradual transition using as an example the first overtone of HCl. With further increase in the molecular density, the spectrum alters again, at first gradually, then on reaching the crystalline state more markedly; the bands now show a smaller half band width and are therefore sharper and more prominent.

Much effort has been expended on both theoretical and experimental

93

Fundamentals of the Theory of Infrared Spectra

investigations to obtain and interpret these alterations in the spectra. This work cannot be discussed in detail. It will only be mentioned here that for a few molecules with a very large rotational splitting, e.g. H_2O, NH_3, and CH_4, a rotational structure of the vibrational band is to be expected, and has sometimes been observed [Buswell *et al.* (*106*), Breneman and Williams (*83*), Greinacher *et al.* (*277*)], for the liquid state for sufficiently dilute solutions in inert solvents. However, the spectra of the pure liquids should in practically all cases be free from any signs of quantized free rotation, and this is in agreement with observation. This also applies to solids, for which the fine structure sometimes found can be explained in other ways, e.g. as the result of an interaction between the molecular vibrations and the lattice vibrations.

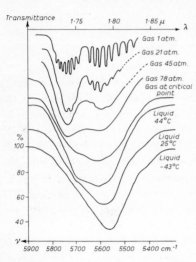

FIG. 20. *Overtone band of the HCl molecule at* $1 \cdot 76$ μ *in the gaseous and liquid states and at different temperatures and pressures [from West, J. Chem. Phys. 7, 795 (1939)]. The individual curves are displaced with respect to the ordinates.*

2. Vibrations of Liquids and Solids

In contrast to the rotational characteristics just discussed, the vibrational part of the spectrum is usually altered little or not at all on going from gas to liquid or to solid. Such alterations as are found are small *displacements* of the spectral position of the bands (for which Fig. 20 is also an example). However, *new* bands can also appear in the spectrum of the liquid. In general, the first effect is at most a 5 per cent alteration in the wave number and is nearly always a decrease in the vibrational frequency. In this, the various vibrations of one and the same molecule can behave quite differently, the reasons for which are not yet clear. Although this difference in the position of the bands of the spectra of the gas and the liquid must be taken into account for theoretical purposes, it is of little or no importance for practical applications.

The appearance of new bands in the spectrum on going from the gas to the liquid is important. This effect can also be considered to be such a pronounced positional displacement of an existing band, that the general interaction between the more closely packed molecules cannot be the reason for it, and that some other basis must be found for its clarification. The cause

94

of this phenomenon is always the formation of molecular *aggregations*, which can be associative (by means of subsidiary valencies), or polymeric, i.e. formed by means of main valencies. The most important example of the first group is *hydrogen bonding* between the oxygen and nitrogen atoms of different hydroxyl and amino-groups or between hydroxyl or amino-groups on the one hand and carbonyl-groups on the other hand; there are also a few other cases, e.g. the association of HF. This effect leads, in general, to a displacement of the OH or NH stretching band towards lower wave numbers, compared with the spectrum of the gas. The magnitude of this displacement gives a measure of the strength of the hydrogen bond. Such a displacement of the OH band can therefore serve for the detection of chelation or for the recognition of the formation of more or less stable polymeric molecules. The best-known example is that of the alcohols. It is known from their infrared spectrum that no molecules exist free in the liquid state; they are associated, in general, to polymeric complexes. However, in very dilute solution in an inert solvent, in addition to the displaced, the undisplaced OH band is found (that shown for the gaseous state); this indicates that now non-associated alcohol molecules also occur. The band intensities offer the possibility of determining the degree of association. This behaviour will be further discussed later.

A decided simplification of the spectrum is found on passing from the liquid to the crystalline state for substances with the possibility of hindered internal rotation, in that some very strong bands completely disappear. It is quite obvious that this is because only one single rotational isomer of those possible exists in the crystal. This effect is of considerable importance. In the first place, with its help the existence of *rotational isomerism* can be definitely demonstrated. Further, this observation of the simplification of the spectrum in the crystal is important for the interpretation of the spectrum of the liquid and for the theoretical assignment of its bands to definite vibrations. It is clear that such a spectrum of a liquid, containing two or even more rotational isomers, can hardly be successfully interpreted using the methods and rules previously discussed. This is because it will contain, in general, the bands of several 'molecules' which usually cannot be disentangled. Only if the molecule in question is symmetrical with respect to the single bond about which the internal rotation takes place, as in, for example, ethane, are the rotational isomers indistinguishable spectroscopically. Otherwise, the spectrum always contains the complete set of fundamental vibrations for every isomer, according to the symmetry and activity properties applying. This is especially disadvantageous for the recognition of the band connected with the vibrations of the CC bonds. For molecules which possess only two rotational isomers, the spectrum of the crystalline state usually affords the necessary simplification for the interpretation of the spectrum. It can be decided on thermodynamic and energetic grounds which isomer exists at low temperatures, and its spectrum can be assigned and interpreted (if possible, the results of Raman spectroscopy should be taken into account). The corresponding bands in the spectrum of the liquid can now be assigned. From the number of bands remaining, a conclusion can be reached about the number of the possible rotational isomers which occurs. For a total of two, the set of fundamental vibrations for the second will usually be recognizable from the remaining bands. Even if this is not completely attainable, the theoretical analysis of the first isomer gives sufficient criteria for a more or less definite analysis of the second isomer. These relationships will later be discussed further, together with practical examples. Another important point is that the temperature dependence of the bands of the various rotational isomers gives values for the energy differences between them.

As has been seen, the vibrational spectrum of the free molecule in the gaseous state depends essentially on the symmetry properties of the molecule, i.e. to the point group to which it belongs as a result of its geometry. In the ordered condensed state, i.e. in the *crystal*, similar but not identical criteria apply. The arrangement of the molecules in the crystal lattice is given by certain *space symmetry groups* together with the knowledge of how many molecules occur in the *unit cell*. These characteristics also determine the nature of the vibrational spectra of crystals.

In general, the motions of the atomic nuclei in crystals can be divided into two types of vibrations: (1) *Lattice vibrations* between the molecules considered as rigid entities. (2) *Molecular vibrations* within the individual molecule. The latter correspond, at least approximately, to those of the free molecules, but are more or less modified by the intermolecular forces. Therefore, the similarity of the spectrum of the crystal to that of the individual molecule depends on the strength of these intermolecular forces (in so far as the latter's rotations may be taken as hindered). The stronger the attractive forces of the crystal lattice, or, in other words, the more strongly polar the molecule, the greater will be the expected spectral difference between the gas or liquid and the crystal.

For the symmetry considerations, the symmetry of the unit cell of the crystal has to be used. Usually, but not necessarily, this is lower than the symmetry of the individual molecule. It is described mathematically in the form of a group, the so-called *space group*, just as the free molecule is described by a point group. The space group determines the selection rules and the activity of the vibrational species allowed. The symmetry of a unit cell, as given by a definite space group, can only be realized for certain definite numbers of molecules in the unit cell. A definite symmetry applies to every part of the crystal, which is described by a site group.† Each space group can only contain certain site groups, which have been given by Halford (*288*). For example, the site groups C_2, C_3, D_3, C_{3i}, and C_i belong to the space group $\boldsymbol{D}_{3d}^6$. For a crystal of this space group, the number of molecules in the unit cell determines the possible site groups; for calcite (CaCO₃) with two molecules in the unit cell, these are the groups $\boldsymbol{D}_3$ (for the CO₃ ion) and $\boldsymbol{C}_i$ (for the Ca ion). The vibrational species allowed for a site group correspond according to rules with the species in other site groups belonging to the same space group. This is shown in Fig. 21 for the space group $\boldsymbol{D}_{3d}^6$. From this, it can be seen, e.g. that each species belonging to the groups $\boldsymbol{C}_2$ and

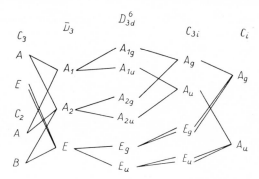

FIG. 21. *Relationships of the vibrational species of the site groups of the space group* $\boldsymbol{D}_{3d}^6$ [*from Hornig (330)*].

$\boldsymbol{C}_3$ contributes to two species of the group $\boldsymbol{D}_3$. Again, every species of $\boldsymbol{D}_3$ contributes to two species of $\boldsymbol{D}_{3d}^6$, etc. Obviously, alterations in the activity can occur, especially for species which belong to a group which is forbidden in the whole vibrational spectrum (Raman and infrared spectrum). As an example, the symmetry classes for gaseous and crystalline naphthalene are given in Fig. 22 according to Pimentel and McClellan (*536*). In addition to these symmetry influences, a splitting of the frequencies in the spectra of crystals is often observed, which depends especially on the number of molecules in the unit cell and also on the symmetry. Finally, the spectra are also influenced by polymorphism. However, the relationships are too complicated to be discussed in detail here. In addition to the references already given, the following are recommended for further study: Carpenter and Halford (*112*), Halford and Schaefer (*289*), Pimentel *et al.* (*537*), Richards and Thompson (*589*), Winston and Halford (*748*), Hexter and Dows (*777*).

† Usually called a 'lattice complex'.

The *lattice vibrations* of the crystals mentioned above usually lie at such small wave numbers that they can be measured either not at all or only very imperfectly with the normal infrared spectroscopic technique. The proper experimental technique for these (in addition to Raman investigations) is the measurement of the long-wave infrared spectrum, for which this application is one of the most important uses. The lattice vibrations can also be measured by means of their interactions with the molecular vibrations which cause absorption in the middle infrared in the form of combination bands. Thus, values for the lattice vibrations can be obtained from the vibrational analysis, but these values are usually rather liable to error, as a result of difficulties inherent in this method of investigation.

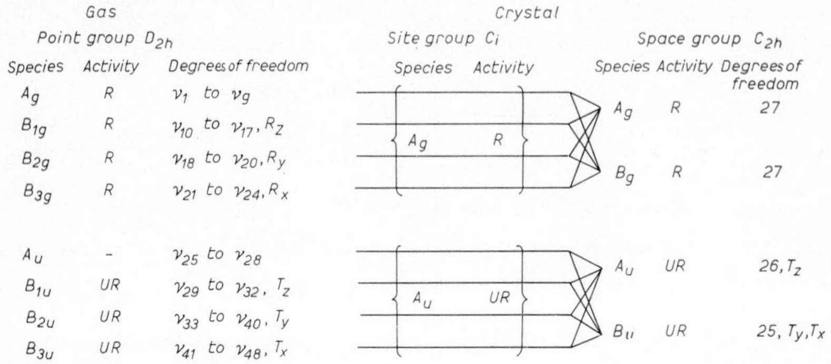

	Gas			Crystal			
Point group D_{2h}				Site group C_i		Space group C_{2h}	
Species	Activity	Degrees of freedom		Species	Activity	Species	Activity Degrees of freedom
A_g	R	ν_1 to ν_9					
B_{1g}	R	ν_{10} to ν_{17}, R_z		A_g	R	A_g	R 27
B_{2g}	R	ν_{18} to ν_{20}, R_y					
B_{3g}	R	ν_{21} to ν_{24}, R_x				B_g	R 27
A_u	–	ν_{25} to ν_{28}					
B_{1u}	UR	ν_{29} to ν_{32}, T_z		A_u	UR	A_u	UR $26, T_z$
B_{2u}	UR	ν_{33} to ν_{40}, T_y					
B_{3u}	UR	ν_{41} to ν_{48}, T_x				B_u	UR $25, T_y, T_x$

FIG. 22. *The vibrational species for gaseous and crystalline naphthalene [from Pimentel and McClellan (536)].*

An interesting complication of the vibrational spectrum of some crystals has been clarified by Haas (*282*). This complication is that, for example, for cubic crystals, in spite of the presence of only a single type of harmonic oscillator, two lines occur in the Raman spectrum. These correspond to one broad, ill-defined band in the infrared spectrum, the limits of which agree with the Raman frequencies. The effect is found only for strongly infrared active vibrations, and can be interpreted on the basis of the classical dispersion theory as an interaction of the vibrating dipoles. Certain effects in the spectra of uniaxial crystals can be explained similarly.

In addition to the alterations caused by symmetry influences in the spectrum of a crystal compared to that of the gas, the fixed *orientation in space*, which is inseparable from the concept of a crystal, must be considered. This effect can result in vibrations, which are highly active in the free molecule, not appearing when the crystal is investigated with polarized radiation. This happens when the orientation in space of the vibrating dipoles is such that no component of the moment lies in the direction of the incident radiation. Accordingly, the spectra of crystals can show dichroism, i.e. a dependence of the absorption on the direction of vibration of the incident radiation. (More exactly, this effect can be seen in the spectra of single crystals and ordered crystal arrays, but obviously not when the crystalline particles are statistically oriented in space.) These effects will be discussed later. Crystals in which the intermolecular forces are much smaller than the intramolecular forces can be treated, to a first approximation, as an 'orientated gas'. This is because the spectra of the gas and of the crystal are rather similar except for effects of orientation. Indeed, this analogy has been very useful for the interpretation of some crystal spectra. Such an application is found, for example, for naphthalene [Pimentel *et al.* (*537*)].

The vibrational spectra of gases and crystals can be interpreted on principles which are perhaps complicated in practice but which are fundamentally simple. However, the prospect of a similar interpretation for liquids is rather hopeless. The liquid state is completely unordered and strong interactional forces occur in it between the individual molecules. Therefore, the selection rules, which would be given by the symmetry either of the free molecule, or of the crystal lattice, are more or less cancelled. Accordingly, liquids should give the most complicated spectra with regard to the number of bands, and this is found in practice.

PART II

Equipment and Experimental Technique of Infrared Spectroscopy

Equipment and Experimental Technique

Equipping a laboratory for infrared spectroscopic investigations was, until quite recently, dependent on the initiative and constructional ability of the experimentalist himself. Until about the middle of the third decade of this century, infrared spectroscopy was merely a part of optics. As such, it interested only physicists and a few investigators of molecular structure, whose principal aim was the inclusion of this spectral region in the general physical concept of the universe and the determination of physical constants. The scientific instrument industry then understandably offered only a very limited range of suitable equipment. Many of the investigations in the infrared region approached the limit of the technically possible; the experimental physicist was engaged in a continuous fight for the sensitivity, degree of precision and reproduceability of his apparatus. This often led to situations which today appear quite grotesque, such as siting the apparatus in thermally controlled, shock-free, blacked-out cellars and working only at night, etc. All this was suddenly changed when developments of infrared spectroscopy made it an indispensable tool, not only of the specialist, but also of all practical chemists, and particularly when the chemical industry began increasingly to use this technique for the solution of its various problems. This development began in the middle of the second decade in the laboratories of the Badische Anilin- und Soda-Fabrik A.G., Ludwigshafen, and reached a climax there in about 1940. It was taken up by the American industry on a tremendous scale during and after the Second World War and from there has spread to all centres of civilization. It is a sign of this still continuing development that today the industry offers a wide range of instruments, some with quite extraordinary properties. These have incorporated many advances connected with infrared spectroscopic measurements. The degree of precision, sensitivity, and reliability of these instruments is often such that they can be used without a detailed knowledge of their construction and method of working, just as for an ordinary electrical instrument. Accordingly, the infrared spectroscopist, who uses his measurements for solving definite chemical problems and not for their own sake, need not necessarily understand every detail of his apparatus. This is primarily the responsibility of the physicists and engineers who are responsible for further fundamental developments and to whom recourse must be made should the apparatus develop a serious fault. The following chapters are therefore limited to a summary of the present state of development of instrumentation, sufficient for the experimentalist's needs. Reference must be made to the textbooks and monographs of physics for many of the more fundamental points.

1. Radiation Sources

1. Introductory

The infrared spectrum of a substance gives information about some of the energy states of its molecules, as has been seen. To differentiate between these energy levels, which frequently do not differ by much, and to characterize them quantitatively, a delicate method of investigation is required. Such is found in electromagnetic radiation of appropriate frequency, i.e. which agrees with the frequency of the vibrations and rotations of the molecule under investigation. As a given molecule always has many energy levels, the positions of which are initially unknown, the radiation used for these measurements must include all the possible frequencies in the desired region. In other words, a radiator is needed with a continuous emission spectrum. Such a spectrum is emitted by solids and liquids according to definite laws; gases are characterized by a selective emission, i.e. one limited to few, usually narrow, wavelength regions. Liquids are eliminated as radiation sources for practical reasons.

The radiation emitted by solid bodies depends on their temperature, type of surface, and chemical composition. The higher the temperature of a body, the more intense the radiation emitted at every wavelength. Therefore, solid glowing bodies, usually electrically heated, are used as sources of radiation for infrared spectroscopy. The larger the radiant energy of a selected wavelength, the easier and more definite is its measurement. Hence the radiation source is usually used at the highest possible temperature. However, certain conditions limit this temperature; as it increases, the useful life of the source falls sharply, so that in practice a compromise must be made. The differences in the chemical composition and the surface type of actual heaters are so great that their emission of radiation can only be defined with reference to a definite standard, even at known temperatures. Such a standard is defined as a *black body*. This is a body that, at given temperature and wavelength, emits maximum energy of radiation; it cannot be surpassed in this property. For purposes of standardization, black bodies can be prepared to any desired degree of approximation, but they are not suitable for routine measurements. The emission of radiation of a black body is described by well-known laws, the most important of which is the Planck *law of radiation.* This states that the radiating efficiency of a black body, for polarized radiation, at the absolute temperature $T°$ A in a wavelength range $\Delta\lambda$, with the mean wavelength λ, is given, per unit surface and per unit solid angle, by[†]

$$J_\lambda = E_\lambda . \Delta\lambda = \frac{c_1}{\lambda^5} (e^{c_2/(\lambda . T)} - 1)^{-1} . \Delta\lambda \qquad [\text{II, 1.1}]$$

where c_1 and c_2 are constants not dependent on T and λ. The graphical representation of this law is called the isotherm or isocromate of black-body radiation according to whether T or λ is taken as constant. It should be so well known that it need not be given here, especially as similar curves appear presently. From this law all the properties of black-body radiation may be derived: the total radiation (Stefan-Boltzmann) law, the law that the emission maximum is displaced with increasing temperature to shorter wavelengths (Wien displacement law), the distribution of the emission for individual wavelength regions, etc. Although the black body is not suitable as a radiation source for infrared spectroscopy in practice, an effort is made to approximate to it, as far as possible, with the sources used. The selection of possible sources is not very large: in addition to a few special types, it is practically limited to the Nernst *glower* and the *globar*.

2. The Nernst Glower

The Nernst glower[‡] [Munday (*490*)] is a solid or hollow rod of zirconium oxide with additions of yttrium oxide and the oxides of other rare earths and sometimes small quantities of other substances. It is a few centimetres long and about 1–3 mm thick. In the cold it is a non-conductor of electricity but on heating to about 800° C it becomes conducting and can then be heated by the electrical energy released; its resistance falls as the temperature further increases. To supply the electrical current, platinum wires are embedded in the ends of the rod. The initial heating is carried out by means of a small gas flame or better electrically; this is cleaner and more protective to the Nernst glower. Electrical heating is by means of one, or better two, platinum filaments on ceramic supports which are situated near the Nernst glower and heat it by radiation and convection to the point of sufficient conductivity. Frequent extinction, by turning off the current, and re-ignition should be avoided in order to lengthen the life of the glower. The electrical data for the use of the glower depend on its dimensions and on the temperature desired or necessary. The power consumption of glowers of the usual size is between

[†] Unpolarized radiation requires an additional factor 2; radiation taken with respect to the hemisphere requires an additional factor π.

[‡] Obtainable from: British Thompson Houston Export Co. Ltd, Rugby (Warwickshire), England; all suppliers of infrared spectrometers equipped with a Nernst glower.

50 and 100 watts. The emission at sufficiently high temperatures approaches fairly closely (50–60 per cent) to that of a black body of the same temperature. At lower temperatures, moderate selectivity is observed for the short wavelength part of the spectrum. However, this has little or no influence on the usefulness of the Nernst glower (*C*). Fig. 23 shows the emission curve of a Nernst glower at about $1900°$ A in the region of $1–12\,\mu$; the deviations from the smooth course of the Planck curve, noticeable principally near $2\cdot7$, $4\cdot25$, and $6–7\,\mu$, result not from any peculiarity of the glower itself but from the absorption of water vapour and carbon dioxide of the atmosphere. Considering the scale of the ordinate, this curve shows very clearly the large

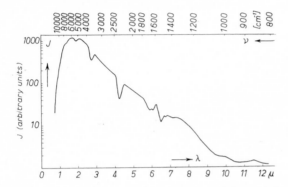

FIG. 23. *Distribution of the spectral energy of the emission of a Nernst glower at approximately $1900°$ A resolved by means of a NaCl prism.*

differences in the emission for different wavelengths. Between the radiation at the maximum and that at 12μ or greater wavelengths, there is a difference of several powers of ten, which is very important in many ways for the construction and use of infrared spectrometers.

Although the Nernst glower is probably the source of radiation most used for the infrared region, it has several serious defects. One of the most important is the disadvantage from the mechanical point of view; the glower is damaged by relatively small mechanical stresses, especially by shearing forces. Therefore its support must allow it free motion and yet guarantee its position in the optical system of the spectrometer. Under the influence of still smaller deforming forces the glower bends easily, which lessens the efficiency of the optical adjustment and so constitutes a serious limitation especially for the modern double-beam instruments with direct registration of the percentage absorption. The emission can vary at different parts of the surface, and can even alter during use (so-called flickering). Further the fact that it is impossible to run the glower at temperatures higher than about 2000° A for any length of time is sometimes an impediment. Even at relatively low temperatures, the life of the glower is measured in weeks or months. There have been many attempts to find a substitute for the Nernst glower but no serious competitor has yet been found except the globar.

3. The Globar

The name globar† denotes a solid or hollow cylindrical rod of silicon carbide, of which a part, usually a few centimetres long and 6 to 8 mm broad, glows. The ends of the rod remain cold either by increase in the cross-section or by an alteration in the conductivity, i.e. in any case by a reduction in the resistance. An advantage compared with the Nernst glower is that SiC conducts in the cold, so that the globar can be heated up simply by switching on the current. However, in contrast to the Nernst glower, which with the introduction of a suitable resistance can be run directly from the mains, the globar needs a transformer or a heavy duty source of direct current at a low voltage, because the globar works at a low voltage and with a high current. The power consumption of the globar is several times that of the Nernst glower; in order to avoid damage to the surroundings by the strong heat radiation, it should only be run in a water-cooled container. The constructional material limits the useful temperature region to values below 1400° C. The variation of resistance with temperature is uneven: at first it falls and then it increases again. After long use, resistance alterations occur as a result of chemical alterations in the material, which must be compensated by suitable alteration of the applied voltage. The emission of the globar corresponds almost exactly to 75 per cent of that of a black body at the same temperature [Brügel (98); Silverman (658)]. The life of the globar is longer than that of the Nernst glower. The globar is also sensitive toward mechanical strains, but on the whole the mechanical stability is better than that of the Nernst glower. Difficulties sometimes arise in leading the current to the globar, variable resistances can form between the electrodes and the material of the rod causing alterations in the emission. However, in general the globar is less sensitive than the Nernst glower towards voltage variations, because of the globar's greater mass. On the whole, these two radiation sources are equally valuable.

4. Special Radiation Sources

The two sources just mentioned are those most frequently used for infrared radiation, but, on occasion, others have been recommended or used for

† Obtainable from: Carborundum Co., Niagara Falls, N.Y., U.S.A., and suppliers of infrared instruments equipped with the globar.

special purposes. Their importance for normal infrared spectroscopy is very small, so that a short summary and references to the literature will suffice for this account.

The increase of the radiant energy with increasing temperature is of practical importance only in the region of the maximum emission and on the side of the Planck curve towards shorter wavelengths. The radiation law [II, 1.1] shows that, in the long-wave region of the spectrum (and for this purpose a wavelength of 10 to 12 μ is considered as long wave), increase of temperature of the radiation source results in a rather small increase in the emission. However, it is just in this long-wave region, of low emission, that the danger of errors from stray light from the energy-rich short-wave region near the maximum emission is greatest. This danger is the greater the more intense the emission is in the short-wave region. Therefore, for investigations in the far infrared above about 50 μ, the Auer burner is used as a source. It consists of a glowing mantle of thorium oxide heated by gas to about 1800°A. It is characterized by a very small emission below about 10 μ, while above this an emission very much like that of the black-body radiation is found (C). In the very long-wave infrared, above 200 to far above 1000 μ, only the suitably filtered radiation of the quartz mercury vapour lamp is useful [McCubbin and Sinton (*455*), Genzel and Eckhardt (*251*)].

For the other end of the infrared spectral region, the short-wave part under about 3 μ, a specially constructed Tungsten lamp with a V-shaped hollow element has been recommended as a source of radiation [Taylor *et al.* (*699*)]. For this region, it is also possible to use normal Tungsten lamps; rather than those with a ribbon filament, it is better to use those with a fine coil, because the radiation is then more like that of a black body.

The emission of the carbon arc has been recommended for the whole infrared spectral region, especially for micro-spectroscopy where exceptionally high radiation intensities are desirable. When this radiation source is used, double resolution of the spectrum is necessary to obtain really useful results, because of the large amount of stray light [Rupert and Strong (*618*); Rupert (*617*)]. Recently, reference has been made to the zirconium point lamp as a very active source of infrared radiation [Hall and Nester (*291*)]. For this application, it is fitted with a sealed-on window, transparent to infrared radiation, which together with the cementing material is cooled by water flowing around the emission chamber (Fig. 24). The emission chamber is filled with argon. The useful life has been given as several thousand hours. Because of the small surface emitting and the high intensity of the radiation, this lamp is also specially suited to micro-spectroscopy [Cloud (*121*)].

A radiation source which is easily constructed has been described by

Equipment and Experimental Technique

Herscher (*321*) and by Genzel and Neuroth (*249*). It consists of a small ceramic rod on which is wound a heating element of platinum, or a platinum alloy of high melting point. This element is embedded in a cement of aluminium oxide, thorium oxide, silicon carbide, etc., and then sintered for a short time. The cement covering is heated to glowing by means of the ele-

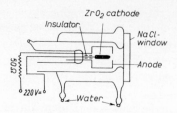

FIG. 24. *Diagram of the zirconium point lamp according to Hall and Nester (291).*

ment. The advantage of this radiation source is its cheapness and its ease of preparation.

Recently, there have been several attempts to substitute electrically-heated metal wires or rods for the normal infrared radiation sources (the Nernst glower or the globar). Nickel-chromium alloys, and also metals of the platinum group (which can be heated to much higher temperatures), have been used; by suitable blackening of the surface, a far higher emission is obtained than for the bare metal.

108

2. Monochromators

In most types of spectrophotometer, the radiation of all wavelengths, emitted by the source, passes through the substance under examination. The alterations which occur thereby are then investigated, quantitatively and qualitatively, as to their dependence on the wavelength. For this, the radiation must be spectrally resolved, i.e. separated into the wavelengths or wave numbers it contains. This is the task of the monochromator, which must be able to select one definite wavelength from a beam of radiation of all possible ones. Fundamentally, every monochromator consists of two parallel slits (the width of which can be altered as desired), an optical system, and a method of resolving the radiation (prisms or diffraction gratings can be used for this).

1. Slits

There is not very much to be said about the monochromator slits. Today, only bilaterally symmetrical slits are used, i.e. for an alteration of the slit width, both sides of the slit move in such a way that the midpoint of the slit-opening remains fixed and coincident with the line on which the sides of the slit touch each other when completely shut. In infrared spectrometers, the slit widths are between 10 and about $1000\,\mu$, according to their construction and purpose. Obviously, a very high mechanical precision, both in respect of their being parallel to each other, and of their motion, must be attained in order to get reproduceable adjustment especially at small slit widths. The jaws of the slit are sharply ground down on one side, at an angle to their line of contact. In order to avoid interfering reflections from these sloping surfaces, the ground side of the slits never faces the oncoming radiation. Thus, this side of the slit faces inwards for the entry slit of the monochromator and outwards for the exit slit. The optical system of the monochromator is generally designed so that the two slits are optically conjugated to each other, i.e. the one is the image of the other in the geometrical optical sense. When prisms are used to resolve the radiation, the image of the entry slit thrown on the exit slit is not straight but curved as a result of the peculiarities of image formation through a prism. The curvature is convex towards the refracting edge of the prism and for a single passage through the prism its radius is given by [Gates (*242*), Bolz (*75*)]:

$$R = \frac{n}{n^2-1} \; \frac{\sqrt{(1-n^2 \cdot \sin^2 \phi/2)}}{\sin \phi/2} \cdot \frac{f}{2} \qquad [\text{II}, 2.1]$$

where *f* is the focal length of the optical system, *n* is the index of refraction, and ϕ is the refractive angle of the prism. In addition to the prism, the rest of the optical system of the monochromator bends the image of the slit; according to circumstances, this can increase or decrease the curvature due to the prism. This curvature interferes with exact measurements of wavelengths. To eliminate it, high-grade prism monochromators have the entry slit bent in the opposite way and to the same amount as the exit slit, or the exit slit adjusted for the curved image. Because the radius of curvature is dependent (via *n*) on the wavelength of the radiation [II, 2.1], any compensation for this curvature only applies exactly for one particular wavelength and increasingly deviates for the others the further they are from this wavelength. A wavelength is selected which is the mean of the region in which the monochromator will be used; or one for which the curvature is especially critical (these are the wavelength regions with small slit widths). For the remaining wavelengths, the deviation from the exact value must be accepted as an error of the second order.

The slit width determines the amount of radiant energy which enters a monochromator and which leaves again in the resolved form (at constant wavelength and for a given source of radiation). The larger the area of the slit-opening, the larger this energy is, and the easier it is to measure exactly. For practical reasons, the slit height cannot be more than 20 to 30 mm, and an increase in the area of the slit must derive by increase in the slit width. On other grounds, to be mentioned later, the smallest possible slit width is desired. Instead of the real width in mm, it is the corresponding spectral width that is important, and this depends also on other properties of the optical system and on the method of resolution. A compromise has to be made between these two non-compatible requirements.

It can be seen from Fig. 23 that the emission of the normal sources of radiation varies strongly with the wavelength. If the slit width were held constant, the radiant energy issuing from the monochromator would also vary correspondingly with the wavelength. However, it is desirable on technical grounds (principally connected with the amplification of the signal) that the intensity of this issuing radiation should be at least approximately constant over the whole spectral region to be measured. Therefore it is necessary to reduce the slit width when the radiation emitted increases, and to increase it when the radiation decreases. This automatic adjustment of the slit width, one of the marked characteristics of modern infrared spectrophotometers, must be exactly coupled with the means (rotation of prism, grating, or mirror) for concentrating radiation of the selected wavelength on the exit slit. The adjustment must also be adapted to the emission curve of the source and to the dispersion curve of the resolving material. The coupling can be a purely mechanical one by means of suitably shaped cams or, more elegantly, by an electrical method. Details will be given with the discussion of the commercially available infrared spectrometers.

2. Optical System

The object of the optical system of a monochromator is threefold: (1) The introduction to the entry slit of the greatest possible amount of the radiation emitted from the source. (2) Geometrical and optical modification of the radiation as required and leading it to the resolving medium and then to the exit slit. (3) Leading the radiation passing out of the exit slit to the detector. On grounds that will become clear later, only those parts of the optical system which perform the tasks listed under 2 and 3 will be discussed here.

Instruments for infrared spectroscopy are characterized by the almost exclusive use of mirrors instead of lenses. In principle, lenses could also be used since the technique of growing crystals has made available suitable materials in sufficient size and quantity; a whole series of materials of sufficient transparency to be used over a very broad spectral region is now known. The contrary indications are: (1) The so-called chromatic aberrations which vary with the wavelength of the radiation and which render impossible the formation of a sharp image over a broad spectral region. (2) The higher energy loss by frequent reflections on the surface of lenses (loss per lens is about 10 per cent compared with about 5 per cent or less per mirror). (3) Mechanical instability of the materials which are often quite soft and which can often be easily deformed by stresses in the cold. (4) The highly hygroscopic nature of just those materials which are transparent to infrared radiation. In contrast, superficially covered hollow mirrors are free from chromatic errors, and are non-hygroscopic. Further, by a proper choice of the materials of the support and the mirror surface, they can easily be protected from mechanical and chemical damage. In general, mirrors of gold or aluminium sublimed on to glass or on to fused quartz in a high vacuum are used. These have a very high degree of reflection (Al up to about 96 per cent, Au up to about 99 per cent) in the infrared, together with sufficient chemical resistance and good mechanical stability. Generally, the form of the mirror is spherical, but for special purposes parabolic, elliptic, or even toroidal forms are used, together with planar mirrors for altering the direction of beams of radiation.

The mirrors inside the monochromator are there to convert the divergent beam of radiation from the entry slit, which can be considered as a secondary radiation source, into a telecentric or, as it is often called, parallel beam of radiation. They must further direct this radiation on to the resolving medium and concentrate each separate wavelength of the radiation, after it has been resolved in a convergent beam of rays, on to the exit slit. On optical grounds, it is necessary that the beam of radiation incident on and passing through the resolving medium must be parallel. The arrangement of the mirrors is different according to whether the radiation after passing through the resolving medium is reflected back (auto-collimation) or not. This will be discussed later, in connection with the insertion of the resolving medium in the path of the radiation.

The radiation passing out of the exit slit of the monochromator must be led to a suitable radiation detector, in order to be measured. In the infrared region, a very high energy concentration is necessary on the small surface of the detector. For this, in general, the exit slit, e.g. $0 \cdot 2 \times 20$ mm in size, is directly reflected on the detector very much reduced in size, in the ratio of about $5:1$ to $10:1$. The use of spherical mirrors necessitates several mirrors for this purpose. It is better to use one elliptic mirror of suitable eccentricity, focus, and aperture; the centre of the exit slit is at one focus and the radiation detector lies at the other.

3. Prisms

The means of resolution most often used in infrared spectroscopy is a prism of a suitable material, where the phenomenon of dispersion (or the wavelength dependence of the refractive index) is used. Sufficient dispersion must be accompanied by sufficient transparency for the spectral region investigated. Therefore, the size of the prism, i.e. the thickness of the layer through which the rays pass, is important. With respect to these two most important properties of the material composing a prism, the following rule applies: the further into the infrared the material stays transparent, the smaller is its dispersion. Also to be taken into account for the selection of the material for a prism are its hygroscopic nature or otherwise, its optical-mechanical properties, and the dimensio s in which it is available.

All the materials that are suitable for use as prisms for infrared spectrometers are listed in Table 7.† The table gives the spectral region for which they can be used (in general for a 60° prism with a 75 mm base) and their

† Obtainable from Harshaw Chemical Corp., Cleveland, Ohio, U.S.A., and suppliers of spectrometers.

TABLE 7. *Substances used for Prisms for the Infrared Spectral Regions*

(The less important are given in brackets)

Material	Limiting wavelength μ	Hygroscopic nature†		Notes
Glass	2·0	0		
Quartzglass	3·5	0		absorption at 2·75 μ
Quartz	3·8	0		absorption at 2·9 μ, doubly refracting, optically active
LiF	6	0·26	(18)	usually absorption at 2·8 μ
CaF$_2$	9	0·0015	(18)	
(NaF)	10	4	(15)	
NaCl	15·5	36	(20)	
(KCl)	20	34	(20)	
KBr	25	53	(20)	
(KI)	29	127	(20)	
CsBr‡	40	124	(25)	
CsI‡	50	44	(0)	
KRS 5‡	40	0		44% TlBr + 56% TlI, very soft, much reflection, poisonous

† Solubility in 100 g water in grams (rounded off) at the temperature in °C given in brackets.

‡ Literature: CsBr: Plyler and Acquista (*548*); CsI: Acquista and Plyler (*3*), Rodney (*605*), Mills *et al.* (*480*); KRS 5; Rodney and Malitson (*786*).

solubility in grams per 100 g water at room temperature, as a measure of their hygroscopic nature. With respect to the dispersion, several different numerical values can be used to measure this: the real dispersion, the angular dispersion (i.e. the angle between two rays of different wavelengths after passing through the prism), or the theoretical dispersion, i.e. that for vanishingly small slit widths, which only depends on the material and size of the prism. For the last two characteristics the refracting angle or the base thickness of the prism is also needed. In Fig. 25, the interval between two wavelengths, and wave numbers, which are just resolved, is given graphically for a base of 75 mm. The smaller the value of $\Delta\lambda$ or $\Delta\nu$ for a given material and given wavelength or wave number, the more favourable is the dispersion or degree of resolution. Together with Table 7, this figure demonstrates the rule previously mentioned connecting the region of transparency and the dispersion.

It can be concluded from this discussion that with respect to the dispersion and the resolution it is not favourable to measure, e.g. the whole region from

Equipment and Experimental Technique

1 to 40 μ with a CsBr prism. To obtain the maximum resolution at every wavelength would require the successive use of almost all the materials enumerated in Table 7, one after another; this would cause too much inconvenience and expense. However, modern infrared spectrometers are all fitted for interchange of prisms in one form or another. Usually this is limited to three prisms: LiF or CaF$_2$ for the short wave, NaCl for the middle wave, and KBr or, recently, CsBr and CsI for the long wave part of that region which is measurable by means of prisms.†

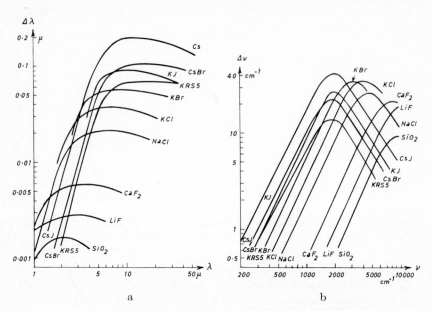

FIG. 25. *The smallest distance between two bands in* (a) *wavelengths and* (b) *wave numbers which is just resolvable by a prism of 75 mm thickness according to the theoretical (Rayleigh) resolving power.*

With respect to the size and shape of the prisms, the angle of refraction and the base thickness are important; these determine the length of the side edge, while the height is chosen accordingly. The angular dispersion is essentially determined by the angle of refraction, but this cannot be made too large because of *inter alia* total reflection in the prism. Usual angles are 60° for materials of medium refractive index (around 1·5), about 72° for

† A monochromator, fitted with three prisms (LiF, NaCl, KBr) any one of which could be selected automatically, was probably described first by Nielsen *et al.* (*500*). Today such instruments are available commercially.

very small refractive indices (around 1·3), and about 35° and smaller for very high refractive indices (over 2). As already mentioned, the base thickness determines the theoretical degree of resolution and is therefore made as large as possible. However, the size of the crystals available free from flaws sets a limit to the possible size. Today, for example, there are NaCl prisms with bases up to 180 mm, but values of 75 to 100 mm are more usual.

The dispersion of the prism depends in a very complicated manner on the wavelength or the wave number. Accordingly, the individual wavelengths do not follow each other on the exit slit after resolution in a linear manner; this circumstance is unfortunate for reading and understanding the spectra. In modern spectrometers, measures are taken to obtain a linear wavelength or wave number scale (or a desired modification, e.g. logarithmic in λ) despite the non-linear dispersion. The correction that has to be made depends, of course, on the material and the angle of refraction of the prism, and for a change in prism the correction must also be modified. This question has been discussed by Brackett (*795*).

4. Diffraction Gratings

Gratings have always had a certain importance as resolving media for infrared radiation because of their great power of resolution and because they are not limited to definite wavelength regions. However, they were formerly used more for fundamental investigations of geometrical or dynamical structure, using the rotational fine structure of the vibration bands, and for the investigation of that long wavelength spectral region for which no suitable prism materials are available. Recent developments have now made them notable competitors of prisms for the regions of the near and middle infrared mainly used in the modern scientific and industrial spectroscopic laboratories. It is significant that, recently, commercial infrared spectrometers have been advertised with gratings as the method of resolution, which have exactly the same application as the prism instruments that up to now have been more usual.

Two big improvements are the principal reason for the appearance of the grating instruments. The advantage of larger dispersion was for long accompanied by the serious disadvantage of very small light intensity. This was caused essentially by the usual gratings (e.g. regular planar parallel arrays of fine lines at constant distance) fanning out the incident radiation into spectra of different orders, the intensity of which decreased with the order in a definite manner. This disadvantage has now been overcome by giving the ridges of a grating a definite form, so that the energy is concentrated in one spectrum, or at least in only a few. Such gratings are known as *echelette* gratings and today they are almost the only type used. However, it was only when an easier and cheaper method of manufacture was found for this type of grating that the way was open for their general use. This is the so-called Merton process [Merton (*471*), Hall and Sayce (*290*), Dew and Sayce (*172*), Dew (*171*)]. In this process, the desired grating is cut on a metal cylinder in the form of a screw thread, which is then coated with a solution of a plastic. When this dries, the plastic coating forms a mould of the screw thread which, after detachment from the screw and coating with metal, serves directly as the grating. It is obvious that in this way an almost unlimited number of true copies of the original grating can be made easily and cheaply. It is advantageous for the process that for the infrared region relatively

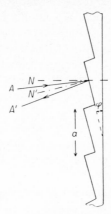

FIG. 26. *Geometrical constants for an echelette grating:* α = *grating constant,* ϕ = *groove angle,* N = *the normal to the plane of the grating,* N' = *the normal to the plane of the groove base,* A = *direction of the incident ray,* A' = *direction of the reflected ray.*

little mechanical exactitude is required; the usual grating constants are 0·01 mm or more, because of the large wavelength.

In Fig. 26, the essential constants of an echelette grating are indicated. They are the grating constant α and the groove angle ϕ. These quantities together determine the distribution of the radiation diffracted by the grating in the spectra of different orders; it is possible for a concentration of energy to occur in a certain order. Calculations on this point have been made by Stamm and Whalen (*681*) and Friedl and Hartenstein (*233*). In the literature describing grating instruments the expression 'blaze' or the phrase 'a grating is blazed for a certain wavelength' is found. This means that a grating of given groove angle is used so that the given wavelength is almost geometrically reflected and therefore the concentration of energy is found in a spectrum of definite diffraction order. Fig. 27 can be considered as an example of the efficiency in practice of an echelette grating of 2400 grooves per inch and a concentration of radiation of wavelength 7 μ in the first-order spectrum [from Cole (*131*)]. This has been obtained by measuring the radiation of the green mercury line at 5461 Å which is diffracted by the grating in the spectra of different orders. The lower abscissa gives the order for the green mercury line and the upper one gives the wavelength of the radiation which appears in the first-order spectrum as a result of the shape of the grating. The ordinate indicates the relative energy. The high concentration of the green mercury line in the spectra of order 12 to 14 can be seen, with the marked prominence of the thirteenth order. Accordingly, the wavelength region 5 to 15 μ is principally found in the first-order spectrum,

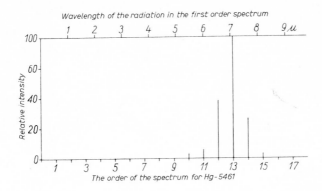

FIG. 27. *Efficiency of an echelette grating according to Cole* (*131*).

3 to 5 μ in the second-order spectrum, and 2·5 to 3 μ in the third order. Therefore, it is possible to obtain with this single grating, by successive changes of the order, the whole infrared spectral region covered by a NaCl prism, and with favourable energy relationship and markedly improved resolution. This result can be improved still further by a somewhat different choice of the grating constants.

However, the use of gratings does not only bring advantages. An important disadvantage is the overlap of the spectra of different order. If a grating monochromator is set for the wavelength λ in a given order, in addition to radiation of this wavelength, there passes through the exit slit also that of the wavelength λ/k where k is any integer. Therefore care must be taken that this radiation of smaller wavelengths is eliminated from the beginning, and that the radiation incident to the grating covers at the most a spectral region of an octave. This can be done by an appropriate preliminary resolution of the radiation by means of a prism or a grating, or by a suitable filter.

In spite of the considerable energy concentration that is possible by an echelette grating with a suitable form for the grooves, it is generally assumed that a grating is inferior to a prism with respect to the light intensity. Jacquinot (*350*) has pointed out that this is a consequence of the fact that the comparison is usually made under circumstances which are disadvantageous to the grating. According to him, when comparing a prism with a grating the following must be borne in mind: (1) The base surface of the prism should be the same as the active surface of the grating, and (2) the comparison should be carried out for the same degree of resolution. Then the grating is shown to be superior to the prism in the question of the light intensity by a factor of 2 to 20 [see also Perry (*526*), Jacquinot (*351*)].

Further details on grating spectrometers for the short-wave infrared region can be found in the work of Rank *et al.* (*575*) and Lord and McCubbin (*428*); for the medium infrared in the work of Cole (*131*), Martin (*448*) and Yates and Buhl (*762*); for the far infrared in the work of Oetjen *et al.* (*508*), McCubbin and Sinton (*455, 456*), Bohn *et al.* (*74*), Genzel and Eckhardt (*251*), and Plyler and Acquista (*549, 550*).

5. Interferometers

The highest resolution yet obtained in the infrared ($0 \cdot 043$ cm^{-1} at $1 \cdot 5\,\mu$ or 6700 cm^{-1}) has been obtained by means of an interferometer [Jaffé *et al.* (*354*)]. Here, as for diffraction gratings, the interference of coincident radiation is used. The order of the interfering rays is low for the gratings but for interferometers very high orders are used and the high degree of resolution follows from this, because it is directly proportional to the order. Interference spectroscopy of this sort is a new infrared technique and is therefore mainly of academic interest. Therefore in this book merely a short summary will be given. The main instrument is the well-known Fabry-Perot etalon, the sheets of which are covered with suitable substances (usually dielectric materials such as ZnS, MgF$_2$, Te, KBr and NaCl) in order to raise its reflective power. The interferometer must also be used in conjunction with preliminary resolution, which in this case must be especially efficient, because of the extremely high dispersion and the overlapping of various orders. Usually quite a refined instrument is used for this, e.g. a grating monochromator with its own preliminary resolution. According to Jacquinot (*350*), the interferometer is superior to the grating in the question of light intensity by a factor of 30 to 400 when the comparison is suitably made (active surfaces equal, and the same degree of resolution). Further details on infrared interferometers and interferometer spectroscopy can be found in the references already given, to which may be added Greenler (*276*) and Jacquinot and Chabbal (*352*).

6. Optical Arrangement

The slits, the optical system, and the means of resolution (prisms will be considered principally, the modifications necessary for gratings will be obvious) can be arranged in quite different ways to form the monochromator. Of the possible arrangements, only one is still in use for infrared spectroscopic purposes, because this has numerous definite practical advantages such as better use of the component parts, more compact construction, etc. A few points of mainly theoretical nature are sacrificed, e.g. maintaining the minimum prismatic deflection.

This is the so-called *Littrow arrangement*, the principle of which can be seen from Fig. 28. The radiation coming from the entry slit Sp_1 is made

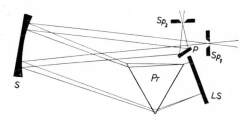

FIG. 28. *The Littrow optical arrangement for an infrared spectrometer.*

parallel by the collimator mirror S. It then passes through the prism Pr, and falls in the resolved state on a planar mirror LS (Littrow mirror), which is perpendicular to the direction of the ray. It is reflected in this mirror and passes through the prism a second time. Somewhat diverted from its initial path, it is then concentrated by the collimator mirror acting in co-ordination with a planar mirror P on to the exit slit Sp_2. The alteration of wavelength is carried out by revolving the Littrow mirror about an axis passing through its plane and perpendicular to the radiation. The prism remains motionless, once it has been adjusted. The advantages are obvious: prism in a fixed position and therefore no rotary motion of heavy parts, a single collimator mirror fulfilling two purposes, small dimensions as a result of the auto-collimation, double use of the prism and therefore doubling the angular dispersion and the resolving power.

In the Littrow arrangement just described, the radiation incident on the collimator mirror approaches it at an angle. Using spherical mirrors this causes fundamental errors, which show up as an unsymmetrical broadening

119

of the image of the entry slit thrown on the exit slit (*coma errors*). This phenomenon is disadvantageous for precise spectral measurements. There are various methods of avoiding or reducing such errors. Still using spherical mirrors, which are inexpensive because of the simplicity of their preparation, the arrangement shown in Fig. 29 allows the collimator mirror to be used so that the radiation is almost normal to it by using a small supplementary planar mirror *N*, the so-called Newton mirror. However, according to the size, a larger or smaller part of the ray is shut off by the Newton mirrors. Lehrer and Luft have proposed a system of two mirrors of suitable dimensions where, in spite of slanting incident radiation, the quality of the

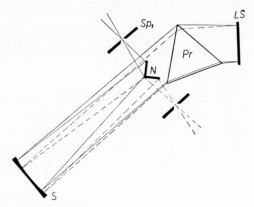

FIG. 29. *Axial use of a concave mirror by means of a so-called Newton mirror N.*

image formation is excellent as a result of the compensation of the error of the first mirror by that of the second which is of the same size but oppositely placed.† The American instruments use a slanting path for the radiation, but frequently incorporate parabolic concave mirrors; this retains the advantages of the original simple arrangement but is more expensive than the use of spherical mirrors.‡

The Littrow mirror, the rotation of which causes the alteration in wavelength, can be directly connected to the prism instead of being placed away from this. For this, it is necessary that the prism (which normally has a refractive angle of 60°) be bisected to halve the angle and that the Littrow mirror be mechanically connected to the plane of bisection. Obviously, the prism must now itself be rotated for selection of the wavelength. The angular

† DRP 737 161.
‡ So-called 'off axis' mirrors are almost always used, i.e. mirrors which are small parts cut out of a large parabolic mirror away from the axis.

dispersion obtainable with such an auto-collimated half prism is naturally only half that which can be obtained by the true Littrow arrangement.

Diffraction gratings are usually inserted into the optical system in the Littrow arrangement with autocollimation. This is possible because reflectional gratings are used. The grating thus replaces the combination of the prism and the Littrow mirror (except in double monochromator instruments); the wavelength at the exit slit is varied by revolving the grating. In addition, the arrangement of constant deviation of Fig. 30 is used, for which, in contrast to the autocollimation arrangement, two collimator mirrors are necessary.

For prisms, the dispersion is known to depend in a complicated manner on the wavelength. For an even rotation of the Littrow mirror or the prism (which is the equivalent), the succession of wavelengths on the exit slit is not linear with respect to the time, but is a complicated function of the

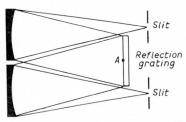

FIG. 30. *Grating arrangement with constant deflection.*

time. The interpretation of spectra is facilitated if the wavelength or wave number scale is linear. Accordingly, the Littrow mirror is not rotated evenly but according to a programme which depends on the dispersion properties and the angle of refraction of the prism. Such a programme is prepared once and for all (mechanical or electrical wavelength linearization). For gratings, the selected wavelength depends on the sine of the angle of rotation. For sufficiently small angles of rotation (up to about 5°) this itself gives a practically linear wavelength scale, because within these limits $\sin \alpha \approx \alpha$ (in radians). For larger angles of rotation, a programme for linearization is needed also for gratings, but one that can be carried out much more simply. A device for this is described, for example, by Genzel and Eckhardt (*251*).

7. The Characteristic Constants of Monochromators

The efficiency of a monochromator is determined principally by four quantities: (1) The useful *spectral region*; (2) the available *monochromatic efficiency* with respect to the spectral slit width; (3) the attainable *degree of resolution*; and (4) the *quality of the monochromatic nature* of the resolved radiation, i.e. the amount of radiation of other wavelengths when set for a given wavelength, the so-called stray light. These four quantities have to be considered when monochromators are compared with each other.

The spectral region that can be used is determined by the prism in

Equipment and Experimental Technique

question. The available monochromatic efficiency S_λ for a selected wavelength λ and calculated on the spectral slit width is given by

$$S_\lambda = \frac{J_\lambda.h.D.A.\alpha}{f} \qquad [\text{II, 2.2}]$$

It is thus dependent on the intensity of radiation J_λ falling on the entry slit, the height of the slit h, the angular dispersion D of the prism, the absolute aperture A of the monochromator, the efficiency α of the monochromator, and the focal length f of the collimator mirror. The quantity J_λ is determined by the type and quality of the source of radiation, taking into consideration the amount of radiation that is directed on to the entry slit. The angular dispersion D is determined by the composition and the refractive angle ϕ of the prism:

$$D = \frac{2\sin^2\phi/2}{\sqrt{(1-n^2\sin^2\phi/2)}}\cdot\frac{dn}{d\lambda} \qquad [\text{II, 2.3}]$$

where n is the refractive index for the wavelength in question. By the absolute aperture A, the size of the projection on the collimator mirror (or, what is equivalent, on a plane perpendicular to the collimator axis) of the irradiated surface of the prism is understood.† The efficiency α takes into account the fact that in the monochromator itself there are losses in the radiation energy by reflections at the mirrors and prism surfaces, absorption in the prism, scattering, etc. If the absorption in the prism is not taken into account (in general this is only important at the long-wave end of the usable region), then a value for α of $0\cdot75$ can be taken.

For this discussion, a more detailed consideration of the conception 'spectral slit-width' is necessary. As a result of the phenomenon of angular dispersion and of the finite width of the slits, the radiation emitted from the exit slit consists not of a single wavelength but always of a whole *interval* of wavelengths, with a characteristic distribution of the energy depending on the wavelength (the so-called *slit function*). For equal width of the entry and exit slits, which almost always holds in practice, this distribution approaches a triangle with the apex (the maximum energy) at the wavelength for which the monochromator is set. This can be visualized in the following way: using monochromatic radiation, as the prism or mirror is revolved the image of the entry slit passes over the exit slit. When there is no overlap at all, no radiation will pass out of the exit slit. The energy passing out begins to increase as soon as the side of the image crosses the side of the exit slit, it reaches a maximum for an exact coincidence, and then the energy again decreases and becomes zero again when the other side of the image

† The absolute aperture determines the illumination relationships in the monochromator. All the remaining factors (mirrors, etc.), which, by shielding, limit the illumination, should be at least as large as the absolute aperture.

crosses the far side of the exit slit. If the two slits are not of equal width, it is obvious that the energy distribution diagram will be a trapezium. The spectral slit-width $\Delta\lambda_s$ or $\Delta\nu_s$ is defined as half the base of the triangle (Fig. 31). For the Littrow arrangement described, it can be calculated [Barnes *et al.* (*41, 42*); Keussler (*381*)] as

$$\Delta\lambda_s = \nu^2 \frac{(1-n^2\sin^2\phi/2)^{1/2}}{8.\sin\phi/2\,\dfrac{dn}{d\lambda}} \cdot \frac{s_1+s_2}{f} + F(s)\frac{\lambda}{2b\dfrac{dn}{d\lambda}} \qquad [\text{II, 2.4}]$$

or

$$\Delta\nu_s = \frac{(1-n^2\sin^2\phi/2)^{1/2}}{8.\sin\phi/2\,\dfrac{dn}{d\lambda}} \cdot \frac{s_1+s_2}{f} + F(s)\frac{\nu}{2b\dfrac{dn}{d\lambda}} \qquad [\text{II, 2.5}]$$

These expressions are sometimes known as Williams's Formulae. In addition to the abbreviations already introduced, s_1 and s_2 are the width of the entry and exit slits, b is the base of the prism, and $F(s)$ is a function of value $0\cdot89$ for $s=0$ and about $0\cdot5$ for such a value of s, that the first term in [II, 2.4] or [II, 2.5] is equal to $\lambda/2b(dn/d\lambda)$ or $\nu/2b(dn/d\lambda)$.

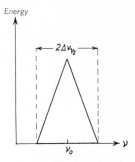

FIG. 31. *Slit function for slits of equal width and image formation of 1 : 1.*

The two terms originate from the two influences which determine the resolution: the first takes into account the influence of the *finite slit-width* on the degree of resolution via the angular dispersion. The other takes into account the *theoretical resolving power* of a prism for infinitely small slit-widths, which, according to Rayleigh, depends only on the dispersion of the material of composition and the base of the prism used. The degree of resolution determines whether two neighbouring bands in the spectrum will be separated or will appear as one single band. For two bands of the same intensity to be seen to be resolved, their absorption maxima must be separated by a definite wavelength interval given by [II, 2.4]. In addition the absorption between them must fall to at least 80 per cent of the maximum value. For bands which are not of equal intensity there must be a definite minimum between them.† The second term in the

† Frequently, if two overlapping bands are not fully resolved, this shows itself by a broadening or by the formation of a so-called shoulder on the side of one of them. Giese and French (*254*) have shown that in such cases it is more favourable to consider the derivative of the transmission or absorption curve, because the separation is more marked in the derivative curve. However, such a treatment must be carried out cautiously, and there are still no instruments which record the derivative instead of the transmission curve [see also Collier and Singleton (*772*), Pemsler (*833*)].

equation is the theoretical limit (Rayleigh) to the possible resolution; naturally, this is not attained in practice, because of the necessary finite slit width. In general, double or treble this quantity has to be reckoned with, or in special cases, where the circumstances are specially favourable, one and a half times. Fig. 25 gives the theoretical Rayleigh resolving power; together with the foregoing approximation an indication may thus be obtained of the actual resolution that is possible with a prism of given material and dimensions.

Every absorption band possesses a definite, so-called natural width, which is independent of the properties of the apparatus with which it is measured, and is determined only by the nature of the substance in question and certain environmental influences, such as temperature and pressure. Obviously, this natural band width can only be experimentally measured when the degree of resolution of the apparatus used is very much greater than this band width. In other cases, the apparatus determines the measured band width to a greater or lesser extent; in certain circumstances it can be almost the only factor which determines it. For example, White and Howard (*738*) found the width of various bands of liquids to be 3 to 5 cm^{-1} by means of a grating spectrometer with a resolution of less than 1 cm^{-1}; the use of prism instruments of lower resolving power had given markedly higher values. It is therefore important to know something about the spectral slit width available, i.e. the practical resolving power of a given monochromator. Such knowledge is most simply obtained by measuring a suitable test spectrum with known small intervals between the bands and observing at what intervals the bands are still just resolved. The test spectrum (e.g. the fine structure of ammonia in the region 8 to 12 μ) must naturally have been previously determined with a superior instrument. However, a suitable test substance is not always available, and there is not one for every wavelength. In such cases, the method of Coates and Hausdorff (*123*), which was originally proposed for the determination of cell thicknesses (see page 190), is helpful. An empty liquid cell of definite width placed in the path of the beam forms a Fabry-Perot etalon, provided the plates are sufficiently planar and parallel. Thus, it does not possess a transmittance which is constant with the wavelength, but one which is overlapped by interference bands according to definite laws (Fig. 32). The distance δ, in cm^{-1}, between successive interference maxima and minima depends on the refractive index of the medium between the plates, and in addition on the distance between the plates (i.e. the cell thickness). The amplitude of the interference bands depends on the slit function and on $\Delta\nu/\delta$, where $\Delta\nu$ is the spectral slit width. It can be shown that the amplitude of the interference bands becomes zero or is at a minimum when $\Delta\nu = \delta$, thus allowing a direct determination of

124

$\Delta \nu$ from the interval between two successive interference maxima. The experimental determination of the spectral slit width is made in the following way: the transmittance of a cell of known thickness is measured in the spectral region in question, together with the interference bands. If the spectrum is linear in wave numbers, the interval between successive interference maxima remains almost constant because the refractive index of the air in the cell varies but little. The amplitude of the interference bands is observed. If the condition mentioned above for the amplitude of the interference bands to become zero is fulfilled somewhere in the spectral region investigated, then at this point the spectral slit width is determined for the mechanical slit width selected. If the condition is not fulfilled, at least it is known that the spectral slit width is smaller than the interval between

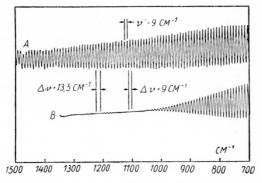

FIG. 32. *The determination of the resolving power of spectrometers according to Coates and Hausdorff (123). Explanation in text.*

the interference bands. Fig. 32 gives examples for both the cases mentioned: for the curve A (slit setting 1) $\Delta \nu$ is always greater than $9 \, cm^{-1}$. For curve B (slit setting 2) the spectral slit width is shown to be $\Delta \nu = 9 \, cm^{-1}$ at the wave number $1100 \, cm^{-1}$. If the measured spectrum is linear in wavelength, the calculation is not quite as simple, because the interval between the interference bands alters with the wavelength. In this case, the interference interval must be extrapolated or interpolated at the place at which the amplitude of the interference bands becomes zero.

The last quantity to be discussed is the *stray light*. Even with the greatest care and the best construction, the presence of a certain amount of short-wave radiation, of which an especially large amount is emitted, cannot be avoided in the regions of longer waves (which are more weakly emitted). This stray light, which interferes with the monochromator and therefore with the accuracy of the spectra, is caused by undesirable, but unavoidable, reflections, refractions, and scattering by dust particles in the air, optical inhomogeneities in the prism, flaws in the prism and mirror surfaces, etc. Naturally, care is taken to construct the parts of the monochromator in such a way as to minimize the stray light. Martin (445) has discussed the various possibilities for the Littrow arrangement. The positioning shown in Fig. 28 has been found to be the best compromise, the non-refracting edge of the prism faces the radiation entering from the entry slit. However, the short-wave stray light in a NaCl monochromator at 15 μ can still represent 20 per cent and more of the radiation energy. High precision instruments must therefore include special precautions to decrease the stray light to a reasonable proportion; these will be described in detail together with the individual instruments. Where such precautions are not taken, the proportion of stray light present can easily be quantitatively determined. For this, the radiation energy is measured once, in order to determine a reference point, with a metal screen which is non-transparent to all wavelengths. A second measurement is made with a screen which is transparent to short-wave radiation but not to the wavelength for which the instrument is set, e.g. a screen of LiF. The difference between these two measurements gives directly the amount of short-wave stray light. The action of stray light on the position of bands and the method of taking it into account in qualitative and quantitative interpretation of spectra will be discussed later (see page 283).

8. Double Monochromators

The resolution obtainable with a monochromator of the type just described does not suffice for some purposes. The simplest method of increasing the dispersion and resolution together with simultaneous decrease of the stray light, is the coupling of two or even more such simple monochromators. This method has been in use for a long time. The exit slit of the first monochromator is then the entry slit of the second and so on. If the dimensions are identical and the monochromators are correctly connected together, the angular dispersion is the sum of those of the individual monochromators. However, a limit is soon set to this increased dispersion due to energy factors, as the increase in resolution is gained at the expense of the available energy, so that eventually the difficulties of its measurement offset the advantages obtained in the resolution. In addition, the expense of such an arrangement is considerable.

Walsh (724) has recently made a proposal which is an important advance in this last-mentioned respect. The essential idea is that the radiation which in a usual Littrow arrangement would fall on the exit slit, i.e. which has already twice passed through the prism, is made to pass through the *same* monochromator again by means of a relatively simple and cheap modification. This arrangement has already been incorporated in commercial instruments. Fig. 33 shows diagrammatically this arrangement, which is known as '*double pass*'. The ray 1 coming from the entry slit passes through the monochromator in the usual way, strikes a mirror at an *angle* (ray 2), which sends it back into the monochromator deflected a little to one side (ray 3). After passing backwards and forwards through the prism a second time, it (ray 4) is now led to the exit slit. The deflection of the ray which occurs on the reflection at the mirror set at an angle is important in view of the addition of the dispersion in rays 2 and 4. It is clear that the radiation

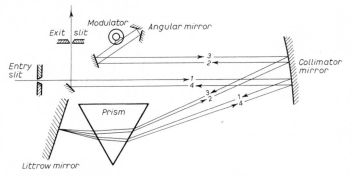

FIG. 33. *Double-pass arrangement for a Littrow monochromator according to Walsh (724). Explanation in text.*

passing through the exit slits consists not only of the desired wavelength which has passed through the prism a total of four times, but also of another wavelength which has passed only twice through the prism. In other words, spectra of different orders occur, where by an order is meant all that radiation which has passed through the prism the same number of times (which is necessarily divisible by two). As a means of separating the different orders, one of them, the highest and that with the largest dispersion, is characterized by the modulator in ray 3, and only this order is then measured. A somewhat different arrangement for a double-pass instrument has been proposed by Rochester and Martin (*604*).

The economy of this principle of repeated use of the same prism is obvious. Naturally the coupling of monochromators with the double monochromator is possible in so far that the available radiant energy in the spectra of higher order suffices for measurements. It can even be arranged that the order of the spectrum can be selected as desired in a multiple-pass monochromator. For this, the two mutually perpendicular parts of the angular mirror are placed symmetrically at the two sides of the entry slit (Fig. 34) [Ham *et al.* (*293*)]. As a result of the continuity of the spectrum, for a definite setting of the monochromator a wavelength of the first order can be selected. This ray strikes the angular mirror near the middle and then, after each new transition through the prism, strikes the mirror nearer the outer edge. Finally, as a ray of order determined by the arrangement, it passes through the exit slit. In order to differentiate it from radiation of other wavelengths, which pass through the exit slit as other orders, a modulator must be suitably placed (e.g. at point A), thus using the device mentioned above. It can be easily arranged that the position of this modulator is altered from outside, so that the spectrum of the desired order can be distinguished

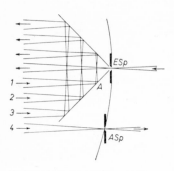

FIG. 34. *Multiple-pass arrangement for the Littrow monochromator according to Ham* et al. (*293*). *Explanation in text.*

from the overlapping radiation. However, in every such multiple-pass arrangement, special care must be taken that the curvature of the slit is correct (see page 110) so that the quality of the image does not deteriorate [see also Walsh and Willis (*726*)].

This double-pass arrangement is not directly applicable in the so-called double-beam spectrometers (see page 145). In these, the radiation which passes through the substance under investigation and some of equal wavelength which passes through a blank cell are alternately led through the monochromator; the rate of alternation is one suitable to the sensitivity of the detector. The modulating frequency in the multiple-pass arrangement has to be made very much larger (> *ca.* 50 c.p.s.) than the frequency of alternation (*ca.* 10 c.p.s.); it all depends on the sensitivity of the detector [Walsh (*725*)]. A double-beam double-pass spectrometer has been described by Cole (*130*) using the Rochester-Martin arrangement. Certain other types of modern spectrometers, in which the investigating and comparison beams pass simultaneously through the monochromator, allow the double-pass arrangement in principle. A limit is always set by the available radiation energy, i.e. by the ratio of signal to noise. In multiple-pass arrangements, this becomes less favourable because of unavoidable losses by reflection, etc. To improve this ratio, the slits are opened wider in double and multiple-pass monochromators than is usual for simple monochromators. In this way, part of the extra resolution is lost again. The limit of such measures to increase the resolution is reached when the improvement effected no longer bears any proportion to the effort needed. The possibility of using gratings remains. When freedom from stray light is especially desired, the best method is to use a double or multiple monochromator, possibly incorporating a grating. They are indispensable for high-temperature radiation sources (because of the high emission at short wavelengths) and for work at long wavelengths especially with CsBr and CsI prisms.

9. Adjustment and Wavelength Standardization

Commercial infrared spectrometers, which are still to be discussed, are in general assembled and adjusted by the makers at the place of manufacture. By adjustment is meant setting the mechanical, optical, and electrical components in such a way that their action is as near as possible to the theoretical expectation. For any further adjustment which may be needed from time to time, the makers supply printed instructions for their instruments: it is beyond the scope of this book to discuss such instructions, which are usually quite detailed. For self-made spectrometers, the final adjustment is the last stage of the construction. It should be based on a profound knowledge of the principles and laws governing the instrument; the detailed discussion also lies outside the scope of this book, and specialized monographs must be consulted for this. The adjustment of an instrument is also a question of experience; the more intensively one has been concerned with it, the greater is the chance that the best adjustment will be reached. An understanding of the mode of action and inter-relationship of all the components is the best set of instructions. If possible, the adjustment should be made using visible light and the very sensitive human eye, perhaps with the help of optical instruments. If this is not possible the detector must be set for a region of well-known bands, e.g. the water vapour bands at $6\,\mu$ or the ammonia bands at $10\,\mu$. By comparison of the results obtained with those from the literature, their accuracy can be estimated. Directions for the especially important adjustment of a Littrow monochromator have recently been given by Kendrick and Schnurmann (*379*).

Special consideration must be given to the wavelength or wave number standardization of the spectrometer. This takes the form of a table or a curve which gives the relationship between the wavelength (wave number) and the rotation of the prism, the Littrow mirror, or the grating. This relationship is either calculated from the known dispersion of the material of the prism and its dimensions, especially the refractive angle (or the corresponding data for the grating), or obtained empirically by measurements of definite bands of known spectral positions [McKinney and Friedel (*458*); Guy and Towler (*281*); Ross and Little (*606*); Martin (*446*); Torkington (*713*); Tilton and Plyler (*710*); Downie *et al.* (*176*)]. Even with the best

K   129 instruments, the standardization curve must be tested from time to time and the momentary deviation from the correct values determined. For this, a large number of bands of definite compounds which have been exactly measured is available, having been specially selected for this purpose with possible errors kept in mind. The rotational-vibrational bands of CO at $2143 \cdot 38$, $4260 \cdot 12$, and $6350 \cdot 39$ cm^{-1} have been found especially useful for this purpose because their positions can be calculated theoretically. However, these cover only a small region of the spectrum and the total number of bands is relatively small. Except for the absolute method, which depends on the direct measurements of wavelengths by means of interference [Cooper and Stroupe (*143*)], recourse must be made in practice to the absorption bands of solids and liquid compounds which cannot be exactly calculated theoretically. Recently it has been customary to use 1,2,4-trichlorobenzene and polystyrene for this purpose, because they possess numerous sharp bands throughout the whole infrared region. In Table 8, a collection of such bands which are used for standardization is given [Plyler and Peters (*551*); Plyler, Blaine, and Nowak (*820*)]; a few emission lines are included. A similar collection of more than 80 useful absorption bands for standardization purposes in the region of 2 to 15 μ and 12 further bands in neighbouring regions up to 33 μ has been given by Jones *et al.* (*808*). Details for the region above 25 μ wavelength, which are not included in Table 8, can be found in the work of Mills *et al.* (*480*); further values for the 15 to 25 μ region have been given by Roberts (*595*). Jones (*358*) has listed exact values for individual lines of the water band at 6 μ. Mecke and Oswald (*463*), Acquista and Plyler (*2*), and Plyler *et al.* (*554*) have given several bands of compounds suitable for the short-wave region. For this region, the situation is more favourable in that the prominent emission lines from various discharge tubes can be used; such lines are sharper than absorption bands and their position is known to an extraordinary degree of accuracy. However, the conversion of the optical system of a monochromator from conditions for absorption measurements to those for emission is generally such a complex and undesirable operation that absorption standards must still be used. Wavelength measurements in the whole of the infrared region used experimentally can be directly correlated with very exactly known wavelength values by observing atom emission lines and rotation-vibration bands of simple molecules in a very high order with a spectral resolution by gratings [Rao *et al.* (*823*, *824*)]. Another method of standardization which gives very exact values for the short-wave infrared using a Fabry-Perot etalon has been described by Rank *et al.* (*574*, *576*, *577*); the sources of error in this method have been discussed by Rank and Bennet (*573*) and Brodersen (*88*). Plyler *et al.* (*555*) and Rank *et al.* (*785*, *822*) have given a very exact method of

TABLE 8. *Wavelength Standards for Prism Spectrometers according to Plyler and Peters (421) and Plyler et al. (820)*

Wavelength μ	Wave number cm^{-1}	Substance	Conditions
0·54607	18307·8	Hg†	emission
1·01398	9859·4	,,	,,
1·12866	8857·7	,,	,,
1·39506	7166·3	,,	,,
1·52952	6536·2	,,	,,
1·6606	6020·3	1,2,4-TCB‡	liquid, 0·5 mm
1·81307	5514·0	Hg	emission
1·97009	5074·5	,,	,,
2·1526	4644·2	1,2,4-TCB	liquid, 0·5 mm
2·24929	4444·6	Hg	emission
2·3126	4322·9	1,2,4-TCB	liquid, 0·5 mm
2·32542	4299·1	Hg	emission
2·4030	4160·3	1,2,4-TCB	liquid, 0·5 mm
2·4374	4101·6	,,	,,
2·4944	4007·9	,,	,,
2·5434	3930·6	,,	,,
2·7144	3683·0	methanol	vapour, 50 mm
3·2204	3104·4	polystyrene	film, 25 μ
3·2432	3082·6	,,	,,
3·2666	3060·5	,,	,,
3·3033	3026·5	,,	,,
3·3101	3020·3	methane	gaseous, 50 mm
3·3293	3002·8	polystyrene	film, 25 μ
3·4188	2924·2	,,	,,
3·5078	2850·0	,,	,,
4·258	2349·3	CO_2	atmospheric
4·866	2054·0	methanol	vapour, 50 mm
5·138	1945·8	polystyrene	film, 50 μ
5·549	1801·6	,,	,,
6·238	1602·7	,,	,,
6·692	1494·6	,,	,,
7·268	1375·5	methylcyclo-hexane	liquid, 0·05 mm
9·672	1034·2	methanol	vapour, 50 mm
9·724	1028·1	polystyrene	film, 50 μ
11·035	906·0	,,	,,
11·475	871·3	methylcyclo-hexane	liquid, 0·05 mm
11·862	842·8	,,	,,
13·883	720·5	CO_2	atmospheric
14·98	667·3	,,	,,
17·40	574·5	1,2,4-TCB	liquid, 0·025 mm
18·16	550·5	,,	,,

† All doublets have been omitted.
‡ TCB = trichlorobenzene.

TABLE 8—*continued*

Wavelength μ	Wave number cm^{-1}	Substance	Conditions
20·56	486·3	1,2,4-TCB saturated in CS$_2$ toluene	liquid, 0·05 mm
21·52	464·6	,,	,,
21·80	458·6	1,2,4-TCB	liquid, 0·025 mm
22·76	439·3	,,	,,
23·85	419·2	H$_2$O-vapour	atmospheric
2·735 ± 0·0002	3655·3	quartz	solid, 0·7 mm
2·9990 ± 0·0005	3333·5	NH$_3$	gas, 1-m cell, 5 mm (Hg)
3·3101	3020·2	methane	gas, 5·0-cm cell, 10 cm (Hg)
3·2204	3104·4	polystyrene	solid, 25-μ film
3·2439 ± 0·0002	3081·9	,,	,,
3·2668 ± 0·0002	3060·3	,,	,,
3·3026 ± 0·0001	3027·1	,,	,,
3·3303 ± 0·0001	3001·9	,,	,,
3·422 ± 0·001	2921·5	,,	,,
3·5070 ± 0·0005	2850·7	,,	,,
5·138 ± 0·005	1945·8	,,	solid, 50-μ film
5·343 ± 0·003	1871·1	,,	,,
5·549 ± 0·002	1801·6	,,	,,
6·238	1602·7	,,	,,
6·692	1493·9	,,	,,
8·662	1154·2	,,	,,
9·724	1028·1	,,	,,
11·035	906·0	,,	,,
3·2393 ± 0·0006	3086·2	toluene	liquid, 0·05-mm cell
3·2908 ± 0·0003	3038·0	,,	,,
3·4220 ± 0·0009	2921·5	,,	,,
3·4840 ± 0·0004	2869·5	,,	,,
3·6566 ± 0·0002	2734·0	,,	,,
5·1472 ± 0·0007	1942·3	,,	,,
5·3818 ± 0·0006	1857·6	,,	,,
5·549 ± 0·002	1801·6	,,	,,
21·52	464·6	,,	,,
3·2434 ± 0·0003	3082·4	1,2,4-TCB	liquid, 0·025-mm cell
3·5564 ± 0·0002	2811·1	,,	,,
3·7067 ± 0·0002	2697·1	,,	,,
5·3034 ± 0·0025	1885·1	,,	,,
17·40	574·6	,,	,,
18·16	550·5	,,	,,
21·80	458·6	,,	,,
22·76	439·3	,,	,,
20·56	486·3	1,2,3-TCB	liquid, 0·05-mm cell (sat. soln. in CS$_2$)
2·7144	3683·0	methanol	vapour, 5·0-cm cell
4·866	2054·5	,,	,,

TABLE 8—*continued*

Wavelength μ	Wave number cm^{-1}	Substance	Conditions
9·672	1033·6	methanol	vapour, 5·0-cm cell
4·258	2347·9	CO_2	gas, atmospheric
13·883	720·1	,,	,,
14·98	667·4	,,	,,
7·268	1375·5	methylcyclo-hexane	liquid, 0·05-mm cell
11·475	871·2	,,	,,
11·862	842·8	,,	,,
23·85	419·2	water	vapour, atmospheric

measurement of absorption wavelengths in the near infrared based on the use of an etalon and the emission lines of atoms.

Spectrometers in use are standardized from time to time using the standards of wavelengths just mentioned. The deviations found are plotted as a function of the wavelength (or wave number) and must be taken into consideration in exact measurements. For instruments which are not fitted with a thermostat, special note should be made of the influence of the temperature on the dispersion of the prism. With the resolution and dispersion now obtainable, this can lead to considerable errors in the wavelength found [Towler and Guy (*715*)]. Some form of temperature compensation is usually available for these instruments, but its efficiency is questionable. The best means of avoiding errors in the wavelength from temperature effects is to keep the temperature constant at the value at which the instrument was standardized. Sometimes small errors can be caused by the thermostat working intermittently; therefore this should be adjusted carefully.

3. Radiation Detectors

1. Introductory

Thermal detectors have always been the most important for the measurements of infrared radiant energy. In instruments of this type, the energy of the radiation to be measured is converted (by absorption) quantitatively into heat energy and some particular effect caused by this heat energy is measured. Only recently have *photoelectric* detectors become important for the short-wave infrared up to about 5 or 6 μ.

The history of thermal radiation detectors is at the same time the history of the development of the infrared spectral region. An advance in one respect was at the same time an advance in the other, whether it was towards longer waves or towards greater accuracy. Of the many forms of thermal detectors evolved in this development, only a few are still in use: *thermocouples, bolometers,* and *pneumatic detectors.* Their theory and technology is extremely complicated, just as that of all thermal radiation detectors. A detailed discussion lies outside the scope of this book. In any case, the selection and construction of the detector must be left to the makers of the commercial instruments, because the detector is an essential part of the instrument and because the selection of a detector necessitates the consideration of many factors dependent on other parts of the spectrometer.

The available radiant energy in the infrared region is extraordinarily small, especially when it is resolved. In order to make the rise in temperature of the measuring instrument (which will be responsible for the effect to be measured) as large as possible, all thermal detectors are built with a very small heat capacity, i.e. with very small dimensions. This necessitates, as already mentioned, a heavy concentration of radiation on the small surface of the detector by means of the optical system. The heat liberated in the detector by the absorption of radiation is not all used in causing the effect which will be measured. A certain amount of this heat is lost unavoidably by conduction through the supports, electrical leads, etc., by convection of the surrounding gases, and also by the heated detector radiating to the colder surroundings. After a short time, an equilibrium is reached between the energy gained by absorption of radiation and the energy lost in the ways just mentioned. The time necessary for reaching the equilibrium (or a state

134

almost corresponding to the equilibrium), the so-called *response time*,† must be kept as short as possible. Every thermal detector is dependent on alterations in the temperature of and the level of radiation in its surroundings. These alterations are in general slow and often occur in one direction for some time; it was this that caused the unfortunate drift in the measuring instruments of early infrared spectroscopy. Therefore, the radiation to be measured is now cut off periodically, in a way depending on the individual instruments. The radiation, and the temperature alterations in the detector which it causes, are thus characterized by a special parameter (in contrast to the non-modulated action of the surroundings). In the succeeding measuring process, it is arranged that only the modulated effects are considered. For example, this can be done by separating the alternating signal transmitted by the detector from the constant signal of the non-modulated effect with an alternating voltage amplifier using electronic methods. The principle of the use of alternating radiation is one of the characteristics of modern infrared spectrometers, together with the variable slit width, the linearization of the wavelength and the wave number scale, and the double beam principle (which is still to be discussed). However, for alternating radiation to be used, the time for attainment of equilibrium by the detector must not be too large. If the thermal inertia is too large, it is obvious that the energy radiated to the detector in a period is not used fully for the measurement; in other words, the sensitivity of the arrangement is low. However, the other advantages of the principle of alternating radiation are so important that some loss in sensitivity is accepted rather than give them up. If the inertia of the detector is large and the frequency of alternation of the radiation is made very low, the point of this method (which is the elimination of the influences of the surroundings) can be lost, or at least the amplification of the signals obtained becomes very difficult. Usually modulation frequencies of about 10 c.p.s. are applied to the radiation when using thermal detectors. The response time of the detector, defined as about 97 per cent of the time needed to attain the energy equilibrium, is in the region of several hundredths of a second or less.

The question of the comparison of different radiation detectors, especially those which work according to different principles such as bolometers and thermocouples, remains very difficult, in spite of much effort [Jones (*359, 360*)]. The sensitivity, i.e. the amount the signal transmitted by the detector alters for a given change of the radiation energy to be measured, is a

---

† In addition to the response time, the *time constant* is frequently used to characterize the slowness of response of a detector. By this is understood the time that is needed for the detector to show $1/e$ or about 37 per cent of the original value when the irradiation is discontinued.

characteristic of the detector as such. However, the sensitivity is only of moderate interest, for with the techniques in use today other factors are also influential, e.g. the amplifier, which produces a signal of measurable size from the minute signals given by the radiation detector itself. The essential quantity is the amount of radiation that can be measured within a given error, independent of the source and taking into consideration the properties of the detector and possibly those of other instruments coupled to it. Every detector has a certain *noise level* by reason of generally valid principles. Essentially this depends on the electrical resistance of the detector arrangement, the temperature, and the frequency band which is allowed through the total measuring arrangement. In addition to the noise of the radiation detector there is also that of the amplifying arrangement. It is usual to consider as the limiting output detectable that output which corresponds to the total noise level. By means of these considerations, comparisons can theoretically be made between different detectors if their properties are known exactly. Because that is not always the case, empirical comparisons under similar conditions retain their value.

2. Thermocouples and Bolometers

The *thermocouple*† is one classical form of the thermal detector. In recent years considerable technical developments have resulted in the dimensions becoming increasingly small, in an increased sensitivity by introduction of new highly thermoelectrical materials, and in modification of the external construction. At the present time it is once again the most popular of all the thermo-radiation detectors. In addition to the wire elements formerly used, recently other types of elements have been constructed; these cannot be discussed in detail in this book. Reference must be made to the review by Geiling (*246*) and references to the original literature given therein. This review also gives details of the efficiency of thermocouples of various origin.

The other classical form of the thermal detector is the *radiation bolometer*. The theory has been summarized recently in a modern form by Jones (*361*); the principle used is the alteration of the resistance of a conductor by temperature changes caused by the absorption of radiation. The importance of the bolometer has decreased recently: during the third decade and also at the beginning of the fourth decade it had been more important than the thermocouple. However, compared with this competitor, it has the disadvantage that a special source of current is necessary for its use. Keeping this current constant sometimes causes difficulties at a very critical point

† Quick response thermocouples with high output, Czerny bolometers, and various photoresistance cells are obtainable from : Physikalische Werkstätten (Prof. Heimann), Wiesbaden-Dotzheim.

in the apparatus, i.e. the entrance to the very sensitive amplifier. The early form of the platinum strip bolometer is now used only in the older instruments. Modern bolometers are prepared by electrolytic methods: very thin nickel bolometers according to Brockman (*85*); the non-blackened and quickly responding bolometers of Czerny (*157*) made by volatilizing suitable metals, usually antimony or bismuth, in a high vacuum on to celluloid films; and the semi-conductor or thermistor bolometers [Becker and Brattain (*50*); Wormser (*754*)]. The short time of response required is achieved in the nickel bolometers by the very small mass of the bolometer filament. In the Czerny bolometers it is attained similarly and also by the elimination of the blackening compound. The thickness of the filament selected is equal to the co-called Woltersdorff thickness where 50 per cent or in the limit 68 per cent of the absorption of the bare metal surface is obtainable. For the semi-conductor bolometers, the large temperature coefficients of the electrical resistance of certain semi-conductors, sometimes several powers of ten greater than those of pure metals, are used. The disadvantage of the semi-conductor bolometers is the relatively high noise level, which originates from the large resistance and from the very nature of semi-conductors. The limiting output measurable for good bolometers appears to be not very much higher than that for thermocouples, so that other circumstances, more of an electronic nature, are responsible for the decrease in the importance of bolometers.

3. Pneumatic Radiation Detectors

Because thermocouples and bolometers have been used for a long time and their construction and mode of action is well understood, the foregoing account has been quite brief. However, somewhat more time must be spent on a discussion of the details of construction of the *pneumatic detectors*, the newest type of thermal instrument for measuring radiant energy. The discussion will be based on the Golay type detector, which has great importance already and which will certainly become still more important in the future [Golay (*261, 262*)]. Fig. 35 depicts the instrument diagrammatically; Fig. 36 gives an actual view of the detector. The radiation, which is to be thought of as coming from the left, passes through a transparent infrared window in a small cell filled with gas and containing an absorbing film (antimony, or more usually aluminium, on collodium) a few square millimetres in size. The heat developed by the absorption of the radiation passes to the gas contained in the cell, in an amount varying with the modulation of the radiation. This causes a periodic change in the pressure of the gas which is followed by the flexible mirror forming the back wall of the cell. This mirror is part of an optical system which also contains a lamp, a condenser lens, and a line grating. The grating is adjusted in this optical system so that its image coincides with the grating itself, and in such a way that the ridges

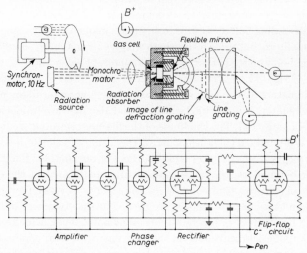

FIG. 35. *Diagram of the Golay detector [according to Golay (262)]. Explanation in text.*

fall on the furrows and the furrows on the ridges. Thus, the light from the lamp does not reach the photoelectric cell by way of the planar mirror.

FIG. 36. *View of the Golay detector. Front, left, the actual detector cell; behind, the amplifier (photograph: The Eppley Laboratory, Inc.).*

When the flexible mirror now bends as a consequence of the rise in pressure in the gas cell, following the absorption of radiation, the image of the grating is altered relative to the grating itself so that the furrows of the

138

grating coincide more or less with the furrows of the image of the grating. The light beam now reaches the photoelectric cell, in an amount which gives a measure of the radiation absorbed. The varying current in the photoelectric cell is amplified and led to a gauge or to a recording instrument. The response time of the Golay detector can be varied at will between 1 and 30 milliseconds by adjustment of the electronics of the instrument. The modulation frequency is 10 c.p.s.; the limiting output measurable has been given as better than 6×10^{-11} watts. A special advantage of this type of detector is that it can be used for spectral regions which lie at longer wavelengths, even those outside the conventional infrared, i.e. in the region of the microwaves. Further advantages are their large region of linearity and a surprising robustness for the complicated construction.†

4. *Photoelectric Infrared Radiation Detectors*

Very recently, it has been possible to apply photoelectric radiation detectors, with their many advantages and possibilities, to the infrared region. They have been used for the short-wave end, i.e. from the visible up to about 6μ wavelength or a little more. The normal gas-filled or evacuated photo cells depend on photo effects from without, e.g. the caesium cell. These are of very little use for the infrared region. Photo elements also have little importance. The situation is different for the *semi-conductor photo-resistances* of the PbS type. In these, the increase of conductivity, i.e. the decrease in resistance, which takes place on irradiation, is used; they must be kept at a potential difference which allows a current, the so-called dark current, to flow. Within a few years they have become the most important radiation detectors for the near infrared region.

At the present time three main types of such semi-conductor cells are available for the detection and measurements of infrared radiation, namely those made from PbS, PbSe, and PbTe. They are differentiated by the spectral region for which they can be used, the upper limit of which increases in this order and which ends in the infrared at about $3 \cdot 4$, $4 \cdot 5$, and 6μ (under usual conditions) (Fig. 37). Their absolute sensitivity decreases in the order given, and their noise level increases. These cells can be produced by 'wet' methods, using certain chemical reactions which precipitate the active layer on to glass, etc., sometimes followed by further treatment at higher temperatures. They can also be prepared by physical methods in which the layer is sublimed on to the carrier in a high vacuum and then activated by treatment at raised temperatures in an oxygen atmosphere. Neither the details of the very complicated method of action nor those of

† Obtainable from: The Eppley Laboratory, Inc., Newport, R.I., U.S.A.; Unicam Instruments Ltd, Cambridge, England.

the preparation of these radiation detectors will be discussed here. Such details can be found in the original literature [summaries have been given by Simpson and Sutherland *(665)* and especially by Smith *(674)*].

The photo resistances possess several great technical advantages. They have superior sensitivity in the wavelength regions to which they are applicable, selective detectors are always superior in the applicable region compared with non-selective ones such as the thermal detectors [Eichhorn and Hettner *(187)*]. Their inertia is very small, allowing alternating radiation frequencies usually up to several thousand c.p.s., which facilitates the solution of the problem of amplification. Their noise level is small compared with their great sensitivity (even though this noise level is raised by the typical semi-conductor effects); 10^{-14} to 10^{-13} watts has been given as the

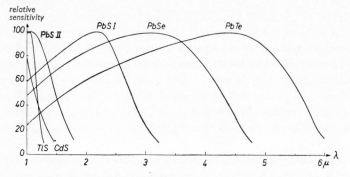

FIG. 37. *Regions in which the photo resistance cells are useful as infrared radiation detectors.*

theoretical limiting output, and apparently 10^{-12} to 10^{-11} watts can be attained in practice [Moss, *(488)*; Müser *(489)*]. Certain disadvantages, which are not very serious, are the non-linearity which sometimes appears when they are used for quantitative measurements (this can be overcome by the use of definite working conditions) and the selectivity of their spectral sensitivity curve. Thus, direct comparison of the readings for different wavelengths is not possible; however, this is usually not necessary in experimental infrared spectroscopy.

Finally, valuable, although not very large, extensions of the spectral region and very marked increases in sensitivity are obtained by cooling the cells with solid carbon dioxide or even with liquid air. These are of great importance experimentally [Nelson *(491)*; Mathis *et al.* *(450)*; Elliot *et al.* *(197)*; Young *(765)*].

For the very short-wave region of the infrared, other types of cells are of importance. One of them, first described by Genzel and Müser *(250)*, is almost identical with the PbS cell already described. It is prepared by a similar chemical method, which differs somewhat in the quantities taken. The main advantage of this cell is that for the attainment of a sufficiently strong photoelectric effect no further activation treatment is necessary. This effect is shown at once; together with a much greater dark resistance than for the normal PbS cell (the R_0 of which is of the order of 10^4 to 10^6 ohms), it shows a much greater sensitivity for a smaller noise level. However, the limit of the spectral region to which it is applicable lies at about $1 \cdot 7$ to $1 \cdot 8$ μ, in addition it is slower in responding than the normal PbS cell. Another very valuable detector for the very short-wave infrared region is the CdS cell, the long-wave limit of which is about $1 \cdot 4$ μ; up to this limit it is superior to all other detectors [Goercke (*260*)]. Up to $1 \cdot 2$ μ the TlS photo resistances and the customary Cs photocells and secondary electron-multipliers may be used, the latter with the advantage of the novel and elegant solution of the problem of amplification. With the increasing importance of the short-wave infrared region for industrial spectroscopy, great utility and correspondingly large application can be prophesied for each of these types of cell.

The *photographic plate* will be merely mentioned here; with suitable sensitizers it can be used in the very short-wave infrared as a radiation detector up to about $1 \cdot 4$ μ, because of its cumulative properties it has certain advantages.

5. Amplification and Registration

Earlier infrared radiation detectors, using the constant radiation method, were fitted with suitable highly sensitive measuring instruments, e.g. galvanometers fitted with mirrors and read optically. In modern industrial infrared laboratories, only instruments with alternating radiation, amplification, and relatively robust gauges or (usually) recording systems are now used. The question of the amplification of the small electric currents or potential differences given by the radiation detectors has always been an acute one. Some of the solutions which were discovered and used experimentally early on (e.g. the Moll thermal relay) have been developed to forms which are still useful. With the introduction of the principle of alternating radiation, the amplification problem became one of the critical questions of the measurement of infrared radiation. The general principles and methods of construction could naturally be borrowed from other techniques but there still remained a series of specific problems to be solved. The development of electronic techniques has helped in a way that could hardly have been foreseen a few years ago. The adaptation of the radiation detectors, which have usually a very low resistance, to the primary circuit of the amplifiers, which is of high resistance, has been made possible by the development of new metal alloys with special magnetic properties. This allowed the construction of transformers possessing the high efficiency needed in the low frequency region. The development of valves and other amplifier components of special quality, especially with low noise level, now permits the use of the detectors nearly to the limit of their natural efficiency. In addition, the electrical industry has made available gauges and registration instruments which possess increased sensitivity, lower response time, and a greater degree of robustness and reproduceability. Thus the unfortunate circumstances which characterized many of the earlier instruments (such as the need for working in a shock-free room and taking readings by means of light beams) are now avoided.

The output of an instrument which records a real line (instead of points at a definite time interval) needs to have undergone massive amplification. The output from the radiation detector usually lies in the region of 10^{-8} watts or less and at best gives a potential difference of a few micro volts across the detector. The large amplification required cannot be obtained in one step without danger of self-excitation in the arrangement. Therefore, in the modern instruments, a preliminary amplifier is used with every detector. Usually this is placed together with, or very near to, the detector and consists of only one amplification stage. Following this there is a multi-stage alternating potential amplification system, which can be placed at some distance away, and which is followed by a rectifier synchronized to the alternating radiation. This rectification serves to make the modulated phases equal. It curtails the frequency band (which has passed as a whole through the proceeding apparatus) as far as desirable and permissible considering the noise level and the other circumstances. The rectified current is not nearly strong enough for driving the usual recording pens and is therefore subjected to further amplification. The ways of accomplishing this are so varied that reference should be made to the specialist literature. The current, now greatly amplified, is recorded as a function of the radiation wavelength or caused to bring about some characteristic manifestation, as will be discussed later. It is characteristic of the state of development of these instruments that all these complicated interwoven electronic stages work almost without maintenance and can be run directly from the main electricity supply.

4. General Discussion of Spectrometers

Radiation source, monochromator, and detector, together with the corresponding amplification, signal, and registration appliances, all connected by the essential optical parts, form the *infrared spectrometer*. Commercially available instruments contain these parts in an arrangement in which the distinction between them is often not clear from outside. Further, the construction usually follows the economical principle in which certain parts are used in several types of instruments of the same firm, thus rendering possible their production in large numbers. In spite of this the acquisition of a modern infrared spectrometer means the outlay of at least several thousand pounds. If one of the large American instruments with their several attachments is needed, this can easily approach ten thousand pounds. However, there is no doubt that it is possible to acquire an infrared spectrometer capable of the solution of numerous experimental problems at much less expense, provided a well-equipped workshop is available and a few technical refinements are dispensed with. Of course, the self-construction of such a machine will only be successful with the help of an experienced physicist and engineer. Such questions cannot be further discussed here. This chapter and the next will be devoted to a discussion of a few points of fundamental importance and then to a treatment of the important commercial infrared spectrometers currently available.

1. Single-beam Instruments

Before beginning the discussion of the individual instruments, a few comments of a general nature will be made. The preceding explanation of the optical system of the monochromator omitted that part which serves to connect the source of radiation and the monochromator. It leads the radiation from the source to the entry slit and usually the beam passes through the substance to be investigated at this stage. This part of the optical system is characteristic of the stage of development of the infrared spectrometer, and its construction is especially important for the efficiency of the instrument.

143

Equipment and Experimental Technique

The simplest method is to focus the source of radiation on to the entry slit by means of a concave mirror. It depends on the ratio of the sizes of the slit and source whether the image of the source should be of the same size, enlarged, or decreased in size. The image of the radiation source must always be larger than the slit, but it should not be too much larger because the part of the image which overlaps the slit, and the corresponding part of the source, makes no contribution to the radiant energy which is eventually measured. The intensity of illumination of the image of the source lying on the slit should be as great as possible because this serves as a secondary radiation source. For this, the solid angle of the emission from the source included by the focusing mirror must be as large as possible. This condition corresponds to a short focus mirror of large size; however, in this respect also, extremes must be avoided on account of inevitable focusing errors. The aperture ratio of the source mirror, i.e. the ratio diameter to focus, must in any case be so selected that the collimator mirror (or to be more exact the image on it of the prismatic face, focused or projected) is completely illuminated. It is clear that, with such arrangements, the substance investigated is placed in a convergent beam, because the substance is usually positioned near to the entry slit or to a real image of the radiation source (when this is not directly focused on the slit). Such an arrangement is quite satisfactory for gases using any desired thickness, but for liquids and solids only if the layer through which the beam passes is sufficiently small. If the layer is too thick, which can happen in the investigation of weakly absorbing substances, e.g. in the short-wave infrared, undesirable and disturbing displacements of the focus result. This is because the optical path becomes larger by the factor $(n-1).d$, where n is the refractive index of the substance and d is the thickness of the layer through which the beam passes. In such cases it is better to investigate the substance with parallel radiation with which these effects do not occur. For this, the substance is placed between two mirrors, one of which makes the radiation coming from the source parallel, and the other then focuses it on to the slit, as described, e.g. by Jenness (357).

If the focusing system is arranged for obtaining an absorption spectrum in the usual form, i.e. percentage transmission or extinction as a function of wavelength or wave number, etc., in principle two measurements at different times are always necessary. The first measurement, made without the absorbing substance, gives the emission spectrum of the radiation source, taking into consideration the effect of the properties of the atmosphere and the apparatus through which the beam passes, as a function of the wavelength. The second measurement, made with the substance, gives the same spectrum but now with the absorption of the substance investigated superimposed upon it. The absorption spectrum of the substance, independent of

144

all the peculiarities of the source of radiation, the apparatus, and the surrounding medium, is obtained as the difference of the two curves measured. An apparatus arranged in this way is called a *single-beam instrument*. If the emission spectrum of the radiation source together with all the other features stemming from the other influences were to be constant for long periods (which is impossible for many reasons), the carrying out of two measurements one after another could be dispensed with; the spectrum of the source could be determined once and for all and then be used for all the later investigations. However, this is impossible; even two measurements following quickly one after another are subject to uncertainties. Between the two measurements there could have been an alteration in the emission of the radiation source, the sensitivity of the detector, the degree of amplification, the absorption of the atmosphere, or any of the many other possible variables. The larger the time interval between two such measurements, the more serious such alterations could be.

This is a fundamental defect of the single beam operation and led to the frequent repetition of measurements and their averaging to eliminate chance errors. In addition, single beam instruments have other disadvantages, especially the greater time needed and the necessity of calculating the required spectrum from the measured curves delivered by the spectrometer. This disadvantage remains, whether the desired curve is obtained by point by point calculation with ruler and slide rule, or by means of a more or less automatic calculating instrument. Such have often been proposed and can probably be obtained in some form or another from the manufacturers of mathematical instruments.

By a suitable slit width mechanism, the Planck curve of the radiation emitted can be evened out for single beam instruments, but the atmospheric absorption bands still cause much interference. They can be avoided by evacuating the whole of the path of the beam or by replacing with an atmosphere of pure dry nitrogen, but both of these methods need constant and careful maintenance. Rientsma and Westerdijk (*594*) have used a better method. They control the slit in the usual way but scan the spectrum of the empty cell for every spectral measurement by means of a photo cell. This is made to control the degree of amplification of the alternating current amplifier so that the absorption spectrum of the substance investigated is recorded as the end result.

2. Double-beam Instruments

Efforts to avoid and overcome the disadvantages of the single-beam instruments just mentioned have led to various alternative forms of construction. Of these, the most important are the *double-beam instruments*, probably first

L

Equipment and Experimental Technique

used for the infrared region by Lehrer (*412*). The fundamental principle is that the radiation emitted by the source is divided into two beams which are energetically, geometrically, and optically similar. At some point before entry into the monochromator, they are brought together again (Fig. 38). In one, the so-called *measuring beam*, is placed the substance to be investigated. The other, the so-called *reference beam*, contains no substance, but usually an empty cell, or for solutions the pure solvent, etc., is placed in it. By means of a so-called optical switch (which in the arrangement shown takes the form of a rotating sector mirror), first the one and then the other beam is caused periodically, at very short time intervals, to be alternatively admitted to the monochromator and thus to the radiation detector. By means of electronic techniques, the radiation detector is caused to measure

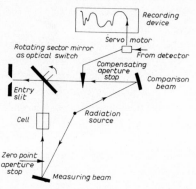

FIG. 38. *Diagram of the photometer unit of a double-beam infrared spectrometer with a wedge or aperture stop.*

the difference in energy between the two beams. By a suitable setting of the zero-point attenuator in the measuring beam it can be arranged that for a certain wavelength both beams are of identical energy. In this case, the detector gives no signal to the amplification and recording systems connected to it. Slow alterations in the energies of the two beams, resulting from anything other than the substance being investigated (slow with respect to the alternating frequency with which first the one and then the other beam is brought to measurement), occur simultaneously in both and are therefore not taken into account by the detector. Hence, if the substance under investigation shows absorption at a particular wavelength, the detector reacts and delivers (via the coupled components) a signal which is a measure of the energy difference of the two beams and thus of the absorption of the substance. Instead of being directly recorded, this signal is used to decrease the energy in the reference beam (e.g. by moving in a suitably formed attenuator) to such an extent that the energy again becomes equal in the two beams, so that the detector no longer gives a signal. The position of the attenuator is thus a quantitative measure for the absorption of the substance and is recorded as such. As a means of weakening the energy in the reference beam, the use of the *optical-wedge attenuator* (Fig. 39) or the

FIG. 39. *Optical wedge attenuator.*

aperture stop (Fig. 40) have become customary. The first-named consists of a number of narrow triangular teeth, the width of which increases from the tip to the base. The further it is pushed into the beam, the greater is the fraction cut off. Any beam attenuator which blocks specific parts of the reference beam (either field or focused image) will cause differences in the distribution of the energy in the sample and reference beams which may affect both beam cancellation and photometric accuracy.

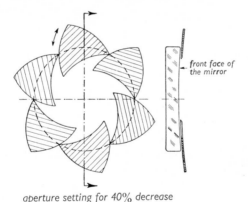

front face of
the mirror

aperture setting for 40% decrease

FIG. 40. *Aperture stop.*

A star-wheel beam attenuator having a number of teeth cut to spiral contour may be used to avoid these difficulties. When such a wheel is spinning there is a linear relation between the radial movements of the assembly and the on/off ratio of the light beam over the working transmission range. If the rate of passage of the teeth past a fixed point is high compared with the response time of the detector used, the detecting system sees the star-wheel as an optical wedge. By using the star-wheel at an image of the slit, errors due to finite slit widths are minimized. An arrangement for weakening

147

radiation by means of movable rotating sectors has been described by Richardson *et al.* (*593*).

Because the arrangement for weakening and compensating in one beam is a characteristic of normal photometry, instruments thus equipped are also called *spectrophotometers*. That part of the optical system, which focuses the radiation source on the entry slit of the monochromator and contains the compensating arrangement, is accordingly often called the *photometer unit*. It can be seen that in this way the disadvantages of the single beam instruments detailed above are avoided. In addition, such double beam instruments correspond to the movement, seen in all forms of measurement, from the old direct methods to compensation or zero-point methods. Obviously, it is not absolutely necessary that the compensation be an optical one involving the decrease in the energy of one of the beams, although this is a natural method for an optical instrument. It is also possible to measure the signal corresponding to an energy difference of the two beams potentiometrically and to perform the necessary calculation for the percentage transmission by electronic means [Zbinden *et al.* (*767*); Savitzky and Halford (*629*); Hornig *et al.* (*334*); Golay (*263*)]. Golay (*264*) has summarized the various systems for spectrophotometers and their advantages and disadvantages.

It is obviously very important for the correct working of the photometer unit that the two beams are of an identical and symmetrical form. In particular, their paths should be of the same length and they should receive the same number of reflections and absorptions, with the exception, of course, of the absorption of the substance under investigation. This can always be accomplished by suitable construction. To compensate any differences still remaining, the investigating beam contains a so-called zero-point attenuator which corresponds to the attenuator in the comparison beam for weakening it. However, in most of the photometer units used experimentally, the two beam paths are not completely symmetrical. As can be seen from the diagram in Fig. 38, by means of similar mirrors an equally large space angle of the emission of the radiation source is used for each beam, but the radiation originates from different portions of the surface of the source. The equality of the emission of different portions of the surface cannot be guaranteed with certainty, as the sources are of a complicated chemical composition and surface type, and errors can thus arise in the energetical symmetry of the two beams. The larger the angle between the two emission directions, the more likely are such deviations. Photometer units have therefore been proposed which take account of this point and are completely symmetrical with respect to the emission, in that radiation emitted in one single direction is used for both the beams. The fundamental

arrangement is given diagrammatically in Fig. 41 [Oetjen and Roess (*506*);
Rugg *et al.* (*612*)]. The radiation source is focused by the first mirror on to a
rotating sector mirror which is placed at an angle of 45° to the direction of
the radiation. This sector mirror alternately contains reflecting and transmitting sectors. For the case of reflection, the radiation is led to the second
concave mirror which again focuses an image of the radiation source on to a
second sector mirror. This second sector mirror is identical with the first, and
their planes of reflection coincide† but the two mirrors act complimentarily:

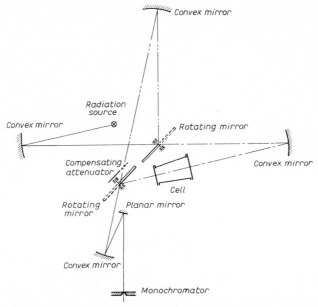

FIG. 41. *Diagram of a completely symmetrical photometer unit
for double-beam spectrometers according to Rugg* et al. (*612*).
Explanation in text.

if one causes reflection, the other causes transmittance. In the case mentioned the radiation passes further to the monochromator. On the other
hand, if the first sector mirror transmits, the radiation passes to a concave
mirror, which is the mirror image with respect to the rotating mirror of the
other concave mirror, and this then forms an image of the radiation source
on the second rotating mirror. In this case, the second rotating mirror is at
reflection for the approaching radiation which is thus led to the monochromator in exactly the same way as the radiation which came from the other

† If a further stationary planar mirror is used in the path of each beam, these
reflection planes need not coincide as long as they are parallel to each other.

concave mirror. Thus, by the co-ordination of the two rotating sector mirrors, first one and then the other beam is led to the monochromator. The paths of both beams are identical from the radiation source to the first rotating mirror and from the second rotating mirror onwards. Between the two rotating mirrors they are split in a way that is completely symmetrical with respect to the plane of the rotating mirrors. The two separate paths thus formed are used as the investigating and comparison beams; the first passes through the absorbing substance and the path of the other contains the attenuator for decreasing its intensity. Obviously, the construction of a completely symmetrical photometer unit necessitates considerable optical and mechanical exactitude, especially in assuring the simultaneous co-ordinated movements of the two sector mirrors and the parallelism of their reflection planes. It appears doubtful whether the advantage gained is in proportion to the effort required. According to all present experience, the use of such photometer units as shown in Fig. 38 does not lead to errors resulting from the 'flickering' of the radiation source which are greater than those from other causes. Hornig *et al.* (*334*) have given another method for the construction of a completely symmetrical photometer unit.

Finally, it should be mentioned that percentage spectra (and obtaining such is one of the main reasons for the use of double-beam instruments) can also be obtained with single-beam instruments. One possibility is the control of the slit width in two successive measurements using a magnetic recording of the value of the first measurement [Hawes *et al.* (*314*)]. This 'memory system' principle is used in the Beckman-IR-3-spectrophotometer and will be explained in connection with the discussion of this instrument. Another possibility has been described by Goulden and Randall (*273*) and Ahlers and Freedmann (*9*). Here the absorption and comparison cells are each placed in a transmitting sector of the radiation modulator which is usually a symmetrical metal disc with two lobes which rotates at a suitable speed in a plane perpendicular to the radiation. With the arrangement of the cells mentioned, the percentage spectrum can be obtained electronically.

3. Optimal Working Conditions

The optical adjustment of a spectrometer to the optimal working conditions for the investigations in view needs some consideration. Three quantities are mainly responsible for the quality of the spectrum obtained; there are also other less important factors. These three quantities are the resolution attained as a measure of the separation of the bands, the exactitude of the absorption measurements as a fundamental indication of their reliability, and the time required for obtaining the spectrum (without considering the preparation of the specimen). The importance of the first two points is obvious. The third point may not be so important in academic investigations, but in any case it plays a part for industrial applications.

The resolution is increased by decreasing the slit width, the accuracy of measurements is increased by raising the ratio of signal to noise, and the time required is shortened by increasing the speed of recording or in other words by lowering the time for recording, e.g. per μ wavelength region. However, all these three quantities cannot be simultaneously altered in the favourable sense. For example, decrease in the slit width, while it increases the resolution, simultaneously decreases the signal-noise ratio. To keep this ratio at the same level, or to raise it, one of the factors determining the noise level must be decreased. The simplest way to do this is to narrow the frequency band that is transmitted by the measuring system (detector, amplifier, and recording apparatus), either by electronic truncation, or by raising the recording time, or by both these measures. This again leads to an increase in the recording time in order to ensure an accurate depiction of the absorption bands.

Luft (*435*) has given the empirical formula

$$g \cdot A^2 \cdot \sqrt{w} = \text{const} \qquad [\text{II}, 4.1]$$

for the connection between the three quantities, resolution A, accuracy of measurements g, and recording time w. This accords with the above discussion. Thus, only two of the three quantities can be selected as desired and the third is then determined. This must on no account be neglected when comparisons are made between spectrometers working under very different conditions and when the requirements of an instrument are being considered.

4. Rapid or Scanning Spectrometers

The time necessary for recording an infrared spectrum from about 2 to 15 μ with a modern instrument is several minutes even in the most favourable cases, and usually lies between 5 and 20 min. This is much too long for the investigation of rapid processes such as combustions, explosions, etc. However, the use of infrared spectroscopy is desired also for these processes. The construction of so-called rapid or scanning spectrometers is now possible; they are so named because they measure spectra in a fraction of a minute or a second [Wheatley *et al.* (*732*); Bullock and Silverman (*101*)]. They have resulted from the development of radiation detectors of very low inertia, the photo resistance cells for the short-wave infrared and the Golay detectors for the whole infrared region, together with registration by means of the cathode ray oscillograph which has almost no inertia. The rapid scanning of a whole spectral region in fractions of a second is made possible by

151

making the Littrow mirror, which alters the wavelength, carry out a very rapid angular vibration of sufficient amplitude instead of the usual slow rotation. This rapid movement of the Littrow mirror can be brought about by connecting it with the vibrational system of a coil through which alternating currents of a suitable frequency is passing. It can be arranged that the spectrum of a flame, etc., can be scanned in a desired spectral region about once every tenth of a second and the successive spectra can be recorded on the phosphorescent screen of a cathode-ray tube. There, they can be photographed or filmed and thus the time variation of significant bands can be observed and evaluated.

Another solution of this problem has been suggested by Agnew *et al.* (*8*). In the focal plane of the collimator mirror in which the whole spectrum is formed, using a motionless Littrow mirror, are placed up to ten exit slits instead of one. These exit slits are placed in the positions where the characteristic bands for the process under investigation occur. Each slit leads to its own detector, each one of which is connected with an independent amplifier. The radiation energy is led to the various detectors by an elegant system of suitably formed and bent silver tubes. The outputs from the amplifiers can either be connected to separate recording instruments or, by means of a rapid electronic switch, be thrown on to a cathode-ray tube at short intervals after one another.

5. Representation of Infrared Spectra

Fundamentally every graphical representation of a spectral property as a function of the frequency, wavelength, or wave number, is a spectrum. This is so whether the transmittance, the absorbance, the power of reflection, the extinction, the extinction coefficient, the absorption coefficient, or the absorption index is recorded. The power of reflection can be expressed by means of the wavelength-dependent refractive index and the absorption coefficient or the absorption index. With this exception, all the quantities are measures of one fundamental property, the wavelength-dependent ability of a substance to convert electromagnetic radiation energy into heat energy. They are connected mathematically in ways that will be discussed in detail later. Which of these quantities is to be preferred is a matter of convenience, and of custom. A mathematical transformation cannot alter the essential relationships, at the most it can allow certain inter-relationships to be seen more clearly.

From the extreme point of view of the theoretical physicist, the representation of the absorption coefficient or absorption index is desirable, because these are most suited to the wave nature of the radiation. Principally because of experimental reasons, such a representation is very seldom

152

found in infrared spectroscopy, especially as such values cannot easily be read off. The extinction coefficient is connected with the experimentally determined quantities (radiation energy or radiation energy density), but the state of development of present instruments hinders their use, especially the insufficient resolution which distorts the bands. Certain relationships, still to be considered, can be written especially simply with the extinction formulation. Many infrared spectroscopists therefore favour the extinction as the characteristic quantity to be represented. This can often be done, but its relationship with the experimental quantities is given by the logarithm of a ratio. The most frequent representations are still the transmittance or absorbance, two quantities which complement each other and which are directly obtained as the ratio of the primary experimental quantities.

With respect to the abscissa for the representation of spectra, the choice is happily not so wide. The frequency, the primary quantity from the physical point of view, is hardly ever used. Wave number and wavelength are used to about an equal extent; recently a general movement in favour of the wave number has been noticeable. The most important disadvantage of the wave number, the fact that wavelength-linear monochromators have a somewhat simpler construction, is not very serious. The drawback of wave number-linear instruments in placing over-emphasis on certain spectral regions (large wave numbers) can be easily removed by a scale alteration at about 2000 cm^{-1}. From an interpretative point of view the wave number representation is certainly advantageous. It represents a direct measure of energy, the band contours are symmetrical, and many regularities in the spectra can be seen directly when it is used. The large influence of the linear representation in wavelength, which earlier was almost always used, can be seen from the fact that even those using the wave number scale usually represent the bands of a spectrum graphically with increasing wave numbers from right to left!

There is no doubt that the appearance of one and the same spectrum can be very different if different methods of representation are used. However, because this is a rather academic question, and because a spectroscopist must in any case become familiar with all the methods of representation, a discussion of this will be omitted and reference made to the examples to be found in (*L*).

For some purposes, e.g. for documentation (see page 269), it is desirable that the spectra should all be of uniform representation. Instruments have been constructed which convert, for example, an ordinate in percentage transmission to one in extinction and an abscissa in wavelength to one in wave number [Bergmann and Kresze (*58*), Bergmann (*771*)].

Equipment and Experimental Technique

Modern self-recording spectrometers usually deliver original spectra of considerable dimensions, e.g. 27 × 85 cm or similar. Sometimes such a format is too large for storing or filing, although it has undeniable advantages for clarity, ease of reading, and interpretation. Various proposals have been made for transforming to a smaller format, by means of a pantograph [Vandenbelt *et al.* (*772*)], recording in the smaller format with the original instrument (certainly the best solution) [Strauss and Thompson (*688*)], reduction by a recording apparatus in parallel to the original recorder [Meakins and Nelson (*816*), Olsen *et al.* (*819*), Stitt and Bailey (*829*)], and reduction by photocopying or photostat, for which a suitable ink must have been used in the original recording [Tanner (*696*)]. A duplicate recorder with a smaller diagram is directly available for some commercial instruments, e.g. the Perkin Elmer Model 21, Unicam SP 100, and Beckman IR-7.

5. Commercial Infrared Spectrometers

The following sections give a short summary of the different types of commercial infrared spectrometers at present available. The description and information are based on specialist publications, in so far as they are available (references to which are given), and descriptions and directions originating from the manufacturers; these are not referred to expressly in every case. The general principles of spectrometer construction and the expected performance of commercial instruments, as well as a suitable basis for assessing them, has been discussed by Williams (743), Hawes and Gallaway (313) and Golay (264). The length of the treatment in this book in no case is to be taken as an evaluation of the instrument in question. Although not inconsiderable differences exist in the technical equipment of the instruments of different manufacturers, especially concerning certain refinements, it is still true that all the spectrometers commercially obtainable at present can be applied to all the usual problems. The technical quality and the price are always related to each other and it is the concern of the user to obtain the maximum efficiency from every instrument.

1. The Baird Spectrometer

The infrared spectrometer Model 4-55 of the firm Baird-Atomic, Inc., Cambridge, Mass., U.S.A. (Fig. 42), is a double-beam spectrophotometer with direct registration of the percentage transmittance against a linear wavelength scale [Baird et al. (29); although this paper deals with an older spectrometer model, much of it applies also to the Model 4-55]. A diagram of the arrangement is given in Fig. 43. The radiation source is a globar working at about 1100° C in a water-cooled container. The source is focused on the entry slit of the monochromator in two ways by means of concave mirrors of 20 cm focal length, according to the system of Fig. 38. In front of the slit is a rotating sector mirror which is at an angle of 45° to both of the beam paths. When reflecting, it throws the measuring beam on to the slit, and when transmitting, the comparison beam; the rotation is at 10 c.p.s. The mono-chromator is of the usual Littrow type, with a collimator mirror of 75 cm

FIG. 42. *View of the Baird infrared spectrophotometer (photograph: Baird Ass., Inc.).*

focal length and a 60° prism of 10 cm base and 8 cm height. The monochromator is normally equipped with a NaCl prism. Further prisms of LiF, CaF_2, KBr, CsBr and KRS 5 are obtainable. The prisms can be changed, together with the necessary alteration in the mechanism for the Littrow mirror and the slits, in about 10 min. The rotation of the Littrow mirror is controlled by a cam disc. The slits are bilateral, 38 mm high and are controlled according to the emission curve of the globar for constant energy. In addition to manual setting as desired, three slit programmes are provided, which for the use of a NaCl prism correspond to spectral slit widths of about $0 \cdot 015$, $0 \cdot 03$ and $0 \cdot 06$ μ. The exit slit is bent. The radiation detector is a blackened platinum foil bolometer of dimensions $0 \cdot 001 \times 0 \cdot 3 \times 7$ mm and is situated in an evacuated container with a KBr window. For the amplification of the 10 c.p.s. signal appearing for inequality of the two beams, a four-stage, resistance-coupled, alternating current amplifier is used. This is set for 10 c.p.s. and has a degree of amplification of 5×10^7, a noise level of 4×10^{-9} volt and a band width of 2 c.p.s. The amplified signal is rectified by a commutator which is placed on the axis of the sector mirror. The resulting direct current is interrupted and again amplified in a three-stage power amplifier. The output of this is used to drive both the recording system and a wedge attenuator in the comparison cell, for the purpose of equalising the energy with

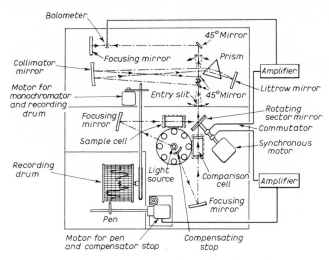

F I G. 43. *Diagram of the Baird infrared spectrophotometer.*

that in the investigating beam. Times between $0 \cdot 5$ and 70 min per μ can be selected for the recording of the region 2 to 16 μ. The accuracy of the measurement of transmission is ± 1 per cent, that of the wavelength $\pm 0 \cdot 02 \, \mu$. The reproducibility is given as $\pm 0 \cdot 5$ per cent for the transmittance, and $\pm 0 \cdot 005 \, \mu$ for the wavelength. The instrument is arranged in several component parts which are accessible individually; the optical parts are kept airtight and are dried chemically. Cells of different types, a slit-programme selector, an extension for single-beam operation, and filters for the decrease of short-wave stray light above $8 \cdot 5 \, \mu$ are available as accessories.

The firm Baird-Atomic, Inc., also produces a smaller self-recording double-beam spectrometer known as Model KM-1. This corresponds to the movement recently noticed to the production of cheaper instruments which are still sufficient for the solution of most problems.

Equipment and Experimental Technique

2. The Beckman Spectrometers

At present, Beckman Instruments Inc., Fullerton, Calif., U.S.A. offer three different instruments, models IR-4, IR-5, and IR-7 which vary in size, price, and performance. The earlier types have been technically superseded and are no longer produced.

FIG. 44. *Appearance of the Beckman IR-3 spectrophotometer* (*photograph: Beckman Instruments, Inc.*).

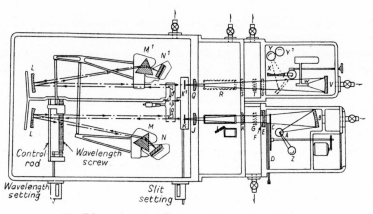

FIG. 45. *Diagram of the Beckman IR-3 spectrophotometer.*

Model IR-4 is the current standard prism instrument with a comprehensive and well-considered range of technical possibilities. It may be applied to all problems except those requiring an extremely high resolution. It is a double-beam instrument which can easily be converted to a single-beam instrument of high photometric accuracy. The appearance is shown in Fig. 46; Fig. 47 gives the block diagram which should be comprehensible from the

FIG. 46. *Appearance of the Beckman IR-4 spectrometer (photograph: Beckman Instr.).*

general notes already given. Optically, the instrument possesses a completely symmetrical photometer unit in the Oetjen-Rugg arrangement with an optical wedge attenuator. This is used with a 60° prism (base 7·5 cm) fitted with a double monochromator of conventional construction (with the middle slit in the Littrow arrangement) by means of which high dispersion is attained and stray light efficiently eliminated. The signal to be measured is obtained by means of a servo motor and optical compensation. It is conveyed electrically to the recorder; special intensity ranges (percent transmittence or extinction) can easily be selected. The recording takes place on a single sheet with preprinted co-ordinates (linear wavelength or wave number scale) on a flat recording table. Thus, the whole of the spectrum already drawn can be seen for the complete course of the recording. By means of a pre-set arrangement for changing the prism, the spectral region

Equipment and Experimental Technique

covered can be quickly altered. In addition to the standard slit programme (for each prism) other slit programmes can be used as desired. The desired recording speed can be selected from a wide range. The time constant of the instrument can be varied between $\frac{1}{2}$ and 32 within a factor of 4. There are eight abscissa scales and it is also possible to record directly on punch cards. Further recorders giving duplicate curves can be connected without difficulty.

Model IR-5 is a simplified and therefore cheaper version of the Model IR-4 just discussed. Optically, it is identical with the former. However, the

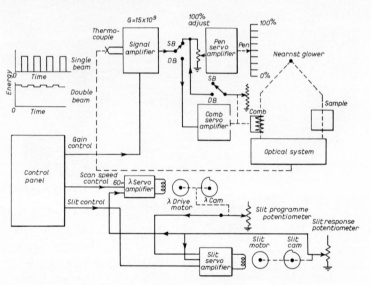

FIG. 47. *Block diagram of the Beckman IR-4 spectrophotometer.*

signal is conveyed to the recorder mechanically and the other electrical and mechanical components are limited to those required for routine operation: single switch control, a single recording velocity of 16 min for the region 2 to 16 μ, one automatic slit programme in addition to manual setting, only an NaCl prism, somewhat smaller size of recording.

In contrast, Model IR-7 is superior to Model IR-4 in resolving power and in photometric accuracy. The higher resolution ($0 \cdot 3$ cm^{-1} at 1000 cm^{-1}) is obtained by the use of a double monochromator, the second portion of which, in contrast to Model IR-4, is fitted with a grating of 64×64 mm surface and 75 grooves per mm (blazed for 12 μ) which is used successively in four orders (with automatic switch-over) for continuous registration from 65 to 4000 cm^{-1}. For the rest, Models IR-4 and IR-7 are largely identical

160

in their optical and mechanical characteristics. The high photometric accuracy of 0·2 per cent (transmittance) for single-beam operation is attained by use of a very accurately constructed optical wedge attenuator and a corresponding system of rigorously linear amplifiers with precision potentiometers at the decisive places. Recording is in per cent transmittance against a linear wave number scale on a desk recorder.

It goes without saying that a wide range of accessories is obtainable for these spectrometers which allow them to be applied to all problems.

2a. The Applied Physics Corporation Spectrometers

The Applied Physics Corporation, Monravia, Calif., U.S.A., has recently offered an instrument under the name Cary-White Model 90 infrared spectrophotometer which is superior in resolution and photometric accuracy (but also more expensive) than all instruments hitherto known. Fig. 47a shows the appearance of the instrument and Fig. 47b the optical path used. Radiation

FIG. 47a. *View of the Applied Physics Corporation spectrometer.*

Equipment and Experimental Technique

from the source (A) is directed by parabolic condensing mirror (BB) to flat (C) where it is carried to auxiliary condenser (D) and reflected to multi-faceted beam splitter (E). The beam splitter divides the beams and directs each half downward to mirrors (F_1) and (F_2) which in turn reflect the beams horizontally to diagonals (G_1) and (G_2). The diagonals turn the beams parallel and through chopper discs (H_1) and (H_2) which chop the sample beam at $13\frac{1}{3}$ c.p.s., the reference beam at $26\frac{2}{3}$ c.p.s. The energy pulses exit the source compartment through windows (I_1) and (I_2) to the sample and reference cells (a_1) and (a_2) respectively. Entering the beam combiner compartment through windows (J_1) and (J_2), the beams fall on diagonals (K_1) and (K_2) which direct them outwardly to (L_1) and (L_2), where they are

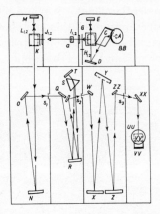

FIG. 47b. *Diagram of the Applied Physics Corporation spectrometer.*

turned upward toward the multifaceted beam combiner (M). The combined beam is directed down to sphere (N) and auxiliary sphere (O) where it falls on the entrance slit (s_1) to the prism monochromator.

Energy from the entrance slit continues to the aberration compensator (Q), then to the prism collimator (R) which directs it to the KBr prism (S) and Littrow mirror (T). After being dispersed a second time by passage through the prism, the radiation is reflected from the prism collimator to aberration compensator (U) and directed to intermediate slit (s_2), where it enters the grating monochromator. Aberration compensator (W) directs the radiation to collimator (X), then to grating (Y), from which it returns to the second collimator (Z) of the Czerny-Turner system and is directed to aberration compensator (ZZ) which carries it to exit slit (s_3).

The monochromatic radiation is directed to the detector by toroidal mirror (XX), by either mirror (VV) or LiF filter (VV′) which are inter-

162

changeable, then to the ellipsoidal mirror (UU) which concentrates the energy on the detector. The first part of the double monochromator is fitted with a KBr-prism of 100 mm base, and the second part with a grating of 100 × 127 mm surface and 75 grooves per mm. The resolution is stated to be 0·3 cm^{-1} or better, the photometric accuracy 0·1 per cent, (transmittance), and the variation of the zero point less than 0·03 per cent (transmittance) per day. Instead of the optical null point compensation by means of a wedge or aperture attenuator which is usually used, the signal is obtained by the formation of a quotient electrically. The recording is carried out from 450 to 4000 cm^{-1} in one operation in which the grating is used in four orders with automatic switch over; no alteration of scale, etc., occurs. The performance of the instrument with respect to slit programme, recording time, time constant, etc., corresponds to the expectation for a high precision instrument.

3. The Grubb Parsons Spectrometers

The firm of Sir Howard Grubb Parsons & Co. Ltd., Newcastle upon Tyne, formerly manufactured prism spectrometers but for some years has produced grating instruments exclusively. Current models may be divided into

FIG. 48. *Appearance of the Grubb Parsons infrared spectrophotometer with the GS2 grating monochromator.*

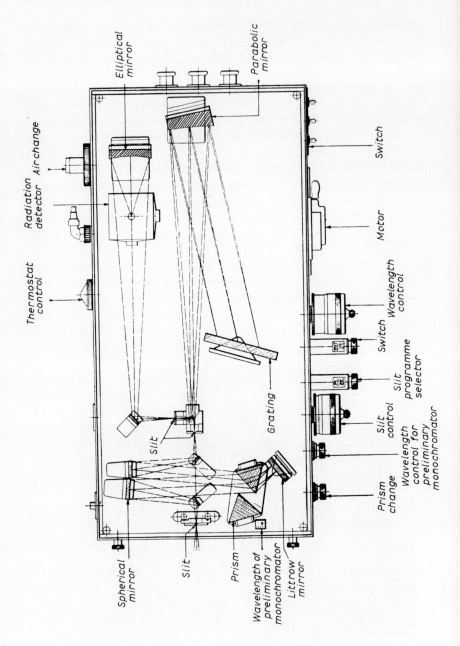

FIG. 49. *Diagram of the Grubb-Parsons grating monochromator GS2.*

two main varieties: high performance wide-range double-beam spectrometers in console form and single-beam inexpensive bench-type instruments. In the former class the grating monochromator employs a single grating (2500 or 1200 lines per inch, cut for maximum energy at about 9 μ in the first-order spectrum) in four alternative orders and is available in two forms: model GS2, 2 to 15 μ with limiting resolution 1 cm^{-1} throughout the range, and model GS2A, 2 to > 20 μ with limiting resolution 1·5 cm^{-1}. The optical system employs a Littrow arrangement with off-axis paraboloid and an ellipsoidal condensing mirror to form a reduced image of the exit slit on a thermocouple detector with receiving area of 2 × 2·0 mm. A fore-prism unit coupled to the mechanism which rotates the grating serves to isolate the spectral order desired, in a similar manner to that described by Cole (*131*) [see page 117; also Martin (*448*)]. The complete spectrum is covered in four sections 2 to 2·5, 2·5 to 3·75, 3·5 to 5 and 5 to 15 or 5 to > 20 μ. A CaF$_2$ fore-prism is used from 2 to 5 μ and for longer wavelengths KBr is employed; the change from one to the other is instantaneous. A double-beam unit with built-in strip chart recorder and reciprocating mirrors for beam switching is attached to the monochromator to form a complete spectrometer, while the pedestals of the console house the electronics units. A servo-operated optical-wedge attenuator automatically maintains balance between the sample and reference beams and sample transmittance is recorded directly. A gas cell up to 10 cm in length can be accommodated.

The less expensive monochromator, Model DM2, is a considerably simplified version of the GS2 and GS2A, important differences being the replacement of off-axis paraboloid by two spherical mirrors (Ebert system) and reduction of focal length from 50 to 15 cm. Model DM2 is available for the following ranges: 1 to 3, 2 to 6, 5 to 15, 15 to 25, and 5 to 25 μ, while a further version, type DM3, is intended to cover the range 2 to > 20 μ with two interchangeable first-order gratings. A Nernst glower is used as infrared source in all spectrometers. All monochromators have slit-programming to keep the energy constant throughout the working range and can be used in single or double-beam equipment.

4. The Hilger Spectrometer

The firm of Hilger and Watts, London, world-famous for the construction of spectrographs for emission spectro-analysis, offer a double-beam recording spectrophotometer (H 800) for infrared spectroscopic purposes.

The H 800 double-beam spectrophotometer (Fig. 50) consists of a simple monochromator of the Littrow type with a 60° prism of base 95 mm and height 55 mm and a parabolic collimator mirror of 66 cm focal length. The instrument is available with either a linear wave number or a linear wavelength scale. Prisms of the usual materials, together with the corresponding calibrated cams and scales, are easily interchangeable. The slit programme is continuously variable to suit any particular measurement which is being taken. A Schwarz thermopile serves as a detector; a condensed image of the exit slit is focused on the detector by means of an elliptical concave mirror. The photometer unit is characterized by the fact that the two optical paths lie one above the other and are alternately directed into the monochromator by means of a plane mirror oscillating at 12·5 c.p.s. in front of the entry slit. The energy of the comparison beam is automatically adjusted by altering

FIG. 50. *View of the Hilger recording infrared spectrophoto- meter H 800 (photograph: Hilger and Watts Ltd.).*

an aperture stop, placed in front of the last focusing mirror in the compari- son path beam. The opening of this stop is a measure of the energy passing and the corresponding recorder deflection is obtained from a potentiometer coupled to the aperture stop mechanism.

5. The Leitz Infrared Spectrograph

The Leitz infrared spectrograph (Fig. 51) was originally a semi-automatic instrument, but it has since been developed to the fully automatic stage. The optical path is shown schematically in Fig. 52. From the electrically started Nernst glower N, radiation is directed in two ways on to the entry slit *S* 1 to the monochromator. The comparison beam with the aperture stop *B* for compensation travels via *Sp* 2, *R* 2 and *Sp* 3, and the investigating beam with the cell *S* via *Sp* 1, *R* 1, and *SS*. The sector mirror *SS*, rotating at 12·5 c.p.s., 'switches in' the two beams alternately. The monochromator is in the usual Littrow arrangement with an axial collimator concave mirror of 1 m focal length, and a 60° NaCl prism of 150 mm base and 100 mm height. The bilateral slits are controlled for constant energy according to the emission curve of the radiation source with respect to the wavelength. A thermo-element serves as detector. An image of the exit slit is thrown on to the

FIG. 51. *View of the Leitz infrared spectrograph (photograph: E. Leitz G.m.b.H.).*

detector via the plane mirror *Sp* 4 (which for wavelengths above 9 μ is automatically replaced by a reflection filter in order to remove short-wave stray light) and a concave mirror *F*. The alternating signal is amplified in the usual way, rectified by a rectifier placed on the axis of the sector mirror, and led to the recording mechanism motor. This controls the aperture stop and the pen of the recording mechanism; the recording is on a drum of 15 cm useful width. The percentage transmittance is recorded against a linear wavelength scale. The spectrum obtained is 49 cm long for a full recording of the NaCl region; limited spectral regions can be extended to three times the normal length. Recording times of 1·5, 3, and 4·5 min per μ are available.

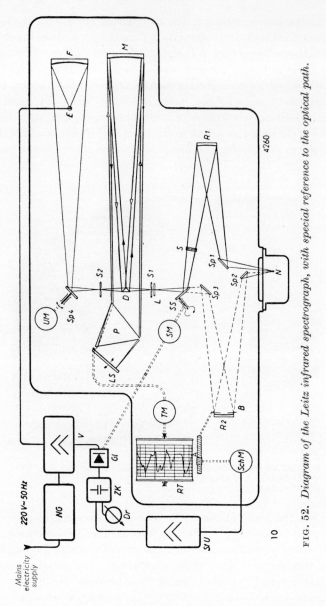

FIG. 52. *Diagram of the Leitz infrared spectrograph, with special reference to the optical path.*

The prism can be changed very simply, parts of the control mechanism of the normal setting (for NaCl) being used again. The spectrometer can also be used for single-beam operation; in addition it can be delivered as a universal instrument including the ultra-violet and visible spectral regions.

6. The Perkin-Elmer Spectrometers

The firm Perkin-Elmer Corp., Norwalk, Conn., U.S.A., with the German subsidiary Perkin-Elmer and Co. G.m.b.H., Überlingen/See, offers at the present time the largest selection of both the larger and smaller instruments available for infrared spectroscopic purposes. The main interest is naturally in the spectrometers themselves. There are two self-recording single-beam instruments (Models 12 C and 112), one of which is a multi-monochromator on the 'double pass' principle. Model 21 is a recording double-beam spectrophotometer with an optical-wedge attenuator. Model 13 can be operated as a single or double-beam instrument with phase differentiation of the two optical paths, potentiometric measurement of the energy ratio, and automatic recording. Model 137 is a smaller double-beam spectrometer with optical compensation. In addition there is a comprehensive range of accessories.

The fundamental construction of the monochromator is the same for all the spectrometer types: it is shown diagrammatically in the left half of Fig. 55. It uses the usual Littrow arrangement of a 60° prism of 75 mm base and 60 mm height and a parabolic collimator mirror of 80 mm diameter and 27 cm focal length. Prisms of all the customary materials are available to cover the whole infrared region at present attainable with prisms; CaF_2, NaCl, and KBr or CsBr are especially recommended. Changing the prisms, together with all the necessary changes in the control mechanism of the Littrow mirror and the slits, can be carried out in a few minutes, retaining the adjustment of the monochromator. A device for compensating the influence of temperature on the dispersion of the prism is included; in addition, the monochromator is kept at a somewhat raised constant temperature by means of internal heating elements. The slits are bilateral, 12 mm high, and can be controlled electrically by means of set programmes or they can be hand-set to any desired value between 0 and 2 mm. The entry slit is bent. A quickly responding high-efficiency thermo-element serves as detector. An image of the exit slit, decreased in the ratio 6:1, is formed on the detector by means of an elliptical concave mirror. To eliminate short-wave stray light, a reflection or transmission filter is used, which is moved automatically into the optical path at about 9 μ or manually at any other desired wavelength. In Model 112 this monochromator is converted into a double monochromator of the Walsh double-pass type.

The individual spectrometer types are essentially differentiated by the type of slit illumination and by the electronic treatment and evaluation of the signal received. For the radiation source, Model 12 C uses a globar at about 1300° C in a water-cooled casing. Its radiation is thrown on to the entry slit by a suitable mirror, being modulated on the way by means of an interrupter disc rotating at 13 c.p.s. The detector signal is amplified several million times in a tuned, resistance-coupled amplifier and then rectified by a commutator placed on the axis of the interrupter disc. The instrument can also be used for measurements with unmodulated radiation. A Brown recorder with a measuring range of 10 mV at 4 sec response time and an accuracy of 0·2 per cent is used for registration. To simplify the precise measurement of wavelength, marks are made automatically on the spectrum

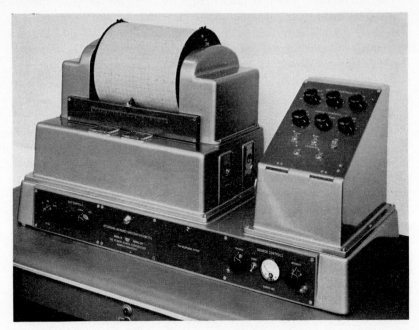

FIG. 53. *View of the Perkin-Elmer double-beam spectrophoto-meter Model 21 (photograph: Perkin-Elmer Corp.).*

by a short displacement of the pen at every one-tenth of a revolution of the wavelength screw driving the Littrow mirror. The exact determination of the radiation zero point can be made, even when stray light is present, by the use of three filters of metal, glass, and LiF, which can be moved into the optical path as desired. Any of four recording times can be selected, namely 1, 2, 4, or 8 min per revolution of the wavelength screw.

Model 112, in most respects similar to the instrument just described, differs from it by the use of a double-path monochromator. Provision is made for three recording times only. A Leeds and Northrup Speedomax Type G is used for recording purposes with a response time of 2 sec.

Recently, the Models 12 C and 112 have been available as grating instruments (then called 12 G and 112 G) with KBr prisms for preliminary resolution and an echelette grating of 75 lines per mm and blazed for 12 μ [Siegler and Huley (*657*)]. In double-beam operation a resolution of 1 cm^{-1} at 2·6 μ, and 0·6 cm^{-1} at 11 μ is obtained. A spectral region from 2 to 17 μ can be covered, and this may be extended at both ends by changing the prism and grating.

Model 21 (view, Fig. 53; block diagram, Fig. 54; optical path, Fig. 55) is probably the best known double-beam spectrophotometer at the present time [White and Liston (*739*)].† The arrangement of the photometer unit

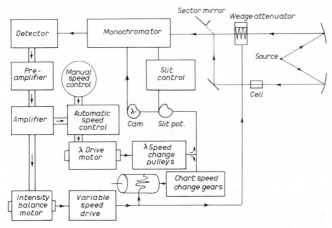

FIG. 54. *Block diagram of the Perkin-Elmer spectrophotometer Model 21.*

can be seen from Fig. 55. The two pairs of illuminating mirrors M 1, M 3, and M 2, M 4 direct the radiation of the Nernst glower (which is started electrically) along two paths which at first lie parallel with one another and then coincide and lead to the entry slit of the monochromator. The optical path furthest from the operator is the comparison beam. In addition to the empty cell, etc., it contains an optical-wedge attenuator which serves to equalize the energy when absorption occurs in the investigating beam. The latter passes through the cell containing the substance under investigation. A second wedge attenuator, which can be adjusted before the measurement and then remains fixed, is placed at the focus of the investigating beam and is used to ensure equal energy in the two beams when there is no absorption.

† Baker *et al.* (*792*) have described an automatic programme control for this spectrometer which allows it to be used economically in routine work because only the necessary spectral regions are recorded slowly and the others quite quickly.

Equipment and Experimental Technique

The two beams are brought together by means of a system of fixed plane mirrors and a sector mirror rotating at 13 c.p.s. which alternately reflects and transmits; thus each beam is subjected to an equal number of reflections. The toroid mirror M 9 actually focuses the beams on to the slit.

The 13 c.p.s. signal occurring for energetic inequality in the two beams is (as can be seen from the block diagram of Fig. 54) amplified in a preliminary amplifier and then in a four-stage main amplifier. It is rectified, its phase compared with a reference signal, thoroughly screened and, after conversion to a 60 c.p.s. signal, again amplified. The resulting output inserts the wedge attenuator, by means of a servomotor, until equal energy results in the two beams; the signal then becomes zero and the movement of the wedge

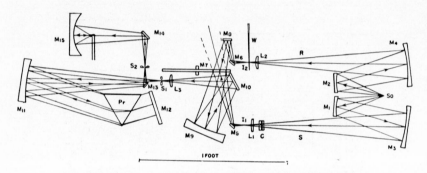

FIG. 55. *Optical path of the Perkin-Elmer spectrophotometer Model 21.*

attenuator stops. The position of the wedge, which is proportional to the absorption to be measured, is registered by the pen-recording mechanism. This consists of a large drum on which spectra are recorded with a linear wavelength or wave number scale and percentage transmission (or the corresponding extinction) scale. The recording speed of the spectrometer can be varied continuously up to a maximal value for certain regions of 3, 10, 40, and 120 min per μ; these times can be increased twelve-fold by means of a special adjustment. An automatic decrease in the drum velocity can be used to increase the recording time when an absorption band occurs, so that an accurate representation of the absorption is recorded. The gear on the recording drum, and therefore the scale of the spectrum, can be varied within very wide limits, namely those corresponding to the values 1, 2, 4, 5, 10, 20, 40, and 50 in. or cm per μ or per 100 cm^{-1} (according to whether the scale is divided into inches or metrically, and whether a linear wavelength or wave number scale is used). The wavelength or wave number and the slit width at which the instrument is set can be directly read off on a counting

172

device. According to the continually changing slit setting in the course of the given programme, very different values of resolution, almost up to the theoretical limit, can be selected. The greater resolution resulting from a decrease of the slit width is accompanied by a deterioration of the signal to noise ratio which is compensated by the truncation of the bands by electrically screening and in the recording instrument. Various tests of the accuracy of the whole construction, such as test signals of defined magnitude, etc., are incorporated and can easily be operated. It is also worth noting that the

FIG. 56. *View of the Perkin-Elmer spectrophotometer Model 13 (photograph: Perkin-Elmer Corp.).*

Model 21 can now also be delivered as a double-pass instrument, the special advantage of which is the elimination of stray light for work in the long-wave region. Model 21 can also be delivered with a grating instead of a prism as the dispersion medium.

As a further development of the well-proven Model 21, the Perkin-Elmer Corporation has produced Model 221. It corresponds externally to Model 21. Indeed, almost all the well-tested parts can be found again in the new model. The automatic control of the amplification to correspond to the requirements of the spectrum being drawn at the time is new. By this means, weak and strong bands are drawn with the same quality and differential analysis is especially improved. In addition an

Equipment and Experimental Technique

ordinate extender (which can also be fitted to Model 21) is built in as a matter of course. Thereby, in any desired region of extinctions, the ordinate can be extended 5-, 10- or 20-fold (in transmittance) which is especially valuable for trace analysis. The recording velocity is automatically controlled to the optimum value so that with larger spectral slit widths the recording time becomes smaller (the mean velocity can of course be selected). Furthermore, the velocity is slowed down in regions of bands and accelerated in regions

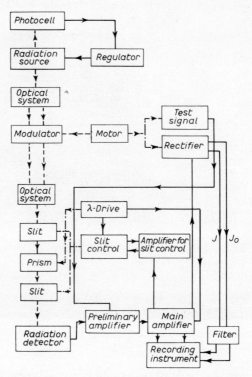

FIG. 57. *Block diagram of the Perkin-Elmer spectrophotometer Model 13.*

free of bands. As a result the recording time is suited to the selected slit programme according to the particular problem. The recording of a spectrum takes about 5 to 6 min and running back to the initial position only 1 min. All the accessories for Model 21 are also applicable to Model 221.

Model 13 (Fig. 56) was first produced as an accessory to an older Perkin-Elmer spectrometer by Savitzky and Halford (*629*). Later it was developed into a complete spectrometer. It is a double-beam instrument which produces percentage spectra by an electronic method and without an optical

174

attenuator. Fig. 57 gives a block diagram of the arrangement. The arrangement and interaction of the individual parts and processes is adequately
indicated by the description in Fig. 57. The arrangement of the photometer
unit is shown in Fig. 58. The Nernst glower N, the emission of which is held
constant by measurement of the visible part with a photocell controlling the
heating current, is placed in the axis of an illuminating mirror $M\,1$ of wide
aperture. The beam reflected on this is divided by the plane mirror $M\,2$ into
two distinct parts which are energetically equal. One of these serves as the
investigating beam and the other as the comparison beam. In order to
ensure full symmetry of the two optical paths (it should be noted that the

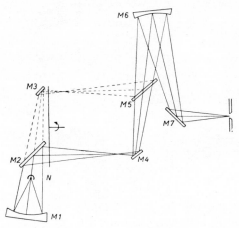

FIG. 58. *Photometer unit of the Perkin-Elmer spectrophoto-*
meter Model 13.

same portion of the radiator surface is used for both) the investigating beam
is also reflected at the plane mirror $M\,3$ before it enters into the cell portion.
Before passing through the cells, both rays are modulated at 13 c.p.s., at a
phase difference of 90° with each other. After passing through the cells, the
two beams are reunited by the two plane mirrors $M\,4$ and $M\,5$ to one single
beam, which is focused on to the entry slit of the monochromator by means
of the concave mirror $M\,6$ and the plane mirror $M\,7$. The monochromator is
the same simple arrangement as in the other Perkin-Elmer instruments;
the use of a double-pass arrangement is not possible. The formation of an
image of the monochromator exit slit on to the detector and the construction
of the detector itself are as already described. The signal relayed by the
detector, composed of parts of both the beams corresponding to the modulation in the two optical paths, is amplified and then each part is separately

175

rectified by mechanical commutators and screened so that the ratio of the two is formed. For reasons of space, details of the complicated electronic parts of the instrument cannot be discussed here. The spectrometer can be used as a single-beam instrument instead of double beam, for this the mirror in the photometer unit is brought into the required position mechanically; the electronic parts must also be suitably modified. Both of these alterations can be carried out by means of simple operations so that the conversion may

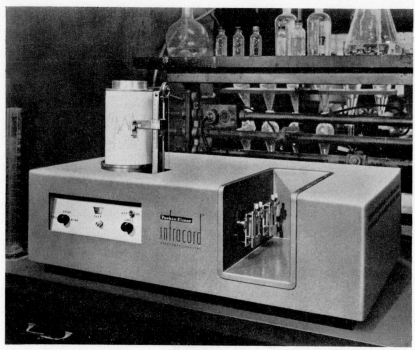

FIG. 59. *View of the Perkin-Elmer 'Infracord' spectrophoto-meter (photograph: Perkin-Elmer Corp.).*

be very quickly effected. This enables the instrument to be adapted to the most diverse problems.

A rapid spectrometer, and an automatic many-component infrared analyser (Model 131) from the range of Perkin-Elmer spectrometers must also be mentioned. These specialist instruments cannot be described in detail. Model 131 uses a modified 12 C spectrometer and measures the radiation energy at ten wavelengths according to a prepared programme. A calculating machine determines the extinction values from the results and makes them available for further interpretation by means of an electric typewriter, a counter mechanism, or a punched card.

The 'Infracord' (or Model 137) is a further spectrometer of the Perkin-Elmer range. It is a robust, simple, self-recording, double-beam spectrophotometer with optical compensation and it is produced at a very much reduced price (less than half that of the Model 21). This has been achieved by means of a simple construction and method of operation; although this gives up many of the refinements of other spectrometers, it still offers everything necessary for a successful application of infrared spectroscopy in laboratories lacking in funds and trained technical assistants. The appearance of the instrument is shown in Fig. 59. An electrically heated, blackened rhodium rod serves as the radiation source. The monochromator contains a 70° NaCl prism included in a refined optical system. The slit control is automatically set to a fixed programme corresponding to a resolution of about $0 \cdot 04 \mu$, but can also be manually set. The recording time for the spectral region attainable ($2 \cdot 5$ to 15μ) is fixed at 12 min, with the speed of recording adapted to the resolution previously mentioned. The percentage transmittance is recorded for a linear wavelength scale on a chart 100 mm wide and 260 mm long. After having been used, the instrument can be manually returned to the initial position without loss of time. Model 137 can also be delivered with KBr optics and under the name Model 137 G as a grating instrument for the wavelength region $0 \cdot 83$ to 7μ.

7. The Unicam Instruments

The SP 100 Infrared Spectrophotometer manufactured by Unicam Instruments Ltd., Cambridge, England, is an elegant and very versatile instrument. It has been designed as a high-precision double-beam spectrophotometer with an evacuable light path, the complete equipment forming a self-contained unit. The appearance is shown in Fig. 60 and the optical arrangement in Fig. 61.

The source is a water-cooled Nernst filament fitted to the outside of the vacuum cover. Radiation from this source is passed alternately at 10 c.p.s. through the sample and reference channels. The two beams come to a focus in the sample cell compartment and on a star-wheel beam attenuator respectively. A prism/grating double monochromator is used in which the grating monochromator has been ganged with a rock-salt monochromator on a linear frequency basis [Tarbet and Daly (835)]. Two replica gratings, each used in a single order, cover the range 650 cm^{-1} to 3650 cm^{-1} in conjunction with a 60° rock-salt prism of 6 cm height, $7 \cdot 5$ cm edge; each monochromator having an aperture of F/6·5. Gratings, cams and other prisms mounted in the thermostatted optical section may be interchanged by operation of

FIG. 60. *View of the Unicam SP 100 infrared spectrometer (photograph: Unicam Ltd.).*

press buttons converting the instrument to a single-prism monochromator as desired and extending the frequency range from $375\,\mathrm{cm^{-1}}$ to $8850\,\mathrm{cm^{-1}}$.

A constant energy background is provided by magnetically operated slits programmed by a tapped potentiometer. A Golay pneumatic detector provides an electrical output signal which is ultimately used to move the star-wheel beam attenuator in the reference beam. The star-wheel has 24 teeth

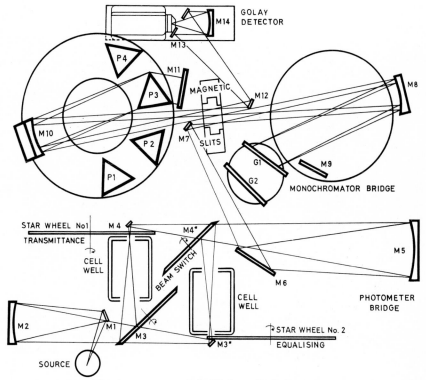

FIG. 61. *Arrangement of the Unicam SP 100 spectrometer.*

and rotates at 3000 r.p.m. It is moved laterally changing the light transmitted in the reference beam according to changes occurring in the sample beam on the normal optical null principle, the teeth being profiled to give linear transmission changes with lateral displacement. The movement of the unit is linked mechanically to the pen-recording transmission. The star-wheel attenuator ensures good beam equality and an accurately linear transmission scale.

The spectra, linear in transmission and wave number, are recorded on a drum recorder, wave number intervals being recorded by pens in both chart

179

Equipment and Experimental Technique

margins. A second proportional output may be used for simultaneous feeding to a second recorder or computer and in this way, scale expansion facilities and absorbance records are obtained.

The whole optical system is evacuable using built-in equipment achieving a pressure of 10^{-3} mm Hg within the vacuum casing, which is considered satisfactory for accurate work under high resolution. Having selected the desired optical configuration, the operator may obtain the optimum recording conditions using the comprehensive gear-box system (the scanning time

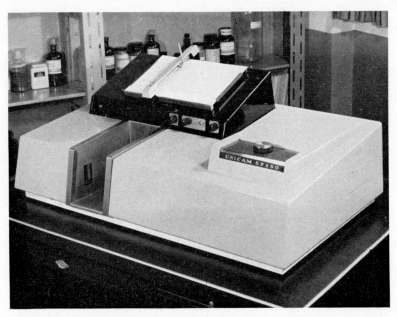

FIG. 61a. *View of the Unicam SP 200 spectrometer.*

per cam is variable from 2 min to $21\frac{1}{3}$ days) in conjunction with electrical gain, damping and servo ratio controls.

Following the recent trend towards low-cost infrared instrumentation for the routine chemical laboratory, Unicam also manufacture the SP 200 spectrophotometer. This instrument is very simple to operate, spectra being recorded over the range 650 cm^{-1} to 5000 cm^{-1} in 10 min on a normal scan. The source used is an air-cooled Nernst and the detector is a Golay pneumatic cell (Fig. 61a).

The wave number scale is linear over the region 650 cm^{-1} to 5000 cm^{-1} with a scale change at 2000 cm^{-1}, the chart being of standard foolscap size for convenient filing.

180

The slits are controlled by a variable mechanical programme. Provision has been made for a second recorder to be used for simultaneous registration of spectra in absorbance or with transmission scale expansion.

8. Zeiss Spectrometer

The firm VEB Carl Zeiss, Jena, has put on the market the infrared spectrometer UR 10 (Fig. 62). This contains a monochromator in the Littrow arrangement fitted with three prisms (LiF, NaCl, and KBr) which are moved automatically and successively into the optical path. Percentage transmission

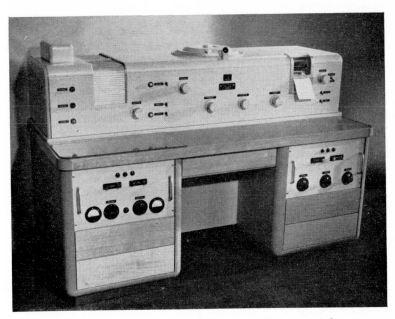

FIG. 62. *View of the Zeiss UR 10 spectrophotometer (photograph: VEB Carl Zeiss, Jena).*

is recorded and this is obtained by the optical compensation method using a wedge attenuator. A noteworthy specialty is that the programme selector (Fig. 63) allows smaller or greater regions of the applicable spectral region of 400 to 5000 cm^{-1} to be selected in sections of 50 cm^{-1}. These can be recorded at a set velocity, which can be varied within wide limits; the other parts of the spectrum being passed through very quickly. The programme selector can also be brought into action while a spectrum is being recorded. The spectrum is drawn on wax paper of 10 cm effective width. A complete system of co-ordinates (percentage transmission and wave number) is drawn by a special mechanism during the registration; discrepancies between printed co-ordinates and spectrometer setting are thus eliminated. Alteration in the abscissa scale or the speed of registration is possible at any time. An automatic decrease in the velocity of recording occurs when passing through very sharp bands in order that a true representation of the absorption is drawn. A thermostat is included as an integral part of the instrument; it keeps the most important and most sensitive parts of the spectrometer at a constant temperature and the apparatus also ensures that the optical path is sufficiently free from water vapour and carbon dioxide.

FIG. 63. *Table of three prisms and programme selector of the Zeiss UR 10 spectrophotometer (photograph: VEB Carl Zeiss, Jena).*

A complete set of gas and liquid cells for the spectrometer is included in the price.

9. Spectrometers for the Near Infrared Region

All the spectrometers discussed or mentioned in the preceding sections can be used in the near infrared region if the dispersion system is suitably selected. This spectral region has recently gained a marked importance for the solution of several special problems. Therefore, spectrometers have been developed and constructed specially for this purpose, in part as complete instruments, in part as extensions to available instruments for the ultraviolet and visible spectral regions [Kaye (*374*); Kaye *et al*. (*376, 377*)]. A list is given in Table 9 (which is not claimed to be comprehensive). Details cannot

TABLE 9. *Spectrometers for the Near Infrared Region*

Manufacturer	Model	Description
Applied Physics Corp., Pasadena, Calif., U.S.A.	Cary 14	Double monochromator, $0 \cdot 19$ to $2 \cdot 8$ μ, fully automatic recording with a double pen writing device, transmittance or extinction against wavelength
Beckman Instr., Fullerton, Calif., U.S.A.	DK	Simple monochromator, $0 \cdot 19$ to $3 \cdot 5$ μ, fully automatic recording transmittance or extinction against wavelength
Mervyn Instr., St. John's Woking, Surrey, England	IRS/A/P	Grating monochromator, 1–4 μ, without recording mechanism
Optica, Mailand	CF 4 NI	Grating monochromator, 1–4 μ, without recording mechanism
Perkin-Elmer Corp., Norwalk, Conn., U.S.A. (Bodenseewerk, Überlingen)	Warren Spectracord 4000	Double monochromator, $0 \cdot 2$ to $2 \cdot 8$ μ, fully automatic recording transmittance or extinction against wavelength
Carl Zeiss, Oberkochen	reg. Spektralphotometer	Simple monochromator, $0 \cdot 21$ to $2 \cdot 7$ μ, fully automatic recording transmittance against wavelength

be given here, but it is common to all these instruments that they use electrically heated metal filament lamps as radiation sources, and photo-resistance cells as radiation detectors. Sometimes registration mechanisms are not delivered by the manufacturers as components of the instruments and these must be obtained separately.

6. Spectrometer Accessories

The infrared spectrometers as described in the preceding sections need certain accessories before they can be used at all and others to extend their range of application. These will now be discussed. The most important are the containers in which the investigated substances are placed for different states of aggregation; they are generally called cells.

1. Cells for Gaseous Substances

The containers used for measuring the absorption of gases consist in general of cylindrical tubes of glass or metal, unless another material is necessary on chemical grounds. The ends are closed by infrared-transparent windows which are cemented to the tubes. Two side tubes, which can be closed by

FIG. 64. *100 mm gas cell (photograph: Perkin-Elmer Corp.).*

stop cocks, are provided for leading the gas in and out, and sometimes a further side tube for holding a little liquid of which the vapour is to be investigated. One set, or for a double-beam instrument two identical sets, of cells of different thickness are needed, e.g. those of 20, 50, 100 mm, and 1, 10, 40 m, the last of which can usually be set for different thicknesses as desired.

184

Fig. 64 shows a 100 mm gas cell, Fig. 65 one of 1 m and Fig. 66 one of 10 m. The use of the different thicknesses will be discussed in the next chapter.

All the materials from which prisms can be constructed come into consideration for the end windows; however, the transmittance of the windows

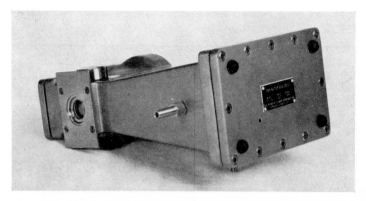

FIG. 65. *1 m gas cell (photograph: Perkin-Elmer Corp.).*

reaches further into the infrared because of their smaller thickness (usually between 5 and 15 mm). Table 10 gives the summary of the materials that can

FIG. 66. *10 m gas cell (photograph: Perkin-Elmer Corp.).*

be used for cell windows. In general, for economy, windows of only one or two materials are used, according to the spectral region in which the investigation will be carried out. The windows, ground planar and polished, are fixed

185

TABLE 10. *Infrared Transparent Materials for Windows*

Material	Limiting wavelength† μ	Material	Limiting wavelength† μ
Glass	*ca.* 2·5	$BaF_2\|$	*ca.* 13
Quartzglass	3·8	NaCl	20
Quartz	4·4	Ge	21
Mica	5·3	KCl	25
Sapphire	6·5	AgCl¶	28
LiF	7	KBr	32
Periclase (MgO)‡	10	KI	37
CaF_2	10	CsBr	50
NaF	10	CsI	60
CdF_2§	10·5	KRS 5	50
As_2S_3	12		

† That wavelength at which the transmittance of a 5-mm plate has fallen to about 50 per cent.

‡ Suitable for temperatures up to about 1500°. Obtainable from The Infrared Development Co., Welwyn Garden City, Hertfordshire, England.

§ Hygroscopic character can be deduced from the solubility: 4·4 g in 100 g H_2O. Literature: Haendler *et al.* (*286*).

‖ Hygroscopic character can be deduced from the solubility: 0·17 g in 100 g H_2O. Literature: Potts and Wright (*821*), Gordon (*803*).

¶ Hygroscopic character can be deduced from the solubility: 9×10^{-5} g in 100 g H_2O; sensitive to light.

to the ends of the cell-tube by means of a cement. This cemented joint can be mechanically self-supporting or merely a means of making the appliance air tight and be supported by means of an outer holder. This depends on the mechanical, thermal and chemical properties of the materials in question, their relationship to one another, and whether it is desired that the cell should be permanent or not. Many thermoplastic substances, e.g. picein, and also cold- or warm-cured synthetic resins, such as Araldite and Glyptal, are available as cements for permanent connections. For cells which can be quickly altered, the usual greases of vacuum technology are used. In each case, the joint must be air-tight and if possible resistant to moderate excess pressures [Smith and Creitz (*676*)]. For larger pressures, special construction is needed as has been described (with simultaneous variable thickness) by Weber and Penner (*729*). For investigations at raised temperatures electrical heating appliances of suitable dimensions are used.

Gas cells of smaller thicknesses (usually up to 100 or 200 mm) can be placed directly into the optical path in front of the monochromator and the beam passes directly through once only. However, the cells with 1, 10, and

40 m thickness mentioned above would be inconvenient or unusable if they were constructed with lengths of this magnitude. Therefore, they are made of a more convenient length and the required absorption thickness obtained by repeated reflection of the radiation within the cell [White (*735*)]. The arrangement of the optical path within a 1 m cell is shown in Fig. 67. The

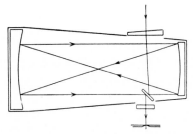

FIG. 67. *Optical path within a 1 m gas cell.*

arrangement of a 10 m cell is shown in Fig. 68. The number of reflections between mirrors M_A and M_B and the mirror M_F, which determines the effective thickness of the cell, is regulated by the inclination of the former two mirrors, which can be altered from outside the cell. For an overall length of the cell container of about 40 cm, optical path lengths of from 1·25 up to

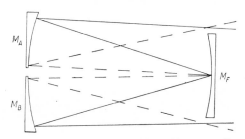

FIG. 68. *Optical path within a 10 m gas cell.*

10 m are obtainable in steps of 1·25 m. Obviously, the coefficient of transmittance of the empty cell is thereby decreased. At 1·25 m it is about 75 per cent, at 7·5 m about 50 per cent, and at 10 m still about 40 per cent [Pilston and Wright (*535*), White and Alpert (*830*)]. Fig. 66 shows how this cell is placed in the spectrometer.

2. Cells for Liquids

In contrast to the gas cells, cells for liquids have only very small path lengths, corresponding to the high extinction coefficients of liquids for the fundamental vibrational bands. A set of cells of the approximate sizes 0·025,

Equipment and Experimental Technique

0·05, 0·1, 0·2, 0·5, 1, 2, and 5 mm is recommended. Cells are of two types, those that are cemented together permanently (fixed cells) and those that can be dismounted; both kinds have their advantages and disadvantages. Details of cell technique can be found in the work of Coggeshall (*125*), Colthup (*135*), Kivenson *et al.* (*391*) and Lord *et al.* (*430*). The production of very small thicknesses is difficult, especially as the usual window materials are rather soft and therefore difficult to grind and keep planar.

The materials already listed (Table 10) can be used as cell windows as long as they are not appreciably attacked chemically or physically by the substance to be investigated. The desired distance is obtained by means of a thin spacer of suitable thickness, usually made of calibrated (and sometimes amalgamated) lead foil. The cell itself is placed in a holder which serves to press the two windows and the spacer between them together, and also to simplify the insertion into the spectrometer. The ease of filling and cleaning the liquid cells is of especial importance experimentally. The use of fixed cells is advantageous for volatile substances because they are completely sealed off. They possess two small holes, outside the area of the radiation path, through which the liquid is introduced, e.g. by means of a syringe. While the spectrum is being measured, these openings are closed by suitable stoppers which prevent volatilization of the liquid. The layer thickness of such a cell is better defined than that of a dismountable cell. However, it must be kept in mind that all window materials suffer a certain amount of wear as a result of their solubility; in time this renders the given layer thickness meaningless. In addition, fixed cells are often very difficult to keep clean. For cleaning purposes, volatile solvents of high solvent power may be used, e.g. hexane or chloroform. If the liquid investigated is subject to polymerization or leaves difficultly soluble residues on volatilization then its complete elimination is difficult and frequently only possible if the cell is taken apart (contrary to normal practice).

Frequently, demountable cells are recommended. These consist simply of two windows pressed together in a holder and separated by a spacer. Such cells are filled by laying the spacer on top of one of the windows, dropping on one or more drops of the desired liquid, and covering with the other window; superfluous liquid overflows between the windows and the spacer. Such a method of filling is not satisfactory, although the cleaning of the cells in the dismantled state is naturally easy. More satisfactory is a type of liquid cell intermediate between the fixed and the demountable cells already mentioned (Fig. 69). This consists of two windows (usually NaCl) which are held in a mounting separated by a spacer and pressed together by means of three screws, but without any cementing material. One of the windows is pierced by two holes of *ca.* 1 mm diameter at 120° to one another. At first, these run

188

from the edge of the window towards the middle and then, as soon as the thickness of the spacer has been exceeded, make a right-angle turn and run parallel to the direction of the beam and end in the hollow space within the cell. On the outside there are two metallic nozzles for filling, which are connected to these two holes by means of a lead or teflon ring. Filling is carried out by means of a pipette or syringe; even for small thicknesses the liquid quickly fills the cell space, displacing the air, by means of capillary force. Emptying is carried out by air pressure or by sucking out by means of a pump. This type of cell can be washed with liquids as desired for cleaning and can easily be taken apart if necessary. Windows that have become cloudy can be repolished and used again with all the other parts. The production of the two boreholes needed is carried out directly in the cell holder through the delivery tubes by means of a suitable hand drill. In this way exact correspondence of the delivery tube and the boreholes is ensured.

Cells for volatile substances need special care in construction and a reservoir from which the cell is automatically refilled with the compound being investigated [Broadley (*84*), Friedel and Pelpetz (*232*), Fry *et al.*

FIG. 69. *Liquid cell (photograph: BASF).*

(*235*), Oetjen *et al.* (*507*), Smith and Miller (*677*), Smith and Creitz (*676*), Pierson and Olsen (*534*)].

For some investigations, a knowledge of the exact value of the layer thickness used is desirable or indispensable. Obviously the thickness of the spacer ring cannot always be relied on because the windows might have been attacked or the material of the ring could have been pressed into that of the windows. Fortunately, the thicknesses of the cells can easily be determined in the infrared spectrometer provided the inner window surfaces are reasonably parallel and planar. When radiation passes through the empty cell (and sometimes even when the cell is full) interference occurs between the rays which have been reflected several times at the surfaces of the cell. The interference bands depend on the wavelength of the radiation and the thickness of the cell used; their influence on the spectrum takes the form of a wave train, the maxima of which become successively further apart (Fig. 70) with increasing wavelength (if the registration is linear in wavelength). Simply by counting the number n of the maxima between two

189

Equipment and Experimental Technique

wave-lengths λ_1 and λ_2 (for registration with linear wavelength scale) or between the frequencies ν_1 and ν_2 (for linear wave number scale) at which interference maxima lie, the thickness follows from the formula†

$$d = \frac{n}{2} \cdot \frac{\lambda_1 \cdot \lambda_2}{\lambda_2 - \lambda_1} \quad \text{or} \quad d = \frac{n}{2} \cdot \frac{1}{\nu_1 - \nu_2} \qquad [\text{II, 6.1}]$$

If this direct method cannot be used, there is a series of other possible methods: determination of the thickness by the weight of a liquid which fills the cell, e.g. mercury [Fox and Martin (*219*)]; from the measured absorption of a substance of known properties, possibly in another spectral region [Nielsen and Smith (*498*), Smith and Miller (*677*)]; from the interference bands which, as a result of multiple reflections, overlap the spectrum of the

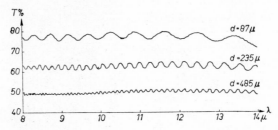

FIG. 70. *Interferometric determination of the thickness of an empty liquid cell.*

substance in the cell at sufficiently small thicknesses [Ellis (*198*), Smith and Miller (*677*), Gordon and Powell (*266*), Sutherland and Willis (*693*)]; from the fact that the optical path is longer in a filled cell than in an empty one and that this can be measured in a Jamin or Rayleigh interferometer [Jaffé and Jaffé (*353*)]. Of all these methods, it has been shown that those depending on the determination of the thickness in question from the strength of absorption of one or several isolated bands of a substance which can be easily obtained pure, e.g. CCl_4, are the best in practice. This applies especially when thicknesses must be measured or tested frequently. For this method, it is necessary that the thickness of one given cell should have been previously standardized, e.g. by the interference method described. Selected bands of the test substance are then measured in this cell under conditions which should henceforth be kept constant. The extinction values thus determined and remeasured at intervals are then used to determine the thicknesses of unknown cells.

† For the maximum at λ_1 the expression $2d = k_1 \cdot \lambda_1$, for that at λ_2 correspondingly $2d = k_2 \cdot \lambda_2$, where k_1, k_2 are whole numbers and $k_1 - k_2 = n$. Solving these expressions for d, the formula given follows.

3. Special Cells

In addition to the normal cells for liquids just described, there are many of special construction for specialized purposes and measurements. The most important will be briefly discussed.

One of the most useful, and also most complicated, of the specialized forms of cell for liquids is that with a thickness that can be altered as desired [White (737); Coates (122); Stuart (690); Adams and Katz (4)]. Usually, the values that can be selected lie between about 0·01 and 5 mm and the accuracy of the thickness is about 0·001 to 0·002 mm. Many types of these cells

FIG. 71. *Cell for liquids with variable thickness (photograph: Perkin-Elmer Corp.).*

exist. Fig. 71 shows the outward appearance, and Fig. 72 a section, of a cell of variable thickness that used to be manufactured by Perkin-Elmer [White (737)]. The cells of this type which are now commercially available are more simply constructed and therefore easier to handle. A variable cell of a wedge shape in which parts of the cell are not displaced with respect to another, but instead the whole cell is pushed into the slit, deserve special mention.† Such cells are especially valuable for the investigation of solutions in double-beam spectrometers where it is desired that the comparison beam should contain an amount of solvent equivalent to that in the measuring beam, in order to compensate for this in the spectrum. The necessary thickness, which of course is smaller than that in the investigating beam, can be correctly selected empirically with the help of this type of cell. These cells are filled by means of a syringe passing through a nozzle which is then shut. Liquids which polymerize, or solutions which leave rubber-like residues

† Manufacturer: Connecticut Instrument Corporation, Wilton, Conn., U.S.A.

191

Equipment and Experimental Technique

when the solvent volatilizes, should not be used in the cell. Understandably, the complicated and sensitive mechanism of these precision instruments makes it necessary that they be handled with care. The manufacturer gives detailed instructions for their use which we shall not discuss here.

Usually, the volume needed to fill the cell is negligible compared with the amount available, provided that the investigation is made in the region of the fundamental vibrations where very small thicknesses are required. Occasionally only very small amounts are available, especially of biochemical substances such as vitamins, ferments, enzymes, etc. In these cases

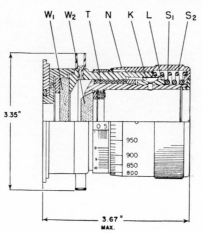

FIG. 72. *Section through a variable thickness cell for liquids shown in Fig. 71. W_1 fixed window, W_2 movable window (mounted in such a manner that it can be made parallel to W_1), T teflon gasket, K and L screw and guide to prevent the movement of the piston with the micrometer screw, S_1 and S_2 protective springs.*

microcells must be constructed and used [Wollman (*749*); Blout *et al.* (*73*); Davison (*164*); Friedel and Pelipetz (*232*)] (Fig. 73). Such cells are placed very near to, or even at, a real image of the radiation source in the optical path and are just large enough to cover the cross-section of the beam used. Such microcells can also be arranged for measurements at different temperatures [Zenchelsky and Showell (*832*)]. The microcells constructed by boring *cavities* in infrared transparent materials are of special interest. These have been described by Jones and Nadeau (*809*) and are manufactured by Connecticut Instrument Corporation, Wilton, Conn., U.S.A. In the overtone region, where layer thicknesses of 10 cm and more are often used, consideration must frequently be given to the amount of material available. Hansen

192

and Mohr (*297*) have shown how these cells can be designed to require the minimum volume.

Cells which allow measurements to be made at other than room temperatures are of great importance for several types of investigation. As long as only moderate temperatures are required, it suffices to introduce the liquid substance into a prewarmed cell. The cell can also be warmed with a stream of hot air during the determination of the spectrum [Mitzner (*418*)]. For temperatures up to about 200° C, a normal cell is slightly modified by surrounding it with an electrically heated mantle which is mounted parallel to the direction of the beam. The resistance to heat of the cement which holds

FIG. 73. *Microcell for liquids with* (left) *the corresponding mounting* (*photograph: Perkin-Elmer Corp.*).

the windows to the rest of the cell is of especial importance. Simard and Stegner (*659*) recommend AgCl for this up to 200° C, but as a result of its different coefficient of expansion from that of glass and NaCl, the cells frequently break on cooling. Therefore Neu (*492*) recommends a silicone, which is useful up to 350° C, while Smith (*671*) suggests Glyptal, but this can be used only up to 200° C. If a very even temperature is desired throughout the whole cross-section of the cell, the heating mantle must be very long (which is often not possible because of space limitations) or the faces of the cell windows must themselves be heated. Such heating is most simply done by means of a very thin platinum wire placed across them. The influence of the radiation emitted by the hot cell on the detector is eliminated by modulating the radiation to be measured before it passes through the heated cell, or by the introduction of similarly heated cells into both the optical paths. Most of the materials to be considered have such a low emission in the usual infrared regions that their own radiation is of little importance.‡ The various

‡ Fischer and Brandes (*211*) point out that it is necessary to consider emission effects of heated samples especially for different relationships in the two optical paths of the spectrophotometer and for materials with marked absorption bands at the wavelengths under investigation.

O

Equipment and Experimental Technique

types of heated cells that have been used differ considerably in their construction, but no one of them has been shown to be generally superior to the others [see, for example, Brown and Holliday (90)]. A simple and very useful type of cell for liquids at moderate temperatures has been described by Funck (236) and is shown in Fig. 74. In this reference details are also given for the construction of heated gas cells and low temperature cells. Fig. 75 shows a heated cell which is horizontally arranged and is especially suitable for solids. It is brought into the optical path by means of two mirrors lying above and below the sample holder; for more efficient heating, a mantle

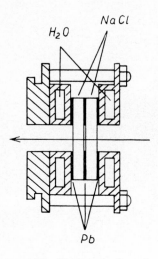

FIG. 74. *Section of a heated cell for liquids according to Funck (166).*

FIG. 75. *Horizontal cell with heating arrangement (without mantle) (photograph: Perkin-Elmer Corp.).*

covers the whole except for suitable openings for the radiation beam. In order to eliminate the elongation of the optical path resulting from this arrangement, and in order to retain the focusing arrangement, one or more of the focusing mirrors must be displaced (unless suitable convex mirrors are used as the cell mirrors). Genzel (247) has described a cell for very high temperatures (above 1000° C). In this, the substance under investigation is placed in a crucible on a temperature resistant plane mirror (platinum); the beam thus passes through the substance twice.

The construction of cells for low temperatures, down to those of liquid hydrogen or helium, is much more complicated. The principle used is to connect the actual cell containing the substance to be investigated by a metal block of large heat conductance to a Dewar vessel containing a refrigerant (Fig. 76). In order to avoid the danger of condensation, the cell itself must be placed within an evacuated container which is transparent to infrared radiation. This avoids condensation on the cell windows and also avoids the windows being warmed by convection. Instead of evacuating the container, displacement of the atmosphere with dry nitrogen frequently suffices. Usually it is necessary to warm the outer windows gently to prevent the condensation which would occur if they were to be cooled down by heat

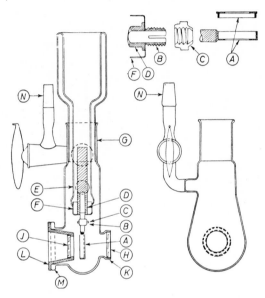

FIG. 76. *Low temperature cell according to Walsh and Willis (726). A, cell in holder; B and C, connection nuts; D, copper cylinder; F, metal connection to E, the refrigerant container; G, evacuated glass vessel with an opening to take E and flanges K and M for fitting the outer NaCl windows H and J; N, union used to evacuate G.*

conduction. A method of cementing the windows of the cell which will keep the cell airtight at all the temperatures of operation has been described by Roberts (596). A specially designed cell window that can be used for the lowest temperatures (liquid helium), and in direct contact with the refrigerant material, has been described by Warschauer and Paul (728). For the details of the different methods of cell construction, reference should be made to the original literature [Walsh and Willis (726); Šimon and McMahon (662); Bovey (79); McMahon et al. (459); Duerig and Mador (178); Roberts (597)]. In order to alter the thickness of the substance under investigation

Equipment and Experimental Technique

quickly and conveniently at low temperatures, a so-called reflection type cell [Holden *et al.* (*326*)] is advantageous and has been mentioned above for investigation at very high temperatures. The substance to be examined lies in a Dewar vessel on a plane mirror of high reflecting power; the radiation approaches from above and passes through the layer of substance on the mirror twice.

A cell with windows constructed (exceptionally) of fluoropolymers is worthy of mention; it is used for special purposes such as the investigation of corrosive substances [Kirby-Smith and Jones (*386*)].

4. Reflectometers and Polarizers

Now and again a knowledge of the infrared reflection spectrum of a substance is of interest, although this technique is seldom used in comparison with absorption measurements. Several supplementary instruments exist for the measurement of reflection spectra; they can be considered to be

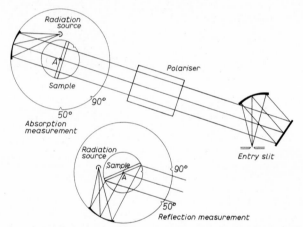

FIG. 77. *Reflectometer accessory according to Oldham (509).*

reflectometers or reflection adaptors for the usual spectrometers. It must be differentiated whether the regular mirror-like reflective power or the diffuse reflection is being measured. In both cases, relative measurements are nearly always used rather than absolute measurements.† This means that the power of reflection is measured with respect to a standard; in the case of the mirror-like reflection, gold or silver mirrors serve as standards because they possess a very high power of reflection, approaching 100 per cent, over

† An apparatus for measurements of absolute spectral power of reflection in the NaCl region has been described by Gier *et al.* (*252*).

196

most of the infrared region. For the measurement of the diffuse reflection, MgO and $MgCO_3$ layers, suitably abrased Al surfaces [Euler (*200*)], or suitably prepared standards of pulverized sulphur, selenium, or NaCl [Agnew and McQuistan (*7*)] are used. The measurement of the mirror-like reflection therefore depends on placing a planar, and often polished, disc of the substance to be investigated and the standard mirror alternately in the optical path; the whole of the rest of the arrangement must remain identical. It is desirable to be able to use different angles of incidence.

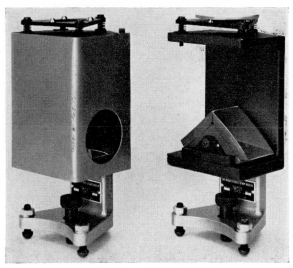

FIG. 78. *Reflection adaptor for the Perkin-Elmer spectrometer with protective casing* (left) *and without protective casing* (right), *with sample in position* (*photograph: Perkin-Elmer Corp.*).

A reflectometer of great adaptability for both reflection and absorption measurements on the same sample has been described by Oldham (*509*). Essentially, it consists of two goniometer tables with a mutual axis and an arrangement of the radiation source, samples, monochromator, and the necessary supplementary mirrors, such that the mutual positions of the radiation source and the sample can be moved from the position for a reflection measurement into that for an absorption one in a single operation (Fig. 77). Reflection adaptors of suitable construction and dimensions are available for several commercial spectrometers. One of these is shown in Fig. 78; the solid sample is placed on the adjustable plate (liquid samples in a container), the radiation strikes the sample from below at an incident angle of 10° and the reflected ray is led off by means of a second mirror at an angle.

Equipment and Experimental Technique

Just as for the heated cell mentioned above, it is necessary to control the focusing.

By comparison, the methods for measuring the diffuse reflection are more complicated. There are two types of instruments, those that measure only a certain angular region of the stray reflection, and those that measure the whole of the reflection deflected backwards. Both types have up to now been used almost exclusively for the short-wave infrared region; such measurements are of technical importance. An example of the first type is the reflection accessory for the Beckman spectrophotometer for ultraviolet and visible radiation. With small modifications, this is usable up to $2 \cdot 7 \, \mu$ wavelength (*374, 376, 377*). In this apparatus, the sample is placed at the focus of an elliptical mirror which uses an angle of reflection of 35 to 55°

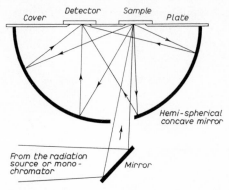

FIG. 79. *Diagram of a hemi-spherical Albedo reflectometer according to Sanderson (626).*

for perpendicular incidence; the radiation detector is sited at the other focus. MgO, $MgCO_3$, or $BaSO_4$ serve as standard substances [Sanders and Middleton (*625*); Jacques *et al.* (*349*)]. Reflectometers, for the measurement of the diffuse radiation reflected in the hemisphere, have been described by Sanderson (*626*) and Derksen and Monahan (*169*). The difficulty in constructing such instruments is the integration of the radiation energy which is spread over the hemisphere. One possible method for this would be to extend the ellipsoid section of the Beckman instrument to a hemi-ellipsoid. Such a solution is hardly possible in practice because of the difficulty of producing such elliptical concave mirrors.† In general, hemispherical concave mirrors are used as integrators and focusing errors are neglected.

† This difficulty has apparently been overcome recently by the use of hemi-ellipsoids obtained by the deep drawing method from thermoplastic materials (DP 947 752).

198

Sample and detector are placed at two points which are conjugate with respect to the surface of the mirror and lie on the limiting plane of the hemisphere near to, but on different sides of, the centre. For reflection at the sample, the radiation is introduced through a small hole in the mirror to the inner space as shown in Fig. 79. The aberrations for this arrangement are obviously larger, the further the sample and the detector are apart. In addition, the usual detectors for the near infrared region of the PbS type do not obey the Lambert cosine law, i.e. their efficiency is dependent on the angle of incidence of the radiation to be measured. Therefore, only the reflections of bodies which cause approximately equal distribution of the reflection over space can be compared. This limitation is removed by the use of an integration sphere (Ulbricht sphere) with a diffusely reflecting surface [Jacquez *et al.* (*349*); Jacquez and Kuppenheim (*348*)]. The preparation of the MgO surface for this has been described by Dimitroff and Swanson (*173*).

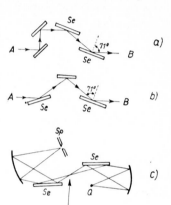

FIG. 80. *Selenium polarizers*, (a) and (b) *with parallel optical path according to Pfund and Schaefer respectively*, (c) *with convergent optical path according to Pfund* (*531*).

Measurements using polarized radiation are of especial importance for the investigation of crystals and certain natural and synthetic products. For such measurements the production of polarized radiation is necessary. In the infrared region, the polarization is always induced by reflection at electrically non-conducting substances, sometimes by using the reflected radiation (when one or two reflections suffice), and sometimes by using the transmitted radiation, for which several elements in series are needed because of the small polarization effect in any one of them. Almost no use is made of polarization caused by means of double refraction. The older reflection polarizers all use compact selenium mirrors as this substance produces a fairly constant degree of polarization throughout the whole infrared region, because of its small dispersion (*R*). As shown in Fig. 80, two selenium mirrors are always used and the sideways displacement of the beam caused by the geometric arrangement is compensated by means of normal plane mirrors [see also Pfund (*531*)]. Recently, more use has been made of transmittance polarizers. For this purpose, thin layers of selenium on cellulose-lacquered films [Elliot and Ambrose (*191*); Elliot *et al.* (*195*)] or on polyvinyl formal films [Ames and Sampson (*18*)] (which are later removed by means of a

199

Equipment and Experimental Technique

suitable solvent) are prepared by sublimation and are used in a series of 4 to 6 units, one after the other. AgCl plates about 1 mm thick in similar number and arrangement can also be used [Wright (759); Newman and Halford (497)]. The angle of incidence of the radiation is about 65–70° for selenium and about 63·5 ° for AgCl, the yield of linearly polarized radiation obtained is about 40–50 per cent of the initial value. The quality of the polarization of the component of the radiation transmitted (which vibrates parallel to the incident plane) depends essentially on the number of plates used, provided other influences of the materials, e.g. absorption and scattering are optically isotropic, i.e. are independent of the direction of vibration of the radiation. Charney (114) has given the transmittance coefficients of the components vibrating perpendicularly as a function of the number of plates (Table 11);

FIG. 81. *AgCl polarizer in container (photograph: Perkin-Elmer Corp.).*

TABLE 11. *Transmittance Coefficients of AgCl Polarizers for the Radiation Component Vibrating perpendicular to the Incident Plane [from (114)]*

Number of plates	t_m
1	0·501
2	0·294
3	0·190
4	0·128
5	0·090
6	0·067
7	0·050
8	0·040
9	0·035
10	0·031

this characterizes the quality of the polarization of a AgCl polarizer. It can be seen that for the usual arrangement of six plates the unwanted component of the radiation is still quite large and that it is only very slowly reduced by increasing the number of plates. Conn and Eaton (142) have described a degree of polarization (for definition, see page 317) experimentally determined as 99·8 per cent for a selenium polarizer of eight films. To avoid actinic effects which can occur as a result of the short-wave radiation emitted by the usual sources acting on the light-sensitive AgCl, the first AgCl plate, i.e. that which faces the radiation source, is frequently protected by means of Ag_2S. This procedure is used especially for the short-wave infrared region; a certain reduction in the transmittance is accepted. The whole polarizer, however constructed, is placed in a casing mounted so that its orientation about the direction of the beam which is passing through it can be altered by 90° on both sides (Fig. 81). Usually, the sample under investigation is placed between the polarizer (which is adjacent to the radiation source) and the monochromator, because it is desirable that the sample should be placed where the beam is most constricted. Makas and Shurcliff (*441*) have shown that this arrangement causes several difficulties of interpretation and technique, when used for AgCl polarizers and in part also for Se polarizers: (1) Because the monochromator is not completely free of polarizing influences (consider the apparatus polarization, see page 317), the dichroic sample lies between two polarizing systems; therefore the Malus law does not hold and the interpretation of the transmittance measured for different polarizer azimuths is difficult. (2) By passing at an

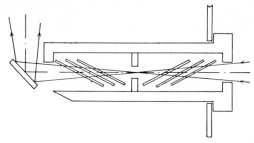

FIG. 82. *Arrangement of a transmittance polarizer according to Makas and Shurcliff (441).*

angle through several plates of a small but not negligible thickness, the beam is displaced sideways; this is especially undesirable when the displacement is perpendicular to the direction of the monochromator slit.† The obvious way to avoid this displacement, by reducing the thickness of the AgCl plates, usually leads to an equally undesirable double refraction. (3) The arrangement of the plates at a small angle to the direction of radiation needs relatively large surfaces which cannot be supported and which are in danger of bending. Accordingly, the authors mentioned suggest an arrangement of the AgCl polarizer as in Fig. 82, which consists of two sets each of three plates of fairly small dimensions. The two sets are arranged at equal and opposite angles to the direction of the radiation so that the correct angle of incidence results. Because of its small dimensions, such an apparatus can easily be placed between the sample and the monochromator.‡ A similar

† This displacement can be compensated, for example, by means of a wedge-shaped NaCl plate of the same size placed in the optical path in the opposite sense.
‡ The arrangement described is that for a fixed azimuth and as used with a Perkin-Elmer spectrophotometer Model 21. Both these limitations can easily be removed.

arrangement has been suggested by Bird and Shurcliff (*793*). They use slightly tapered plates which are arranged with gradually decreasing inclination to the incident radiation so that the angle of taper and the angle of incidence are directed towards the next focal point of the radiation.

In addition to those named, any substance which can be obtained in a suitable form and without absorption and with low dispersion (in order that the Brewster angle, taken as the angle of incidence, only alters slightly within the spectral region in question), but with a high refractive index, can be used for transmittance polarizers, e.g. KRS 5 [Lagemann and Miller (*403*)]. It is advantageous to use grating polarizers in the far infrared region [Mertz (*472*)]. Harrick (*805*) recommends the use of metal-coated germanium plates where the difference in the reflective power of the two polarization components is used for angles of incidence less than the main angle. Mercury has been shown to be a specially advantageous metal. Such a polarizer is useful from the near infrared region up to 100 or 200 μ wavelength. Edwards and Bruemmer (*799*) use only germanium. For the near infrared region, foils of polyvinyl alcohol and similar materials, dyed dichroically, are available [Dreschsler (*798*)]. In general they possess two regions of useful transmittance at $0 \cdot 6$ to $2 \cdot 8$ μ and $3 \cdot 5$ to $6 \cdot 5$ μ with a degree of polarization of over 99 per cent.

5. Equipment for Microspectroscopy

It has already been mentioned that, when biochemical substances are being investigated, the amounts available are frequently insufficient for an infrared determination by means of the usual instruments and cells. This certainly applies to the spectroscopic investigation of single fibres and small crystals. The importance of infrared spectra, and the deductions that can be made from them, is so large that they are also highly desirable in these cases. Therefore instruments and methods have been developed, using in part the techniques taken from other fields, to obtain useful spectra from the smallest amounts of substances.

The problem is that the sample, whether it be a liquid in a cell or a solid, must fully cover the monochromator slit. It was an obvious move to place the sample at a position in the optical path where a sufficiently small real image of the slit could be formed to satisfy this limitation. Accordingly, the monochromator remains unaltered and the modification is made in the system for illuminating the slit; care must be taken that the aperture relationships do not appreciably deteriorate. The so-called microcells are the first step in this direction. In them the cell cavity and the amount of substance needed are reduced by placing the cell close to the slit or at a position optically equivalent to this (Fig. 73). In the usual instruments, the surface which requires to be illuminated is still at least $1 \times 12 - 15$ mm. The next step is to reduce the cross-section of the beam by means of a stop. In double-beam spectrometers this is advantageously placed where the measuring and comparison beams are superimposed rather than in the individual beams [Wood (*751*)]. The same result can be obtained by an optical decrease in the cross-section of the beam (instead of a mechanical one), where the availability of water resistant, infrared transmittent materials of low dispersion but high refraction (such as AgCl) is very important. Using such materials, lenses of short focal distance and large relative aperture with accurate radii of curvature can be manufactured and are available for use as condenser lenses. Anderson and Miller (*22*) have described such a AgCl lens system,† which is also available commercially; it consists of two plano-convex lenses of 24 mm diameter and 22 mm focal length. In the NaCl region, the transmittance is about 50 per cent of the incident radiation; the useful cross-section of the beam is $1 \cdot 5 \times 3 \cdot 5$ mm area, and the amount of substance necessary for solutions is between 20 and 50 micrograms. The system designed by White *et al.* (*831*) for the Beckman IR-5 spectrometer functions similarly. It uses three KBr lenses and gives a useful area of $0 \cdot 5 \times 4$ mm.

Such methods are still not sufficient for fibres and very small crystals. This limitation has been overcome by the use in infrared spectroscopy of the mirror microscope‡ originally designed by Burch (*104*) for ultraviolet spectroscopic purposes. Essentially, the microscope (see Fig. 83) consists of three parts, the condenser, the objective holder, and the objective. The condenser consists of two concave mirrors (sometimes not spherical) of which the lower has a hole at the centre. The condenser forms an image, much reduced in size, of the radiation source on the object under investigation. The objective holder consists of a very small stop, adjustable in every direction, on which the sample is laid. The objective, the optical complement to the condenser, forms an image of this stop, if necessary via an intermediate image and with the help of further mirrors, on the entry slit of the monochromator. The optical construction of the mirrors allows the use of visible light for the adjustment, because chromatic aberrations occur. This is a valuable advantage compared to lenses of infrared-transmittent substances, although these have constructional advantages. Now that the utility of the mirror microscope has been demonstrated, constant efforts are being made to develop systems of greater and greater aperture, i.e. with

† Obtainable from Eastman Kodak Co., Rochester, N.Y., U.S.A.; Research and Industrial Instrument Co., 116 Lordship Lane, London, S.E.2, England.

‡ According to its use, this instrument is not a microscope but a micro-illuminator; however, the customary designation 'microscope' will be retained.

Equipment and Experimental Technique

increased efficiency, so that smaller and smaller samples may be investigated. However, the development of preparative techniques has not kept up with the development in instrumentation; today tiny crystals of about 10^{-6} g would be capable of spectroscopic investigation. From the large literature on this subject, the following references are given: Wilkins (742); Wood (750); Norris and Wilkins (502); Norris et al. (503); Grey (278); Cole and Jones (133); Barer (36); Blout and Bird (67); Blout et al. (72); Mellors (470); Loofbourow (425); Fraser (227); Coates et al. (124); Thornburg (709).

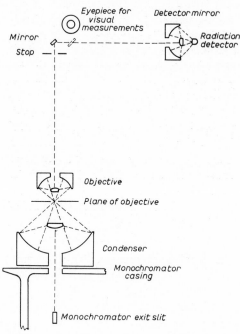

FIG. 83. Diagram of a infrared microscope.

The principal disadvantage in placing the mirror-microscope between the radiation source and the monochromator is that the sample is constantly subjected to the action of the radiation of all the wavelengths emitted and because of the decrease in the surface this irradiation is very intense. Many substances of biochemical interest are very sensitive to ultraviolet radiation (because of photochemical processes) and to infrared radiation (because of thermal effects). Therefore, the mirror-microscope is sometimes placed between the exit slit and the radiation detector. Here, the exit slit takes the place of the radiation source in the arrangement described above and the sample is exposed only to radiation which has already been resolved. For example, the commercial mirror-microscope manufactured by Perkin-Elmer for use with their spectrometers (except Models 21 and 137) is arranged in this way. The appearance of this instrument is shown in Fig. 84 and a diagram in Fig. 83. The monochromatic radiation leaving the mono-chromator exit slit is led into the microscope lying above by a plane mirror. The condenser consists of two spherical mirrors with a mutual centre and in the Schwarzschild arrangement; it forms an image of the exit slit decreased 8·4-fold at the site of the sample. The objective is identically constructed

FIG. 84. *Commercial mirror-microscope coupled to a commer-cial infrared spectrometer (photograph: Perkin-Elmer Corp.).*

and it throws an enlarged image in the plane of the stop; the latter consists of a square stop adjustable from every side. A mirror deflects the beam hori-zontally and a system of two mirrors equivalent to the objective forms a diminished image of the stop on the radiation detector. Behind the deflecting mirror, a further mirror can be moved into the optical path which allows the ray to be led to an eyepiece while the monochromator exit slit is simul-taneously illuminated with visible light. The sample can thus be exactly adjusted with respect to the image of the slit and at the same time the stop can also be correctly adjusted. The available space between the condenser and objective is 32 mm long. A revolvable mechanical stage of normal con-struction holds the sample; this is advantageously supplemented by means of a small goniometer table which orients solid samples in space as desired.

205

Equipment and Experimental Technique

The maximum field of vision is $0 \cdot 22 \times 0 \cdot 65$ mm, and the transmittance is about 45 per cent of the radiation energy leaving the monochromator. Solid substances can be investigated simply by placing them on a suitable support. For liquids, the micro-cell described by Blout *et al.* (*73*) is especially suitable, it consists essentially of a capillary $0 \cdot 075$ mm wide and 1 to 2 mm long in a AgCl plate. For qualitative investigations, Black *et al.* (*794*) placed a solution of a non-volatile or little volatile substance in a volatile solvent in such a capillary and then evaporated the solvent. For volatile substances they used two of the usual KBr discs (see page 231) which are pressed together using a tiny ring of the thinnest platinum wire as a spacer.

The instrument described, and all similarly constructed ones, are generally only applicable to single beam spectrometers, because it is difficult to compensate optically for the extension of the optical path by the microscope. Blout and Abbate (*70*) have recently overcome these difficulties by placing a somewhat modified microscope in the optical path of a Perkin-Elmer spectrophotometer Model 21 and a 1 m gas cell in the path of the comparison beam. By altering the pressure or the composition of the air in this it was possible to compensate for the atmospheric absorption sufficiently. Another arrangement has been described by Haggis (*287*). This is based on a normal single-beam spectrometer, in which, by means of a movable mirror, the radiation emitted from the exit slit can be directed successively through the micro sample and through a comparison cell placed a little to one side. Each of the optical paths thus formed, which lie very close to each other, possesses its own mirror for deflection on to the radiation detector. Ford *et al.* (*800*) use a similar method in which the sample to be investigated is mounted on a vibrating holder which moves the sample and a comparison substance (or the empty shutter) successively into the optical path.

6. Integrators

Up to now, most quantitative interpretations of infrared spectra have been based on the use of band extinctions at the position of the largest absorption, the so-called maximum extinction. As has already been noticed and as will be discussed in detail later, this quantity is strongly dependent on instrumental influences and is therefore an unreliable basis, especially for comparing the results obtained at different places. It has been shown theoretically and experimentally that the use of the integrated absorption, i.e. the area under a band, is more suitable for quantitative purposes. Therefore, it is probable that ever-increasing use will be made of the integral absorption in the future. Of course, every spectral curve can be integrated after it has been obtained by one of the usual instruments. However, such an integration carried out later causes delay and requires further effort. Fortunately, the scientific instrument industry has provided accessories to the commercial spectrometers which perform this task as the spectra are drawn and give the results in the form of numerical values either on a counter or written down. Fig. 85 shows such an accessory that used to be manufactured by

FIG. 85. *View of the automatic integrator for the Perkin-Elmer spectrometer Model 21 (photograph: Perkin-Elmer Corp.).*

Perkin-Elmer for their spectrophotometer Model 21. This piece of apparatus needs two quantities to perform its task; one is provided by the change in the wavelength of the spectrum and the other by the transmittance curve (as a logarithm, for only the extinction is of interest here). Space does not allow details of the mode of operation to be given here. Perkin-Elmer now have an electronic instrument in preparation to replace this accessory.

Equipment and Experimental Technique

It should be noted that, in order to obtain the integrated absorption of a single band in practice, two spectral measurements are necessary. One gives the background absorption (e.g. the solvent) the other gives the sum of the background and the absorption desired; the spectrum is thus obtained as the difference between the two measurements. It is advantageous to extend the spectral region in question as far as possible in order to make the wing errors small with respect to the total area of the band. It is obvious that the two measurements must be made under the same conditions of resolution, stray light, wavelength or wave number limit, etc.

7. Accessories for Single-beam Spectrometers

Although the majority of modern spectrometers are constructed as double-beam instruments or deliver the percentage transmittance or absorbance spectrum by some other means, there are still numerous single-beam instruments in use which record the energy of radiation transmitted. For many purposes this fully suffices, but at times a percentage spectrum is necessary or desirable. In principle, with such instruments, this can be obtained only by means of two successive measurements of the energy transmitted, one with and one without the absorbing substance, followed by a point by point evaluation of the curves obtained. Obviously, it would be advantageous to make this additional time-consuming calculation at least partly automatic and independent of the operator. Details of the different instruments, which have been made available by the manufacturers of mathematical instruments, for the evaluation of the two curves cannot be given, but a simple semi-automatic accessory will be briefly described [Fuoss and Mead (237); Zerwekh (769)]. In this the two curves of the energy transmitted J with the substance, and J_0 without the substance are recorded on a strip of paper over the same wavelength scale. The amplitudes of J and J_0 are then converted into electrical quantities so that J is proportional to a potential difference u and J_0 to a resistance R. The potential difference u applied to the resistance R gives a current i which is obviously proportional to the required transmittance. Fig. 86 shows a diagram of the instrument. By means of the battery B, the potentiometer P_1 is kept at a constant potential gradient from which

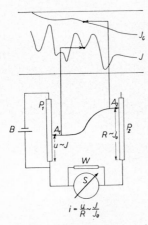

FIG. 86. *Diagram of an instrument for converting recorded energy curves into a percentage transmittance spectrum according to Fuoss and Mead (237).*

208

any desired fraction u can be selected by means of the movable connection A_1. This fraction is led to the connection A_2 of the variable resistance P_2; the resistance R selected is in series with the recorder S. If P_2 has a sufficiently high resistance, the current i flowing in the recording circuit is almost completely determined by u and R. The two movable contacts A_1 and A_2 are each mechanically coupled with a pointer and these are simultaneously traced over the J and J_0 curves respectively. P_1 and P_2 have to be arranged so that the contact zero corresponds to zero energy and the maximum contact to the maximum peak of the energy curve. The resistance W, placed parallel with the recorder, allows any desirable or necessary alteration of the region that can be measured. Obviously, this instrument which works point by point can be converted by suitable modifications, e.g. automatic scanning of the two curves by means of photocells, to one that is fully automatically recording; however, these questions will not be discussed here [Janeschitz-Kriegl (*356*)]. An accessory has been described by Paige (*521*) which gives the absorption coefficient directly.

8. Infrared Filters

Occasionally, in experimental infrared spectroscopy it is necessary to cut out a broader or narrower region from the emission of a radiation source. This can occur when a monochromator is not being used, and sometimes it is required in addition to the action of a monochromator. Such a rough spectral resolution is generally carried out by means of suitable filters of the required transmittance characteristics. Every infrared laboratory needs a wide range of suitable materials in order that filters are always available, e.g. for the elimination of short-wave stray light, etc. The plates usually used can consist of all the possible alkali halides, glass, quartz (crystalline and fused), mica, etc. In addition, filters useful for many purposes are obtained by subliming metals at moderate pressures on to transparent supports. Suitably selected gratings, layers of liquids, etc., are also used. Almost every older infrared paper contains some information of this type. Several recent advances in this field should be noted: especially interference filters for the near infrared,† for the middle infrared [Greenler (*776*), Heavens et al. (*806*) and Smith and Heavens (*826*)] and coloured alkali halide crystals [Gaunt (*243*); Burns and Gaunt (*105*)]. Further information on these filters can be found in (*C*), (*G*), (*M*), (*R*); also in the work of White (*736*); Plyler and Ball (*552*); Banning (*33*); Billings (*63*); Billings and Pittman (*64*); Blout et al. (*71*); McGovern and Friedel (*457*); Henry (*318*); Mohler and Loofbourow (*485*); Braithwaite (*81*).

† Obtainable from: Carl Zeiss, Oberkochen, Württemberg.

7. Non-resolving Infrared Instruments

In addition to the types of spectrometer already enumerated and described in detail, another type of instrument exists. These solve problems similar in principle to those solved by the spectrometers themselves, i.e. the recognition and measurement of substances by means of their spectra, but do this without recourse to spectral resolution of the radiation. These instruments are principally used for continuous control of production processes and atmospheres and for similar purposes, especially those of technical gas analysis. They are characterized by their robustness and by needing little attention, together with reliability, great sensitivity, and selectivity. The most important example is the URAS (Infrared absorption recorder) of the Badische Anilin- and Soda-Fabrik. This type of instrument originated in a patent of Schmick in 1928, but its useful development came only in 1938 with the URAS machine mentioned. The multifold applications of this instrument will not be given here in detail. For this, reference should be made to the specialist texts of process control. The fundamental principles of its construction and mode of action and the most important examples will be briefly discussed [Jamison *et al.* (*355*); Kivenson (*390*); Koppius (*394*); Fastie and Pfund (*205*); Fowler (*217*); Rouir (*608*); Siebert (*656*); (*E*)].

1. Fundamental Principles

All the principal selective non-resolving infrared instruments are one of two fundamental arrangements which are now differentiated as *negative* or *positive filtering*; they are manufactured in many different modifications. Negative filtering will be discussed first. In this type two completely identical radiation detectors E_1, E_2 of any desired type (Fig. 87) are so connected that the corresponding amplification and recording system shows only the difference between their signals. They are illuminated by means of two identical sources S_1, S_2; one source can suffice if its emission is split into two identical beams. Because of the arrangement of the two detectors, this identical illumination is not recorded. In front of each of the detectors, absorption tubes A_1, A_2 (sensitizing cells) are arranged in such a way that the

210

symmetry of the radiation is not disturbed. The gas to be detected, e.g. CO_2, is placed in one while the other remains evacuated or is filled with a gas that does not absorb, e.g. dry N_2 free from CO_2. Corresponding to the absorption of the CO_2, the energy in one of the beams is lowered and this is recorded. Each optical path also contains a further absorption tube B_1, B_2 (measuring and comparison cells) through which a mixture of gases which contains CO_2 is passed. Now the reading is decreased according to the concentration of the CO_2 and the thickness of the layer through which the radiation passes. This is because absorption now occurs in the optical path which was originally free from absorption, while the optical path which was already absorbing is affected to a lesser extent or not at all. Any other gas in the mixture which absorbs has no effect on the reading because its action is the same in both the optical paths. However, this only holds if its bands do not overlap those of the gas to be detected. When such overlapping occurs, resource can be made to another pair of absorption tubes C_1, C_2 (filter cells), one in each optical path, containing the interfering gas. By this means the two optical paths are simultaneously decreased by the radiation energy which is absorbed by the disturbing gas so that any further effect of this gas is of no importance. Briefly, this is the principle of negative filtering, i.e. two identical instruments are used, one measures all the absorbing gases including that to be determined and the other only that to be determined. The selectivity is attained by means of suitable filters.

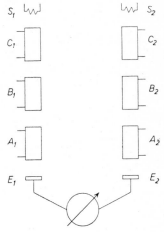

FIG. 87. *Diagram of a non-resolving infrared gas analyzer. S_1, S_2, radiation sources; A_1, A_2, sensitizing cells; B_1, B_2, analysis cells; C_1, C_2, filter cells; E_1, E_2, radiation detectors.*

The principle of positive filtering is to obtain the selectivity within the detector, e.g. by making the gas to be determined act as the detector by means of its absorption and measuring the rise in temperature or some other effect connected with it. If the gas mixture to be investigated is now led through the cell A_1 while the other tubes are empty or filled with N_2, then the absorption of the CO_2 gas weakens one of the beams resulting in a reading. Any other gas in the mixture which causes absorption has no effect as long as its absorption bands do not overlap the bands of the gas to be determined. If such overlapping does occur, recourse can again be made to filling each of the filter tubes C in each of the optical paths with the interfering gas or gases. Obviously the principle of positive filtering is superior with respect to selectivity and sensitivity to that of negative filtering and this is reflected in the practical applications of the instruments.

In addition to positive and negative filtration, there are two other possible principles by means of which sensitization might be achieved; they can be called selective emission and selective modulation. In the case of selective emission, the selectivity is obtained by the use of a radiation source which emits only those wavelengths which the substance to be detected absorbs. Such a radiation source would be the substance to be detected itself at sufficiently high temperatures. However, the technical difficulties and difficulties of interpretation of this method are such that it is only of academic interest. Selective modulation has been described by Luft (*436*). It consists of using an arrangement to modulate the continuous radiation of the source at the wavelength at which the substance to be detected absorbs and at no others. This can be achieved, either by means of suitable filters, or by placing the substance to be detected itself in the modulation apparatus. This method could become important especially for the measurement of liquids with non-resolving infrared instruments.

It is fundamental to the method that only substances possessing an infrared absorption spectrum can be detected and estimated with the instruments described; this eliminates the homopolar gases H_2, N_2, and O_2. Another limitation with respect to the position of the bands which can be used is that these must lie in the region between *ca.* 2 to *ca.* 6 μ, because outside these limits the energy emitted by the usual sources is too small. Understandably, for such instruments a linear relationship between the reading, i.e. the absorption occurring, and the concentration of the gas to be detected is desirable. As will become clear from formulae to be given later, the absorption of a gas is proportional to the product $\zeta.c.d$, provided this product, the extinction, is small compared to unity (which can be taken to be true for absorptions up to about 25 per cent). ζ is the extinction coefficient characteristic of the gas in question, c is the concentration of the gas, and d is the thickness of the layer irradiated. The layer thickness, i.e. the length of the absorption tube, must be selected according to the maximum concentration to be measured, i.e. the area of measurement desired. In practice it is not possible to obtain fully linear standardization, which is not surprising in view of the approximations made; every instrument must be empirically standardized and tested from time to time.

Table 12 gives a summary of the commercial non-resolving infrared instruments available to date.

TABLE 12. *Commercial Non-resolving Infrared Instruments*

Manufacturer	Principle	Year of appearance	Remarks
Bad. Anilin- & Soda-Fabrik A.G., Ludwigshafen	pos. filt.	1938	'URAS'
Leeds & Northrup, Philadelphia, U.S.A.	neg. filt.	1943	improved 1952
Baird Atomic, Cambridge, Mass., U.S.A.	neg. filt.	1946	
Grubb, Parsons & Co., Newcastle, England	pos. filt.	1948	'IRGA'
Infrared Development Comp., Welwyn Garden City, Engl.	pos. filt.	1948	
,,	pos. filt. compensation	1950	
Le Controlle de Chauffe, Bagneaux (Seine), France	pos. filt.	1949	
Mine Safety Appliance, Pittsburgh, U.S.A.	pos. filt.	1952	'LIRA'
Liston & Becker, Stamford, Conn., U.S.A.†	pos. filt.	1952	
Perkin-Elmer Corp., Norwalk, Conn., U.S.A. (Bodenseewerk Überlingen/See, Germany)	neg. + pos. filt. compensation 3 optical paths	1952	'Tri-Non'
Hartmann & Braun A.G., Frankfurt/Main, Germany	pos. filt.	1954	'URAS'
Applied Physics Corp., Pasadena, Calif., U.S.A.	pos. filt.	1954	
Consolidated Engineering Corp., Pasadena, Calif., U.S.A.	neg. filt.	1955	

† Recently obtainable also from Beckman Instr., Fullerton, Calif., or Munich, Germany.

2. The URAS

The URAS (Fig. 88) [Luft (*434*)] was the first non-resolving infrared instrument working on the principle of positive filtering. As radiation sources, it contains two nickel-chromium spirals electrically heated to about 700° C and placed in containers of the form of concave mirrors covered by plates of mica. The radiation emitted from the two sources is modulated in-phase by an interrupter rotating at 6·25 c.p.s. Two chambers containing the gas to be detected (usually diluted with N_2) and fitted with NaCl windows serve as

213

radiation detectors. They are connected with the measuring instrument itself, a membrane condenser. This consists of a rigid metal plate, pierced with holes, and a plate of very thin aluminium foil mounted in an insulated frame so that it can vibrate. Between the two plates of this condenser a constant potential difference of some 20 volts is applied. If the pressure of the gas is constant in both the detectors, which occurs for zero or equal absorption, then the plates are at a fixed distance with respect to one another. If the absorption in one of the detectors is altered the relative pressure changes; then, as a result of the movement of one of the condenser plates, the capacity of that condenser is altered. Because the radiation is being modulated regularly and the detectors are of sufficiently small inertia, the alteration of the pressure in the detector chambers is periodic so that the aluminium foil vibrates and the capacity of the condenser changes periodically. The alternating potential difference resulting from this is amplified, rectified, and recorded by means of a 5 mV pin-point recorder. The two optical paths can be compensated to equal energy by means of an adjustable stop before the instrument is used. The sensitivity can be varied by altering the constant potential difference applied to the condenser. As mentioned above, the area of measurement can be altered by varying the length of the analysis tubes. Filter cells can be used as necessary. Usually, all the cells are glass tubes gilded internally and with ends consisting of NaCl, mica, LiF discs, etc. It is possible to make measurements at several different points consecutively and alternately with the same instrument using a device to 'switch in' the

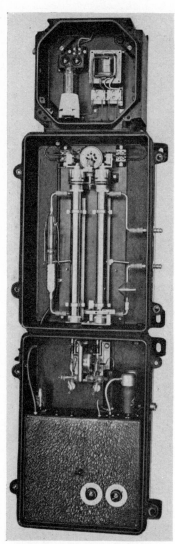

FIG. 88. *Infrared absorption recorder (URAS) of the Badische Anilin-and Soda-Fabrik A.G. (photograph: BASF).*

214

gas from these different points. Similarly it is possible to use two regions for measuring the same gas in the same instrument and in special cases to carry out the simultaneous measurement of two different gases. For the diverse uses of the additional accessories, reference must be made to the manufacturer's prospectus. Table 13 gives a summary of the performance details.

TABLE 13. *Summary of the Performance of the URAS. The Gas Concentration given is the smallest which can be measured in the Normal Manner with an Error of 2 per cent*

Gas	Concentration in vol.-%	Gas	Concentration in vol.-%
CO	0 to 0·05	C_2H_4	0·05
CO_2	0·005	C_2H_6	0·1
CH_4	0·1	vapours of indi-	
C_2H_2	0·1	vidual solvents	about 0·1

3. Perkin-Elmer Tri-Non Analyser

This new instrument, which works on the principle of the null point method, is a further development of the URAS and combines the principles of positive and negative filtration. Its appearance is shown in Fig. 89, and the principle of its construction in Fig. 90. The instrument possesses only one radiation source the emission of which is split by optical means into three beams P_1, P_2, and P_3. The detector is a single cell D filled with the gas to be detected, of which one wall is flexible and is used as the plate of a membrane condenser. A revolving interrupter modulates the radiation so that P_1, P_3 and P_2 in this order are directed on to the detector. The optical path P_1 contains an analytical cell Sa and a sensitizing cell Se. The optical path P_2 contains an analytical cell Sa and a compensation cell Co. The optical path P_3 contains an arrangement to weaken the radiation intensity A. In addition, each optical path contains a trimmer T (adjustable stop) and a filter cell F. For example, if CO_2 is the gas to be measured, the detector D and the sensitizing cell Se are filled with it. Then the optical path P_3 is compensated by means of the trimmer T_3 in such a way that the sum of the radiation intensity P_1 and P_3 is equal to that of P_2. When the interrupter is rotating a certain null point reading is recorded and this is taken as the null point for the purposes of the measurement. If the gas mixture to be investigated now flows through the two analysis cells Sa, the energies of two optical paths P_1 and P_2 are decreased with respect to the radiation which effects the detector.

215

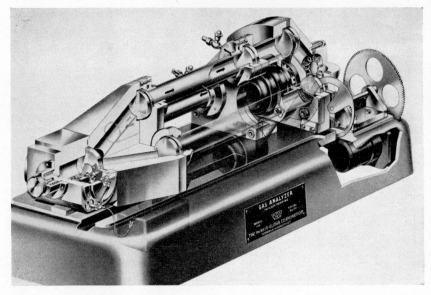

FIG. 89. *Perkin-Elmer tri-non gas analyser (photograph: Perkin-Elmer Corp.).*

This is because of the absorption by the CO_2 in the analytical tubes; this absorption is however of greater importance in P_2 than in P_1 as the latter already contains CO_2 in Se. Accordingly a different signal is now recorded

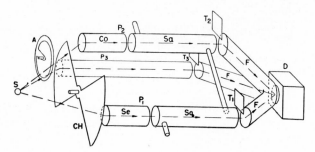

FIG. 90. *Diagram of the Perkin-Elmer tri-non gas analyser. Explanation in text.*

and is used to work the weakening apparatus in P_3 in such a manner that the energy relationship previously selected, i.e. $P_2 = P_1 + P_3$, is restored. The position of the stop necessary for this state is a measure of the absorption by the CO_2 in P_2 and is recorded by electrical means. Other constituents in the mixture of gases do not affect the selective detector unless they possess bands which overlap those of CO_2. In this case their influence can be eliminated by filling the compensation cell and the three filter cells with the gas in question. Obviously, as a result of the use of several cells, the lengths of which can be selected within certain limits, an especially large number of possible combinations is available. This enables the instrument to be applied to the solution of complicated problems [Woodhull *et al.* (*752*)].

Another instrument of the same firm will be briefly mentioned: the bichromator (Savitzky and Atwood (*630*); Savitzky and Bresky (*631*)], the

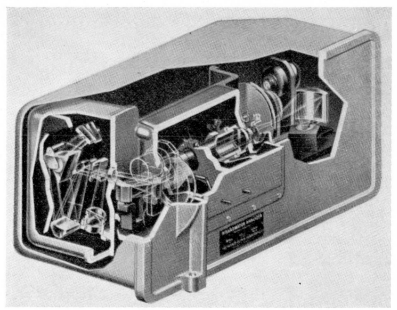

FIG. 91. *Perkin-Elmer bichromator analyser* (*photograph: Perkin-Elmer Corp.*).

appearance of which is shown in Fig. 91. In construction this is a spectrometer, but it is used as a process control instrument in a way similar to those without spectral resolution. It consists of a small robust monochromator of the Littrow type with the peculiarity that the Littrow mirror, the position of which determines the wavelength directed onto the exit slit, is bisected horizontally and each half is adjusted separately. By this means it is possible (for details reference must be made to the manufacturer's prospectus) to determine the ratio of the absorption of two bands of very different wavelengths and thus, for example, the ratio of concentrations of two substances in a mixture. The advantage, compared with the instruments without spectral resolution, is that for the latter the usefulness, sensitivity, and range of measurement cannot be directly deduced from available spectra and that their action is limited to a certain wavelength region. Neither of these limitations applies to the bichromator. A similar instrument has been described by Herscher and Wright (*322*). Another instrument in which (instead of two) six, and recently as many as ten, wavelengths can be measured albeit successively, has been described [White *et al.* (*740*) and (*525*)]; the method is similar in principle, but the technique is different.

8. Preparation of Samples

1. Introduction

The preparation of the sample to be investigated is next in importance to the spectroscopic apparatus in determining the success of the investigation. Although certain conclusions can be drawn from any spectrum, finer points can only be deduced from a spectrum suitable to the particular problem. In addition to the preparation and adjustment of the apparatus, e.g. with respect to the recording time and the degree of resolution, some precautions are necessary in the preparation of the samples. The multitude of possibilities in this respect allows only general fundamental principles and directions to be given here. The application of these to any particular problem must be left to the investigator.

The two characteristics of the sample (apart from its composition), essentially influencing the spectrum, are the concentration and the layer thickness. Without anticipating a more detailed discussion (see page 307), it will be mentioned here that these quantities must be selected for a given substance so that the bands to be measured in the investigation fall within the area of transmittance of about 20 to 60 per cent. The amount of substance which is usually necessary for an infrared spectroscopic investigation (at constant area the cell thickness is a direct measure of this) is so small that it is usually negligible compared with the amount available. This is almost always true except in biochemical investigations and here the application of microspectroscopic methods and instruments can be helpful. In a large number of cases the substance used in the spectroscopic investigation is not lost but can be recovered after the measurement has been completed. In this sense, infrared spectroscopy is a non-destructive method of investigation. This can be an important advantage. Other advantages of the method are the specificity of the information obtained, and the short time required, especially in comparison with chemical methods which often cause fundamental changes in the substance to be investigated. However, there are also cases in which the substance to be examined spectroscopically must be subjected to a preparation which although it does not directly effect any chemical change does cause some physical alterations, e.g. a change in the state of aggregation. This can result in certain undesirable and unintentional

219

Equipment and Experimental Technique

chemical changes. An example is slight thermal oxidation, etc.; the most certain way of detecting this is the infrared spectrum itself, together with experience and care. Such possibilities must certainly be taken into consideration, especially for the interpretation and assignment of weaker bands which appear contrary to expectation. Repetition of the experiments using somewhat altered conditions can be instructive in these circumstances.

With respect to the chemical preparation of the sample, the fundamental principle applies that it should always be separated as far as convenient

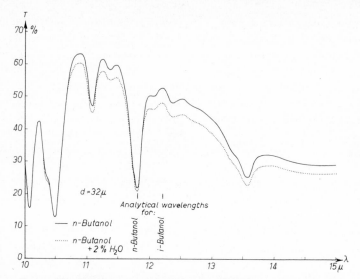

FIG. 92. *Influence of small amounts of water on the spectra of liquids. Full line: anhydrous normal butanol; dotted line: normal butanol plus 2 per cent water.*

into its components by chemical or physical-chemical means. The comprehensibility of the infrared spectrum of a substance decreases rapidly with the number of components present. Although in some cases infrared spectroscopic analysis of a many-component system can be carried out, the effort required in calculation and interpretation is often so much greater that it is no longer economical. The removal or minimization of free water in the samples is especially important. Although infrared spectroscopic investigations in aqueous phases are now possible as a result of developments of technique [Potts and Wright (558), Sternglanz (788)], they are not suitable for routine use because of the large influence of small water concentrations on the spectral transmittance at certain wavelengths. This means that they are more troublesome, less exact, and need special standardization curves (Fig. 92).

220

In addition the sensitivity to moisture of most of the materials used for cell windows is important. Every infrared spectroscopic laboratory should therefore work in close collaboration with a physical-chemical laboratory which takes care of the chemical preparation of the samples. It is a question of organization and of local conditions whether these two laboratories which must work in collaboration should be under unified direction. The actual preparation of the samples themselves, usually only a physical and technical one, should be carried out in the spectroscopic laboratory as a preliminary to the spectroscopic measurement itself.

2. Gaseous Substances

The peculiarities of the infrared spectrum of a gaseous substance, i.e. strong absorption at few wavelengths and almost complete freedom from absorption between them, together with the existence of easily available gases without an infrared spectrum (the non-polar gases such as O_2, H_2, N_2, and the rare gases), simplifies enormously the process of obtaining a spectrum suitable for the problem in question. The possible variation in layer thickness is determined by the cells available. As has been seen, they cover a very wide range and their thicknesses are easily and exactly measurable because of their sizes. It is clear that the smallest layer thicknesses must be used for strongly absorbing gases, especially within the regions of the fundamental vibrations. For mixtures of gases, and preliminary and comparison spectra, it is advantageous to use medium layer thicknesses about 100 or 200 mm. The 1 m and 10 m cells are used for the determination of the purity of weakly absorbing gases. Still longer absorption paths are used for detecting traces and for investigations in the overtone regions; here absorption paths up to 5·5 km have been constructed by means of multiple reflections [Herzberg and Herzberg (323)].

In addition to the layer thickness, the pressure of the gas under investigation, or more exactly its partial pressure, which is a measure of its concentration, can be altered as a means of influencing the spectrum. Besides investigating the gaseous substance at different pressures, it is sometimes advantageous to use a constant partial pressure of the gas in question and to alter the total pressure by means of another gas (naturally one without an infrared spectrum is selected). By these means it is possible to investigate the influence of pressure and of other gases on the absorption over wide ranges and to draw conclusions concerning intermolecular interaction, etc. An arrangement is required for filling the cell so that any pressure of a single gas, or any desired mixture of gases, may be obtained. When the gas under investigation is altered, certain precautions are necessary to avoid contamination from residues of the gases used earlier in the greased taps of the apparatus and the connection to the cell itself. If the connections are made by means of rubber tubing, etc., the danger is always present that the gas previously used will have been absorbed by these and will appear later as a contaminant. The situation for the taps is similar; the gases dissolve in the usual tap-greases and are set free again when the gas is changed. Thorough washing by a gas stream or a complete cleaning and re-greasing of the apparatus are then usually the only remedies. Recourse can also be made to specially constructed apparatus and special precautions such as those developed for similar situations in other techniques, e.g. mass spectroscopy.

3. *Liquids*

Compared with gases, it is more difficult to obtain suitable preparations of liquid substances which give easily interpreted spectra in all cases. Here also, variation of the layer thickness is the first measure to be taken to improve a deficient spectrum. Because of the very much stronger absorption of liquids and solids compared with gases, a result of the denser packing of the molecules, the useful layer thicknesses are quite small and therefore can be less exactly determined. In addition, the appearance of the spectrum as a whole is very much more complicated because of the numerous and often intense overtones and combination bands and because of the general background absorption which is a result of strong interaction between neighbouring molecules and the overlap of the wings of many bands. Layer thicknesses below $0 \cdot 02$ mm can hardly be produced with the necessary precision and reproducibility and are therefore only useful for qualitative investigations. But many substances (especially if they contain numerous and stronger polar groups) in the pure state still produce spectral bands of such strength that the lower transmittance limit of about 20 per cent mentioned above is not attained at $0 \cdot 02$ mm. In such cases, the use of capillary films, e.g. between two discs of NaCl pressed against each other without a spacer or as a layer on a flat surface, can be made for qualitative investigations. This method is obviously not applicable to quantitative work. The only remaining possibility is to reduce the concentration, i.e. to dissolve the liquid in question in a suitable solvent. However, the spectrum then also contains the bands of this solvent and is therefore falsified in ways which are not always obvious. In addition the possibility of direct chemical interaction between the solvent and the solute must be considered. Every solvent possesses a characteristic spectrum of its own. The more complicated the structure of the solvent molecule and the more polar it is, the more complicated and intense will be its infrared spectrum. The infrared spectra of a whole series of available solvents are given in Fig. 93 (for $0 \cdot 1$ mm

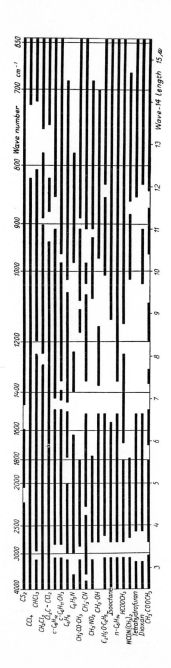

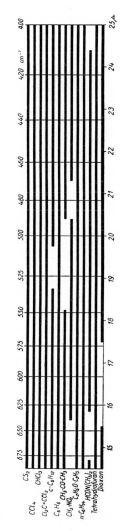

FIG. 93. *A summary of the useful regions (black) of some common solvents for infrared spectroscopic purposes.*

thicknesses) for the region from 2 to 25 μ [see also Torkington and Thompson (714); Anderson and Stewart (21)]. It can be seen from this that CCl_4 and CS_2 are the most advantageous, as would be expected; they are followed by chloroform, cyclohexane, acetone and a few others which are quite useful. Most solvents are useful only for one or for a few wavelength regions which are often quite limited in extent. In this context, 'useful' means the regions where the absorption of the solvent is not more than about 30 per cent at the thickness selected.

In this situation, if the spectrum of a pure substance is not satisfactory and the possibilities of alteration of the thickness have been fully investigated, the use of mixtures or solutions successively in different solvents cannot be avoided. The spectra of the solvents must be carefully considered and the influence of the layer thickness. Double beam instruments have special advantages for this method. In principle, they allow the direct compensation of the spectrum of the solvent by introducing a suitable amount of the solvent into the comparison optical path. However, some subtle precautions are necessary. In the comparison beam, there should be only the same amount of the solvent as there is in the investigating beam, i.e. the cell thickness in the comparison beam is always smaller than that in the investigating beam and the difference is the greater the more concentrated the sample that is being measured. Here the cells with variable thickness mentioned above find their most suitable application. They should be introduced into the comparison beam of the spectrometer filled with the solvent in question and with a thickness estimated as that required. The spectrometer is set at a wavelength at which the solvent possesses moderate absorption but at which the substance to be measured is almost transparent. With care and patience it is possible, by altering the thickness, to compensate the absorption of the solvent at this and thus at all places on the spectrum. When the spectrum is now run through the wavelength regions of interest, only the required spectrum of the dissolved compound is obtained. It is a limitation of the method that the spectrum of the solute should not be altered by the influences of the solvent in the positions and strengths of the bands; such alterations could occur by complex formation, etc. With regard to this, it is often advantageous to work with large thicknesses and small concentrations instead of vice versa because intermolecular solute-solute interactions are usually only noticeable at higher concentrations. Of course, the regions of strong absorption of the solvent are then no longer available for interpretation. It must be remembered that in regions of complete absorption indefinite readings are obtained as a result of the relationship between the radiation energy incident on the detector and the position of the compensation mechanism. For this reason, the recording speed must be lowered at regions of stronger absorption to allow the recording apparatus to draw bands which really correspond to the spectrum. If the vapour pressures of the solvent and solute are very different, care must be taken, especially for lengthy experiments, that the concentration remains constant [Meeks *et al.* (*469*)]. The use of triethylamine is recommended if the solubility is not sufficient [Ard and Fontaine (*25*)].

The thickness of the comparison cell cannot always be set at the correct value by means of the compensation technique. The spectra then obtained are falsified to a certain extent. The extent of this falsification has been investigated by Bayliss and Brackenbridge (*48*), whose results have been amplified by Lippert (*419*). Surprisingly, it is found that the cell thickness and the concentration of the solution are without influence. Only the absorption coefficient of the solvent and solute and their molecular volumes are important. The influence is expressed in the appearance of maxima and minima in the compensation curve which simulate absorption bands.

The use of absorption in the comparison beam is of interest not only for true compensation of the solvent but also for intensifying bands which would otherwise be too weak to appear. Wavelength-independent absorption in the comparison beam is especially suitable for this application. In the simplest case, it consists of wire gauze of suitable mesh; the transmittance always being greater than that of the bands that are to be intensified. The various possible types of difference measurements have been discussed by Kaiser (*370*); because of the possibilities of simulations just discussed, they must be used with care.

Some, but not all, of the difficulties inherent in dilution techniques can be avoided by working in the region of the overtones instead of that of the fundamental vibrations. As a result of the lower intensities of the overtone bands, larger thicknesses can be used in that region. These bands can be measured more exactly and allow greater variation so that interpretable spectra of pure substances can be obtained without dilution. Unfortunately, some questions are not so easily solved by the use of overtones, compared with the fundamental vibrations. On the whole, the technique of measurements in the overtone region is the simpler one.

The infrared spectroscopy of aqueous solutions will be briefly discussed. In general water is the mortal enemy of infrared spectroscopy, because of its own spectrum [Gore *et al.* (*272*); Curcio and Petty (*154*); Genzel (*248*); Giguere and Harvey (*255*)] and also because of the damage to cell windows. Thus, if possible, water should certainly be avoided as a solvent. However, sometimes it has to be used. Methods of avoiding attack on the cell windows by the solvent will be discussed in detail later; alternatively water resistant substances can be used for the windows.† The spectral investigation of aqueous solutions can be carried out in two ways. In one, the usual cell technique is used with relatively small thicknesses (on no account above $0 \cdot 02$ mm); normal water and heavy water complement each other admirably with respect to their spectra, as shown in Fig. 94. The other method is to dispense completely with the use of cells by adding to the liquid a trace of a wetting agent (surface activator) which gives it the properties of a soap solution. This can now be measured as a capillary film on a wire frame. By means of refined techniques, a film of constant thickness can be obtained

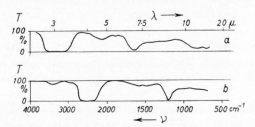

FIG. 94. *The spectra of normal and heavy water in thin layers according to Gore (272); see also Blout and Lenormant (69).*

as desired. This is attained by means of a regulated flow from a reservoir and perpetual removal from the film.‡ The decrease in the transmittance of aqueous solutions above 7μ which is often troublesome can be compensated for by means of the method mentioned above, i.e. by the use of wire gauze, etc., in the comparison beam [Potts and Wright (558)].

4. Solid Substances

The salient facts discussed in the use of liquid samples are of still greater importance to the preparation of solid samples. In only a very few cases can the necessary thin layers be obtained easily, as, for example, for mica. Most solid substances are obtained as more or less crystalline powders, or as crystalline or amorphous compact pieces. Compared with the preparation of substances in the other states of aggregation, the preparation of solids for spectroscopy is therefore of very great importance. The success of a spectroscopic investigation often depends on the development of a suitable method of preparation.

† If the investigation of aqueous samples in water sensitive cells cannot be avoided, and provided there are no chemical reservations, the sample can be saturated with NaCl, or the window material in question, without any special treatment of the cell windows.

‡ Thanks are offered to H. Hausdorff (Perkin-Elmer Corp.) for information concerning this technique.

The simplest approach, and one that is very often successful, is to investigate the solid in solution. Everything of importance about this method has already been given under the treatment of liquids. This method is almost always applicable for purely analytical purposes, but it cannot be applied to the study of phenomena which are connected with the solid state itself. The next most usual approach is to obtain a sufficiently thin film of the substance in question *via* a solution, provided such a film can be formed. For this purpose the solid substance is dissolved in a solvent at a known concentration, a definite amount of this solution is poured on to a suitable support held exactly perpendicularly, and the solvent is removed by evaporation. If possible, the material for the base should be so chosen that the film can be removed from it and thus be measured spectroscopically as such, e.g. held in a frame. Suitable bases are carefully cleaned and polished metal and glass surfaces and also certain liquid surfaces such as mercury and sometimes water, but only if the substance to be investigated neither dissolves in nor absorbs the liquid. If for any reason such a base cannot be used, or if the film cannot be removed when formed, it is sometimes possible to pour the solution and allow the film to form directly on an infrared-transparent support such as those used for cell windows. The thickness of films that have been detached can be measured by the usual methods or by the use of an interference effect [Tolansky (*711*); Bovey (*80*)]; for a film coating the surface-profile microscope [Tolansky (*712*)] comes into consideration [Rahbek and Omar (*567*); Green and Hadley (*275*)]. The thickness can also be calculated from the concentrations, amounts of solutions used, and surface areas, but possible losses must be borne in mind. Provided the necessary apparatus and standardization curves are available, another method is to measure the transparency to radioactive β-rays [Peterson and Downing (*529*)]. Obviously, the spectral characteristics of the supporting material will be incorporated into the final spectrum and must be compensated or allowed for.

This process of obtaining a film *via* a solution is applicable for all substances provided a solvent is available and that a coherent film can be obtained from the solution. The solvent must meet a series of conditions: (1) The solute be sufficiently soluble in it at moderate temperatures. (2) Chemical inertness towards the solute. (3) High purity; however, it is still more important that the solvent should not polymerize under the drying conditions selected and not leave a non-volatile residue [for purification, see Pestemer (*527, 528*)]. (4) Weakest possible absorption bands in order that remaining traces do not cause falsification of the spectra; on the other hand when no comparison is available, the spectrum itself is the surest test that the solvent has been sufficiently well eliminated. (5) The solvent should

Equipment and Experimental Technique

form clear films with a smooth surface. Several of these conditions are obviously required. The fourth condition is academic with respect to the infrared region, because there is no solvent that does not possess one or more strong bands somewhere in this region. As already mentioned, this can serve to test whether the solvent has been completely removed. In those cases where the film itself shows (not too strong) absorption at the same wavelength, the solvent can be detected by comparing the spectra of films which have been dried to different extents and considering the alteration in individual bands, i.e. the decrease in absorption with drying. Finally, the method of drying has a certain influence on the quality of the film formed. Low-boiling solvents are advantageous because complete evaporation is attained more easily, but they tend to give, as a result of too rapid drying, films containing bubbles and of uneven thickness and high scattering power. Slow drying can be obtained in a drying cabinet at a moderate temperature suitable to the solvent. If the substance in question is temperature sensitive, evaporation in a vacuum is advantageous, sometimes with simultaneous heating. In both cases, solidification occurs initially at the surface and the film there formed is pierced by the escape of solvent from the inside and this results in an unevenness of the surface. In this respect, advantageous use can be made of the novel radiant-driers, with their more even drying, including the more deeply lying parts. Because of the depth of action of their radiation highly loaded electrical bulbs are especially suitable.

Those materials which crystallize, instead of separating from solution as a coherent film, are treated by means of the so-called infusion or powder technique on an infrared transparent support. The process is the same as for the formation of films from the solutions but requires suitable drying conditions to keep the size of the particles formed as small as possible and indeed (corresponding to the well known dispersion law) smaller than the wavelength of the spectral region in question. For this it is advantageous to heat the support, usually a NaCl disc, above the boiling point of the solvent (but naturally under the decomposition temperature of the substance) [Hacskaylo (283)]. The thickness of the layer is regulated by means of the concentration and the amount of solution used. A suspension of a finely-powdered insoluble substance in an easily evaporated medium can be similarly treated [Hunt et al. (343)]. At the best this method is semi-quantitative, but it produces good results, especially for the regions of wavelength above about 5μ. If the solid material is already in the form of a sufficiently fine powder or can be pulverized, the use of a solution may be avoided by dusting a thin layer of the powder directly on to a support, if necessary with the use of a very thin adhesive layer of paraffin oil [Lecomte (410)]. Aqueous solutions

228

are evaporated on insoluble, infrared-transparent materials, e.g. AgCl or CaF_2 [Stevenson and Levine (*686*)].

For solid substances, for which no suitable solvent is either known or available, or for which the processes already mentioned are not suitable for some other reason, there is a series of further methods of preparation. Provided the sample can be melted and is not altered in constitution by the necessary heating, thin layers can be produced by melting on a support. The influence of foreign bodies such as solvents, etc., is thus eliminated, but greater consideration must be given to the possibility of decomposition, polymerization, or oxidation. Oxidation can be avoided, or at least minimised, by melting the sample between two plates of the supporting material or in an oxygen free atmosphere. With suitable control of the temperature, films consisting of a single crystal can sometimes be obtained from melts or solutions. The well-known Kofler block is very suitable for obtaining the necessary temperature gradient.

If the last process cannot be carried out, and if the material is available as lumps, an attempt can be made to obtain the film, by purely mechanical means, i.e. by cutting the compact material with a microtome or by drawing and pressing in a calender. In some cases the material must first be pressed into suitable plates or blocks. The hardness and brittleness of some substances are an impediment to this process which has several advantages because structural alteration is avoided. Thermoplastic materials can be prepared by first forming them into a small balloon which is then suitably warmed and blown out with compressed air until the required wall thickness is obtained.

Two general methods of preparation remain for substances which cannot be brought into a useful form for spectroscopic purposes by any of the methods already mentioned. These are the suspension technique and the disc technique; the first is also known as the Mull technique. Both processes are concerned with embedding the substance, to be spectroscopically investigated, in a finely pulverized form, in a suitable liquid or solid medium possessing the least possible absorption of its own. The fundamental basis of these processes is to avoid, by means of the embedding, the numerous atmosphere-solid surfaces which cause diffraction. Diffraction at the embedding material-substance particle surfaces is minimized by making the two refractive indexes close to one another. In both processes it is important that the substance should be pulverized so that the mean particle size is below that of the infrared region wavelengths. This can be attained for sufficiently hard substances by grinding in suitable mills and further treatment in agate or glazed porcelain mortars, but it frequently requires patience. For soft and greasy substances, a sharp decrease in temperature is often necessary in order to carry out this process. With respect to the embedding material, it is obvious that it should possess few and only moderately strong bands in the infrared region, because its spectrum will overlap that of the substance to be investigated.

Nujol is generally used in the suspension technique; it is the pharmaceutical paraffin oil, i.e. a mixture of long chain liquid paraffins. It possesses the bands characteristic for CH_2 and CH_3 groups near $3\cdot4$, $6\cdot8$, $7\cdot2$, and $13\cdot8\,\mu$ (Fig. 95). At these points the spectrum of the substance investigated is masked by that of the Nujol; compensation is difficult and unsatisfactory. In general, the bands do not appear as sharply and prominently as is customary. Especially in the short-wave part of the spectrum, the course of the

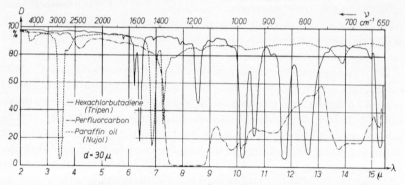

FIG. 95. *Spectra of the usual suspension compounds.*

transmittance curve is levelled out unless the correspondence of the refractive indices mentioned above is very good over the whole region and not only for one wavelength. If it is desired to obtain valid bands for the masked portions of the spectrum, the embedding material must be changed, just as different solvents must be used when it is desired to cover the whole spectrum. In this sense, long-chain liquid compounds similar to the paraffins, but in which all the hydrogen atoms have been replaced by halogen atoms, particularly by fluorine or chlorine, are complementary to Nujol. This is because the bands corresponding to the various CH vibrations then lie at other (in fact increased) wavelengths as a result of the greater masses and altered bond strength. Perfluorkerosine, also called perfluorocarbon, and perchlorobutadiene [Ard (*24*)] are used; the spectra of both these compounds are given in Fig. 95.† The substance to be investigated and an amount of the embedding material (which must be determined empirically) are intimately rubbed and mixed in a mortar. The mixture is placed in a

† Perchlorobutadiene (hexachlorobutadiene) is obtainable under the name 'Tripen' from Wacker Chemie G.m.b.H., Munich.

dismountable liquid cell, using a spacer of suitable thickness, and submitted to spectroscopic investigation. This technique is at best only semi-quantitative because it is not possible to transfer the quantity weighed from the mortar quantitatively into the cell. Efforts have been made to remove this limitation by introduction of a so-called inner standard. Here a third component of known band intensity is added to the mixture and from its spectrum conclusions are drawn as to the thickness and concentration; however, the results are not very convincing [Barnes *et al.* (*40*); Pirlot (*539*); Kuentzel (*397*); Bradley and Potts (*796*)].

The disc technique is currently used in two modifications. In one, the finely pulverized substance is mixed with powdered silver chloride or some other powdered plastic substance of good infrared transmittance, rolled out into a plate of about 1 mm thickness or less, and the spectrum is then measured [Sands and Turner (*627*)]. The second modification uses finely powdered KBr as the dispersion medium [Schiedt and Reinwein (*633*); Stimson and O'Donnell (*687*)]. The substance and the dispersion medium are intimately and evenly mixed. About 1 g or less is then placed in a press (Fig. 96) and compressed to a tablet a few tenths of a millimetre thick, using quite high pressures (over 5000 kg/cm^2).† Potassium bromide shows the phenomenon of so-called cold flow and at high pressures it becomes plastic and encloses the substance in a compact disc which is quite transparent. However, for this to occur, the very hygroscopic KBr must be dried. In addition, the air between the particles of KBr must be removed by evacuation during the compression; otherwise, after lowering the external pressure, the tablet splits again and becomes non-transparent because dispersion of the radiation occurs. The quality of the spectra thus obtained is superior to that using the suspension technique; also, KBr possesses no absorption bands of its own in the usual infrared region. However, to obtain sharp bands a marked correspondence of the indices of refraction of the substance and of the embedding compound is again needed. Quite satisfactory results can be attained by intimate mixing in a mortar. Still better results are obtained from water soluble substances by dissolving them in water and dropping this solution on the KBr while mixing and finally drying the mixture. The simultaneous solution of substance and KBr in water followed by lyophilization appears to be even more promising.‡ Schwartz *et al.* (*825*) have successfully used the lyophilization technique for substances which are insoluble in water but soluble in organic solvents. Good results have also

---

† Ryason (*620*) has used this process to prepare useful cell windows from pure powdered alkali halides. This method appears to be suitable in cases where sufficiently large crystals are not available or their price is too high.

‡ A lyophilization apparatus suitable for this has been described by Holzman (*328*).

(a)

(b)

FIG. 96. *KBr press according to Schiedt (632).† Above fitted together, below in pieces. 1, the outer part of the press with connection for evacuation; 2, inner part of the press; 3, lower part of the plunger; 4, base plate; 5, plate; 6, upper part of the plunger with plate; 7, hollow rubber spring; 8, ring to remove the inner part of the press; 9, rubber gasket (photograph: P. Weber, Stuttgart-Uhlbach). For a similar construction see Franck (225).*

† Obtainable from: P. Weber, Stuttgart-Uhlbach. A micro-accessory for further reducing the size of the discs is available from the same firm.

been obtained by pulverizing and mixing the substance to be embedded and the KBr simultaneously in a small vibrator-mill,† but it appears to be advantageous to remove the heat produced [Fahr and Neumann (*203*)]. A further advantage of the compression technique, compared with the Nujol technique, is that the former is easily extended to become a completely quantitative process. This is based on the fact that the mass can be quantitatively transferred from the mortar, etc., into the press and the other conditions can easily be kept constant‡ [Browning *et al.* (*97*); Kirkland (*387*); Wiberley *et al.* (*790*)]. Special note should be made of the fact that the appearance of the spectra obtained, i.e. the sharpness of the bands, depends considerably on the degree of fineness of the powder. Fig. 97 gives an example of this. The spectrum of glycine ethyl ester hydrochloride is given for different grinding times (under given conditions) and after lyophilization. The differences in the spectra are quite apparent. In quantitative work it must be ensured that the degree of fineness of the substance is kept constant.

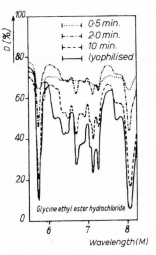

FIG. 97. *Influence of the particle size on the absorption spectra of a substance embedded and pressed in KBr (after the original work of U. Schiedt).*

Other alkali halides have been investigated and recommended as embedding materials. Ford *et al.* (*215*) have summarized the results. The principal qualities required are: (1) Good transmittance in the infrared up to 25 μ, or, better, to 40 μ, which restricts the substances to be considered to ionic crystals. (2) Clear discs should be obtained at low pressures; this points to Cs and K salts (except the fluorides). (3) Constant availability in a pure, non-hygroscopic, or but little hygroscopic form; this eliminates many compounds corresponding to those indicated by (1) and (2). In addition to KBr, KCl has been shown to be very useful. In the very long-wave infrared region, above about 30 μ, paraffin or polyethylene has been recommended as an embedding material [Yoshinaga and Oetjen (*764*)].

The spectra of substances which are obtained by an embedding technique frequently have the appearance shown diagrammatically in Fig. 98. Price

† For construction, see Ardenne (*26*). Obtainable from: P. Weber, Stuttgart-Uhlbach, Germany.

‡ A further advantage of the KBr compression technique is the very small amount of substance needed, about 1 milligram. Using the micro accessory mentioned in the notes to Fig. 96, this amount can be reduced still further.

Equipment and Experimental Technique

and Tetlow (*562*), Duyckaerts (*180*), and Lejeune and Duyckaerts (*413*) have been concerned with the clarification of certain effects observed. Primas and Günthard (*563*) have studied the fundamental basis of general diffraction theory. The effects that have been studied are the increase in the transmittance on the high frequency side of a band (the so-called Christiansen effect), the displacement of the band maxima compared with the corresponding solution spectra, and the decrease in the general transmittance in the short-wave infrared region approaching the visible region. The conditions for a good spectrum from a disc, already mentioned above, follow from the work given in these references. It is found that the course of the extinction curve depends on the product $(n-1).R$, where R is the radius of the suspended particle and n is the (complex) relative refractive index of the

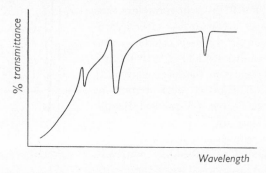

FIG. 98. *Typical appearance of a disc spectrum (diagrammatic).*

embedding material and the embedded substance. The decrease in transmittance on approaching the visible region disappears when one of the two factors of this product becomes zero, i.e. either when R becomes very small (compared with the wavelength of the radiation) or when the two refractive indices become equal. At sufficiently small values of R, provided the extinction is not too large, the contours of the absorption bands are normal in every embedding material. In other cases the product given above determines the distortion of the bands. At not too large differences between the refractive indices, the extinction maximum is displaced towards longer wavelengths if the refractive index of the embedded substance is the larger, and otherwise to shorter wavelengths. In the cases which occur experimentally this displacement is not larger than the half-band width of a band. It is of interest that the form of the band passes through a cycle on increasing the difference between the refractive indices. For not too strong absorption or particle radius, a normal absorption curve is found for $\Delta n = 0$. At

234

$\Delta n = 2 \cdot 1$ it is the corresponding dispersion curve (derivative of the absorption curve), for $\Delta n = 4 \cdot 1$ the emission curve (negative absorption curve), for $\Delta n = 5 \cdot 9$ again a dispersion curve (derivative of the emission curve) and at $\Delta n = 7 \cdot 6$ again an absorption curve. In practice, most cases have $\Delta n, < 2$, so that the extinction curve is formed additively from an absorption and a dispersion curve.

Criteria for the optimum preparation of samples to be subjected to spectroscopy in suspension are given by the preceding considerations. The number of investigations concerned with this subject is very large and is constantly increasing; it is doubtful whether many of the special procedures are really of any importance. These are mainly concerned with minute differences in the technique of preparation. Apparently of greater significance are indications that the embedding material, especially KBr, and the compression technique are certainly not without influence on the chemical constitution of the embedded substance [Bak and Christensen (*30*); Farmer (*204*); Lane *et al.* (*404*); Pliskin and Eischens (*542*); Brockmann and Musso (*86*); Jones and Chamberlain (*779*); Meloche and Kalbus (*817*)]. Although the KBr disc technique is a very valuable method for polar solids, the spectra thereby obtained must be critically considered in order to avoid incorrect conclusions. Wherever possible, solutions should be used.

The preparation of substances in the form of *fibres* need special discussion. Usually these substances also form films *via* solutions. When this is not the case, or is not desirable because the fibres themselves are to be investigated, placing a bundle consisting of many unordered single fibres in the optical path is not recommended, because unwanted diffraction results. Holliday (*327*) has described a better, but somewhat laborious, process. Individual fibres of equal size are wound closely together onto a hollow frame parallel to the spectrometer slit and cemented at each end on to the frame; one of the two layers is then cut away and the grating thus formed is brought into the optical path. If the thickness of the fibres is correctly selected (somewhere in the region of 10 to 30 μ) then this arrangement automatically gives a correct value for the active spectroscopic layer thickness. Diffraction is decreased by the regularity of the arrangement so that sharp and easily-interpretable spectra are obtained. Errors do occur as a result of the variation of the layer thickness from the centre of the fibre to the outside. Their correction has been calculated by Fraser (*801*) as a function of the observed extinction. Micro-spectroscopic methods can also be applied to the investigation of individual fibres.

The most important preparative methods for solids have now been mentioned and their characteristics and application compared. Naturally, there are other specialized methods which are only occasionally applicable. Examples are the precipitation of a film from the gaseous phase and investigation of the products of thermal decomposition [Tyler and Ehrhardt (*719*); Harms (*300*)]. The discovery and success of a suitable method of preparation always depends on a knowledge of the properties of the substance and the information desired from the spectrum. This knowledge, together with care and patience, always leads to a satisfactory result. Therefore, further discussion will be limited to the special preparation of samples for investigation by polarized radiation for the clarification of space lattices or the mutual position of different parts of the same molecule. This concerns crystals of the non-cubic systems and the high molecular weight solids of natural or artificial occurrence such as rubber or polyethylene. Crystal lattices are present in the first group of substances, and crystallography shows how the crystal should be oriented in order that a proposed structure can be recognized in the spectrum. High polymers can also possess a crystalline structure which at times can be recognized from its external features but is usually more or less divided up by amorphous areas. Artificial order, enforced by external influences, in long molecular chains or similar structures can be very important for the solution of many problems, as will be discussed later. The simplest way to attain any marked degree of order is by means of definite mechanical extension or pressure, as, for example, stretching a rubber band. Obviously this gives only a linear order such as the parallel arrangement of long chains. For some substances, e.g. polyvinyl alcohol and polyamide films, a so-called double orientation, i.e. a planar arrangement of the constructional units, can be obtained by cold or hot rolling (Elliot *et al.* (*196*)]. The orientation of the radiation beam plays a part in obtaining and interpreting the spectra of such oriented films, and if this is not taken into account errors can arise.

Obviously such orientation effects occur in the preparation of films from suitable materials to a greater or lesser extent by almost all the processes described above. Also, the radiation emitted from the monochromator and the other optical parts of the spectrometer is always more or less polarized even when this is not desired. This phenomenon is known as *apparatus polarization* and is a result of various reflections at the non-metallic surfaces in the radiation path, e.g. on the prism, which lead to partial polarization according to well known laws. Therefore, it is quite possible, without using polarized radiation as such, for the spectra of such spontaneously oriented films to show typical features of orientation, e.g. varying band intensity according to the orientation of the film with respect to the polarization relationships of the apparatus. Hence, the spectra of such substances must be interpreted with care.

236

5. The Preparation of Cell Windows and Plates

The chapter on the preparative methods of infrared spectroscopy cannot be closed without discussing the preparation of windows, etc., of the typical infrared transparent materials (cf. Table 10) [Lord *et al.* (*430*)]. Although this technique does not belong directly to the preparation of the samples, it is of such great importance and it is so frequently required that every infrared spectroscopist should know the technique and the equipment required. The most important raw materials for infrared optics differ considerably from the usual optical material, glass. Rock-salt, or the synthetic form of this, NaCl, can be considered to be a typical example of the useful materials. There are three essential differences, compared with glass, which make their use in part more difficult and in part easier: (1) Easy cleavage at certain planes determined by the crystal type; (2) considerably softer; (3) more or less marked hygroscopic character.

The first property enables pieces of given dimensions to be easily obtained as long as the form of these pieces accords with the characteristics of the crystal system. Thus, thin square windows of quite large surface can be obtained by simple cleavage of a crystal block along the lattice plane. A thin sharp knife is used for this cleavage; it is placed on in the desired cleavage direction and gently hit with a light hammer. The rule applies that the piece to be cleaved off should always have about the same thickness as the piece remaining; however, after a little practice this rule need not be strictly followed. Most of the alkali halide crystals used can be turned and tapped on the lathe. Round windows can thus easily be obtained without using an instrument like a cork borer which is often recommended but is very tiring to operate. Cuts against the natural cleavage planes must be made with saws; for this fret saws or thin metalled saw blades are used, sometimes with alcohol or water as lubricant.

The second property enables easier finishing and polishing compared with glass, but also causes greater sensitivity towards mechanical action by pressure and scratching, and a lower quality of the final polish. The preparation of refined optical parts, such as prisms, from rock-salt crystals with very high requirements with respect to planarity, accuracy of the angles, etc., should be left as a matter of principle to optical specialists who are specially equipped and who have had many years of experience.† On the other hand, the preparation or renovation of smaller parts, for which the optical requirements are not too high, is a routine process which should be carried out in every infrared spectroscopial laboratory. The equipment needed for this is quite simple: one or more glass grinding plates, about

† E.g. Dr Steeg and Reuter, Bad Homburg v.d.H., Germany; B. Halle Nachf., Berlin-Steglitz.

Equipment and Experimental Technique

20–30 cm in diameter and 2–3 cm thick, ground planar, which can be purchased at any optical firm; abrasive powder of varying fineness; a polishing powder of especial fineness (jeweller's rouge); a polishing plate of glass with fine silk drawn over it; soft chamois, etc.; and a suitable non-aqueous liquid as lubricant (alcohol, olive oil, etc.). The various types of abrasive powder must be carefully elutriated and if necessary winnowed in order to obtain an even particle size. Great care must be taken that a finer type is not contaminated by a rougher type because even a single large particle will spoil all the previous work. If the surface in question is not a good quality cleavage surface, it is first rubbed fairly smooth with emery paper. The grinding process itself is then carried out by placing a little emery powder, of not too coarse a quality, on the grinding plate with the lubricant and then rubbing the component to be polished with a gentle even pressure and small circular movements gradually over the whole surface of the grinding plate (which in time becomes hollowed out). This process is repeated according to requirements with one, two, or sometimes even three types of emery of increasing fineness until a good flat surface is obtained with the appearance of a ground glass plate and so that after working on both sides the component has the desired thickness. The flatness of the surface is tested by means of a spirit level or for more refined cases by means of mechanical precision instruments or even (after polishing) by optical methods. Further treatment, i.e. polishing, is carried out with the constant use of rubber gloves, or at least finger-stalls, in order to avoid the influence of perspiration on the surface; breathing on the component should also be avoided. For polishing, at first the same process is carried out with the polishing material and the lubricant, as previously described for the grinding, until the dull surface gradually vanishes and a smooth mirror-like surface appears. The final polishing is carried out on the silk cloth by dropping a little dry alcohol on it and polishing the component until it is completely dry. Experienced workers often use water as a lubricant for grinding and polishing rock-salt crystals and thus use solution effects in addition to mechanical and heat effects; however, this technique requires long experience. Cell windows, etc., the polish of which is only moderately damaged, can be renovated by dry polishing with a silk cloth or a wad of cotton wool. Softer materials, such as AgCl, require a somewhat different polishing technique. Directions for this have been given by Mitzner (*818*).

Boring holes in salt crystals should also be practised by infrared spectroscopists, because this is needed now and again, e.g. in the preparation of the liquid cells with filling connections as already described. The softness of the material makes this work easier but it still requires patience and a steady hand. As already mentioned, a spiral hand drill of suitable size is used. It

238

should be used in such a way that the direction is suitably controlled, e.g. by boring directly through the connections that will be later used. A certain number of losses as a result of the cell windows breaking must be reckoned with.

The hygroscopic character of most of the salt crystals is especially awkward and often necessitates special precautions. The constant attack by atmospheric water vapour, especially when combined with sudden changes of temperature, in time dulls even prisms, etc., composed of only moderately hygroscopic materials. It is known that natural cleavage surfaces are more resistant in this respect than artificially ground and polished surfaces. It is also known that crystals prepared synthetically are less hygroscopic than natural ones because of the greater purity of the former. For this and other reasons, single crystals synthetically obtained are now almost exclusively used for infrared optics. However, a sufficiently low humidity over long periods can only be attained in air-conditioned rooms. As these are not always available, recent commercial instruments all possess a thermostat which keeps the water-sensitive parts at a somewhat higher temperature than that of the surroundings, in order to decrease the danger of condensation. For liquid cells, it is virtually impossible to ensure complete absence of water from all the substances investigated. Occasional use of liquids containing up to 2 or 3 per cent water will not seriously damage NaCl windows, but if such are constantly used they cause a degree of dullness which cannot be tolerated, as it increases the amount of dispersion (especially in the short-wave region) and thus lowers the transmittance. It is often not realized that the transport of optical components made of salt crystals can be dangerous. Even if this is done in a desiccator, which has cooled down to below room temperature and is then opened without having come back to temperature equilibrium with the atmosphere, water condensation can take place on the surfaces and thus cause dulling. Similarly for emptying cells; if this is done by sucking out, the windows cool down as a result of the latent heat of evaporation, etc., and the humidity of the atmosphere condenses on the surfaces; here, blowing warm dry nitrogen through the cell is advantageous. It need hardly be mentioned that prisms, cells, etc., of hygroscopic materials should be kept in a desiccator when not in use.

There have been many attempts to protect optical components, made from salt crystals, against the attack of humidity by protective coatings. For example, a condensed layer of SiO, which is converted to SiO_2 in the air, does this; naturally the spectrum of SiO_2 then appears, unless arrangements are made for its compensation, to an extent depending upon the thickness of the sublimed film. Another process uses sublimed films of selenium; these have the advantage of no absorption of their own, but the disadvantage of a very high power of reflection [Anderson *et al.* (*23*)]. Recently, a germanium film (less than 1 μ thickness) has also been recommended [Sternglanz (*788*)]. Gatehouse (*802*) uses thin coatings of polystyrene which are prepared by dipping the plates in dilute polystyrene solutions. These prevent reactions between the contents and the window materials as well as protecting the plates against humidity. In some cases polyethylene is preferable for this purpose because of its simple spectrum. With all such protective coatings it should be noted that their preparation must take place in several steps which alternate with a cleaning treatment of the surface to be protected. In this way the probability of unavoidable dust particles falling on top of one another for successive steps of the sublimation is decreased; thus the possibility of the protective film being pierced is also lessened. Another process, to improve the resistance to water, depends upon the observation that cleavage surfaces are less attacked than ground and polished surfaces; obviously, this is because the order of the crystal lattice is retained for the former while the latter are amorphous. The finished surfaces are therefore treated with a solvent or subjected to changes of temperature so that a recrystallisation process takes place.† Practical details are not available for this method.

Among the materials which, because of their transmittance, may be considered for use in infrared spectroscopy, there are fortunately some of low or moderate hygroscopic character, such as AgCl and CaF_2, or the mixed crystals of TlBr and TlI, which are referred to as KRS 5. The last material is especially important for the long-wave part of the infrared spectrum; however, it has some serious disadvantages: excessive softness and therefore low resistance to deformation, very high reflective power, and a poisonous nature which should not be underestimated.

† DRP 760 375.

PART III

The Methods of Practical Infrared Spectroscopy

1. The Interpretation of Chemical Constitution

1. The Characteristic Group Frequencies

The first part of this book was devoted to the theory of the infrared spectra of molecules in the gaseous state. The discussion there showed that it was possible to derive a series of data concerning the *molecular structure* from the coarser and finer structure of a spectrum. These data were all concerned with *geometric* properties of the molecule investigated, such as atomic distances and valency angles, or with *dynamic* relationships, such as force constants, and centrifugal or Coriolis force influences, etc. Therefore, they will be briefly referred to as *geometric-dynamic structural elements*. An advanced experimental technique with high spectral resolution must be used to obtain them, such as is generally only available with grating instruments. In spite of the large prisms which are now obtainable from many different materials, such resolution is usually obtained only with grating instruments. Accordingly, these investigations are principally devoted to academic questions for which the effort demanded does not have to be strictly limited. For the infrared spectroscopy used in industry, where the expense must be related to the result obtained, use has been limited to the prism instruments which, although themselves not cheap, are characterized by a marked versatility of application. Here the high spectral resolution usually necessary for the geometrical-dynamical elucidation of structure, in the way described above, is wanting and investigations are limited to obtaining vibrational spectra with no or very little rotational fine structure (exceptions do occur, especially for gases). These spectra still give significant information which is extremely valuable from a practical standpoint. In addition to the geometrical and dynamic structure, the chemical structure of a molecule, i.e. its composition from different atomic groupings, is important.

It follows from the theory that important quantities, such as force constants, can be obtained from vibrational spectra. If the infrared spectrum is compared (e.g. in the region of 2 to 25 μ) for as large as possible a number of substances of definite structure and known constitution, it is quickly seen that for the same or related chemical constitution bands occur similar with

243

respect to their position and appearance. For example, it is found that all molecules containing CH_2- and CH_3-groups show a band or a group of bands at about $3\cdot3$–$3\cdot6\,\mu$, that the existence of OH groups causes a band at $2\cdot7$–$3\cdot0\,\mu$, and that C=O groups cause a band at $5\cdot5$–$6\cdot5\,\mu$. In such cases the influence of the rest of the molecule, i.e. the other atomic groupings and bonds it contains, is often only apparent after a more detailed study. This necessitates the assumption of a *correlation between certain groups of atoms as components of molecules and bands at definite positions in the spectrum*. This conclusion, based on experience, necessitates the further conclusion that such a band (as an expression of a vibration taking place in the molecule) is essentially determined only by the components of the particular atomic grouping and the valency forces between them. The rest of the molecule has relatively little influence. This conclusion is at first surprising and certainly cannot apply generally and rigidly because it contradicts the earlier finding (see page 61) according to which *all* the atoms of a molecule always take part in a vibration.

It is understandable that the force constant between two atoms should always be approximately the same whatever their environments. This is because the force constant is an expression of a certain electronic structure, the nature of which is essentially determined only by the atoms taking part in it. From this point of view, the above result no longer appears so unexpected. The empirically determined fact of the existence of characteristic bands, i.e. vibrational frequencies for definite atomic groups, is thus based on the circumstance that, to a first approximation, the force constant between given atoms is constant. A more detailed treatment shows that the rest of a molecule is certainly not completely without effect on the vibrations of any group of atoms considered in isolation. This is shown by the fact that the value of the force constant between two atoms depends somewhat on the circumstances and on the environment of the bond in question (Table 14). Similarly, this is shown by the spectral position of the vibrations and bands in question, as will be discussed further.

Questions of nomenclature will now be discussed. The division by Mecke of vibrations into *stretching vibrations* and *deformation vibrations* has already been mentioned. This classification can easily be visualized although of course it is schematic to a certain extent. However, such a classification is certainly not sufficient to deal with all the details which are significant experimentally and theoretically. Especially as a result of the investigations of Anglo-American workers, this nomenclature has been considerably extended. Because of the preponderance of the use of English in scientific writing, the English names have become more or less internationalized. Thus the following are now differentiated:

1. *Stretching vibrations* (abbreviated st) with movement along the bond between the atoms in question;
2. *Bending vibrations* (abbreviated b) motion perpendicular to the bond between the atoms.

The latter term (frequently these vibrations are also referred to as deformation vibrations) is only used when the structure of the vibrating group in

TABLE 14. *Force Constants for a few similar Chemical Bonds in different Molecules (in 10^5 dyne/cm)*

Molecule	C—H	C—C	C=C	C≡C	C≡N
Paraffins	4·8				
Olefins	5·1				
Acetylenes	5·9				
Aromatic compounds	5·0				
Ethane	4·79	4·60			
Propane		3·78			
Hexachloroethane		4·00			
Ethylene			10·8		
Propadiene			9·45		
Acetylene				14·9	
Propyne		5·30		14·7	
HCN	5·4				17·9
ClCN					16·6
BrCN					16·8
ICN					16·7
Cyanamide					17·0
Acetonitrile					17·3

question is so simple that further specification is unnecessary, or when further specification is impossible because deeper insight into the situation is lacking. In other cases the following terms are used:

2a. *Scissor vibration* (abbreviated s), sometimes simply called deformation vibration (abbr. d), especially in early work, for a bending vibration in which an alteration takes place in the *angle* of the group in question;

2b. *Wagging vibration* (abbr. w) for a bending vibration in which no alteration takes place in the valency angle of the group in question but in which this group vibrates as a whole in a plane *perpendicular* to the plane of the group;

2c. *Rocking vibration* (abbr. r), for a bending vibration corresponding to the wagging type, in which however the vibration of the group takes place as a whole in the plane of the group;

2d. *Twisting vibration* (abbr. t), or *torsion vibration*, for a bending vibration similar to the wagging and rocking modes but in which the group in question oscillates as a whole about the bond which connects the group with the rest of the molecule.

Finally, there is a completely symmetrical stretching vibration, found only in ring compounds (and tetrahedral molecules), which is referred to as the *breathing vibration* (abbr. br) and consists of a periodic extension and contraction of all the ring-bonds simultaneously. It can be visualized as a 'breathing' of the ring. These types of vibration, with the exception of the last are shown diagrammatically for the CH_2 groups in Fig. 99.

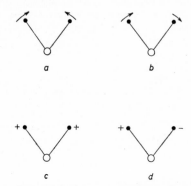

FIG. 99. *The forms of the motion of different bending vibrations of the CH_2 groups.* (a) *Scissor vibration;* (b) *rocking vibration;* (c) *wagging vibration;* (d) *twisting vibration. Arrows denote motion in the plane of the paper, + and − motion perpendicular to this;*

○ *C-atom,* ● *H-atom.*

According to the preceding discussion it should follow that a definite frequency (and thus wavelength or wave number of the corresponding band) should apply to every stretching or deformation vibration (these expressions will be retained in general discussion) for every conceivable group of atoms. This is indeed the case for the individual vibrations if influences, such as pressure broadening, which lead to the frequencies being less well defined, are neglected. If the position of the frequency of a definite vibration for a definite group, e.g. the stretching vibration of the $C=O$ group, is considered for molecules of differing construction, it is seen that this frequency is generally displaced a little by the influence of the different environments, but it almost always falls in one definite region. A group of atoms in a molecule can never be considered to be completely isolated from the remainder of the molecule, and the latter will always have a certain influence on the former,.the size of which will depend on the particular circumstances. As a result of this influence, the force constant between any two atoms, and therefore the frequency of the corresponding vibration, will be somewhat altered. The deviation of the force constant of identical groups of atoms in different molecules is generally small compared with the mean value of the force constant itself. Accordingly, for a given group, there is always a definite region in which the frequency and therefore the wavelength or wave number of a definite vibration is to be expected. The position of the band within this region for any particular molecule (itself sharply defined) will depend on the particular structure of this molecule.

In this sense, and with these limitations, the terms *characteristic* or *group frequencies* are used in infrared spectroscopy for certain frequently occurring groups of atoms. There are many ways in which they can be represented and tabulated. This can be done in the form of tables in which the usual wavelength or wave number region of the vibrations corresponding to every group is given. It can also be shown in the form of a graphical representation by using wavelength or wave number as abscissa and the symbol for the atomic grouping as ordinate. Other authors reject the enumeration of the individual groups of atoms and use in their place the classification of substances according to chemical composition. Here, instead of the vibrations characteristic of definite groups, the spectrum is shown which is characteristic of a particular class of substance as a whole. As each system has its advantages, all three are used in this book: Table 15,† Figs. 100 to 108

TABLE 15. *Spectral Position of the Characteristic Absorption Bands of Groups and Bonds*

Group	Type of vibration	Wave number cm^{-1} ca.	Wavelength μ ca.
OH	st	2500–3700	2·7– 4·0
NH, NH_2	st	3200–3500	2·8– 3·2
$\equiv$CH	st	3300	3·0
CH_{arom}	st	3030	3·3
CH_{olef}	st	3010–3100	3·2– 3·3
CH_{aliph}	st	2840–2970	3·4– 3·5
SH	st	2550–2600	3·8– 3·9
PH	st	2350–2440	4·1– 4·4
SiH	st	2100–2300	4·3– 4·8
N=C=O	st	2270	4·4

† A similar, very detailed summary is found in (*P*), but this must be considered in part to be out of date and superseded. (*P*) also contains two further very comprehensive tables of assignments of definite molecules, one for the so-called 6 μ region (4·8 to 6·9 μ or 2080 to 1450 cm^{-1}), the other for the whole infrared region from 2·66 to 68·9 μ or 3760 to 145 cm^{-1}.

TABLE 15—continued

Group	Type of vibration	Wave number cm^{-1} ca.	Wavelength μ ca.
C≡N	st	2040–2270	4·4– 4·9
C≡C	st	2100–2270	4·4– 4·8
N≡N=N	st–a	2130–2170	4·6– 4·7
C≡O	st	2000–2100	4·8– 5·0
C=C=C	st–a	1960	5·1
C=O	st	1540–1870	5·3– 6·5
C=C$_{olef}$	st	1590–1690	5·9– 6·3
C=C$_{arom}$	st	1500–1610	6·2– 6·7
N=N	st	1570–1640	6·1– 6·4
C=N	st	1470–1690	5·9– 6·8
NH	d	1510–1640	6·1– 6·6
COO$^-$	st–a	1540–1620	6·2– 6·5
O—NO	st	1610–1670	6·0– 6·2
NO$_2$	st–a	1500–1650	6·1– 6·7
CH	d	1340–1490	6·7– 7·5
P—C	st	1430–1450	6·9– 7·0
CH$_{olef}$	r	1280–1430	7·0– 7·8
CH$_3$	d	1360–1390	7·2– 7·4
OH	d	1210–1450	6·9– 8·3
COO$^-$	st–s	1300–1420	7·1– 7·7
C—N	st	1020–1420	7·1– 9·8
C—O	st	1050–1430	7·0– 9·5
N≡N=N	st–s	1180–1340	7·5– 8·5
C—NO	st	1310–1410	7·1– 7·6
N—NO	st	1310–1410	7·1– 7·6
C—F	st	1000–1400	7·2–10·0
SO$_2$	st–a	1300–1350	7·4– 7·7
NO$_2$	st–s	1250–1350	7·4– 8·0
P=O	st	1200–1300	7·7– 8·3
C—O—C	st	1050–1200	8·3– 9·5
=C—O—C	st	1230–1270	7·9– 8·2
C=C=C	st–s	1060	9·4
C—O—C$_{cycl}$	st	1050–1150	8·7– 9·5
P—O—C	st	1000–1230	8·1–10·0
Si—O	st	1000–1100	9·1–10·0
SO$_2$	st–s	1140–1160	8·6– 8·8
S=O	st	1040–1060	9·4– 9·6
C—O—O—C	st	820– 890	11·2–12·2
OH	w	850– 900	11·1–11·8
CH$_{arom}$	w	690– 900	11·1–14·5
CH$_{olef}$	w	690–1000	10·0–14·5
P—O—P	st	930– 970	10·3–10·7
P—F	st	800– 900	11·1–12·5
P—N	st	720	13·9
C—S	st	600– 700	14·3–16·7
P=S	st	600– 650	15·4–16·7
C—Cl	st	600– 800	12·5–16·7
P—Cl	st	500– 600	16·7–20·0
C—Br	st	490– 600	16·7–20·4
S—S	st	400– 500	20·0–25·0

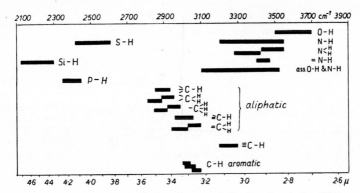

FIG. 100. *Position of the characteristic bands of the XH stretching vibrations [according to Seidel (M)].*

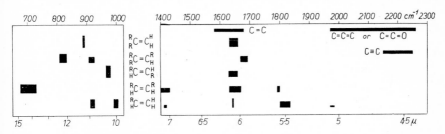

FIG. 101. *Position of the characteristic bands of various CC bonds [according to Seidel (M)].*

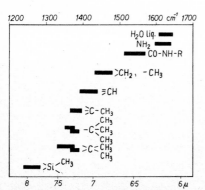

FIG. 102. *Position of the characteristic bands of hydrogen deformation vibrations [according to Seidel (M)].*

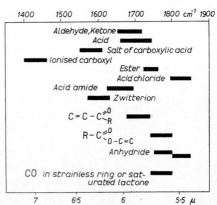

FIG. 103. *Position of the characteristic bands of the C═O stretching vibration [according to Seidel (M)].*

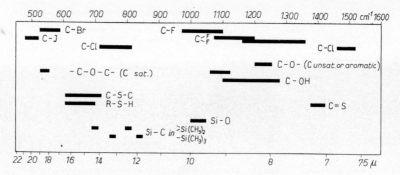

FIG. 104. *Position of the characteristic bands of carbon-halogen, carbon-oxygen, carbon-sulphur, and carbon-silicon bonds [according to Seidel (M)].*

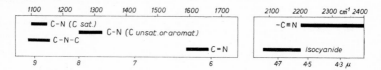

FIG. 105. *Position of the characteristic bands of carbon-nitrogen bonds [according to Seidel (M)].*

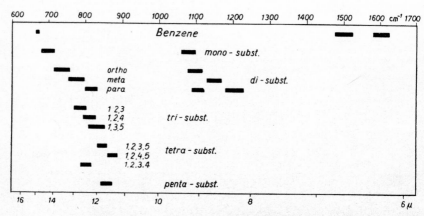

FIG. 106. *Position of the characteristic bands of benzene and benzene derivatives [according to Seidel (M)].*

250

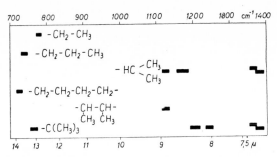

FIG. 107. *Position of the characteristic bands of alkyl groups* [*according to Seidel* (*M*)].

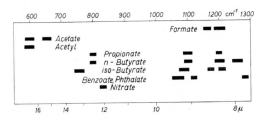

FIG. 108. *Position of the characteristic bands of acyl and acyl-oxy groups* [*according to Seidel* (*M*)].

essentially according to Seidel (*M*), and Fig. 109 according to Colthup (*136*). If these summaries are closely studied, it can be seen that there is a certain natural order in the characteristic frequencies. This does not always apply, but it can serve as a first approximation. Approaching the infrared from the visible region, the group frequencies consisting of stretching vibrations of single bonds connecting hydrogen to other atoms are found in the region from about $2 \cdot 5$ to $4\ \mu$ or correspondingly 4000 to 2500 cm^{-1}. Stretching vibrations of triple bonds are then found to about $5 \cdot 5\ \mu$ or 1800 cm^{-1}, and this is followed by the region of the stretching vibrations of all types of double bonds to about $7\ \mu$ or 1400 cm^{-1}. From about $6 \cdot 8$ to $7 \cdot 7\ \mu$ or 1470 to 1300 cm^{-1} are found the deformation vibrations of the XH bonds, and at still smaller wavelengths the so-called skeletal vibrations of long carbon chains. Of course, this division is only a rule of thumb; many cases of overlapping occur, and, for example, the stretching vibrations of carbon-halogen bonds do not fit into this scheme at all.

A table of assignments for the near infrared has been given by Kaye (*375*). It includes principally the fundamental and overtones as well as the strongest combination bands of the vibrations of the XH bond (Fig. 110). The number of spectra on which it is based is not comparable with that

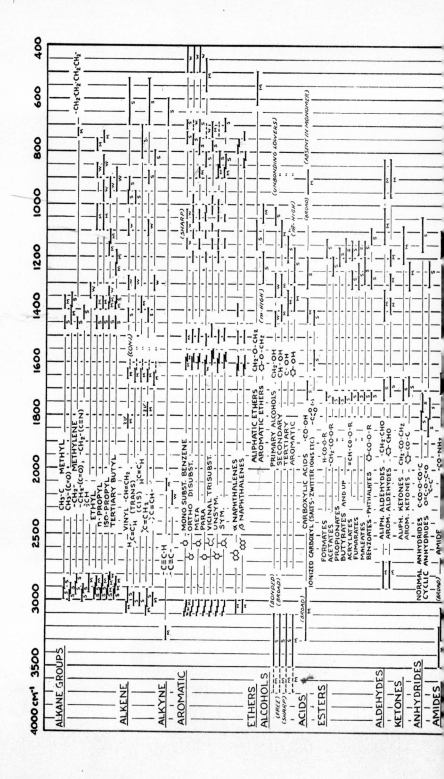

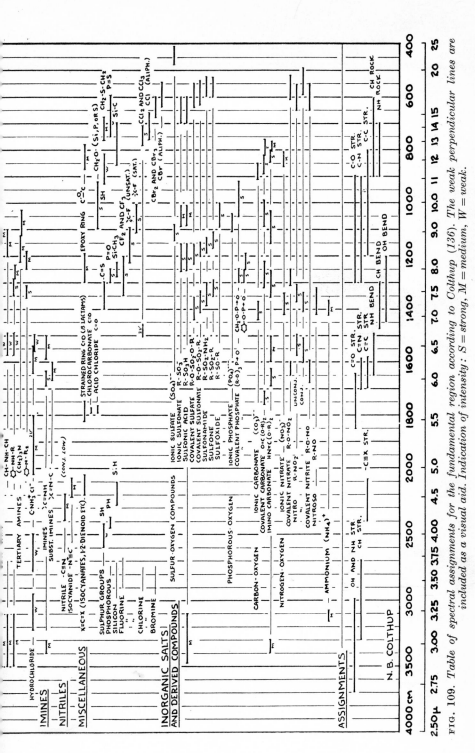

FIG. 109. *Table of spectral assignments for the fundamental region according to Colthup (136). The weak perpendicular lines are included as a visual aid. Indication of intensity: S = strong, M = medium, W = weak.*

(By courtesy of American Cyanamid Co., Stanford, Conn., U.S.A.)

of the Colthup table, so that the assignments are probably only of limited validity. However, it is certainly helpful for a preliminary interpretation.

To conclude the discussion of the characteristic frequencies it should be mentioned that a frequency is the less characteristic, the more complicated the corresponding motion of the atoms in question, and the more the remainder of the molecule takes part in this motion. In this sense, the stretching vibrations are generally 'more characteristic' than the deformation

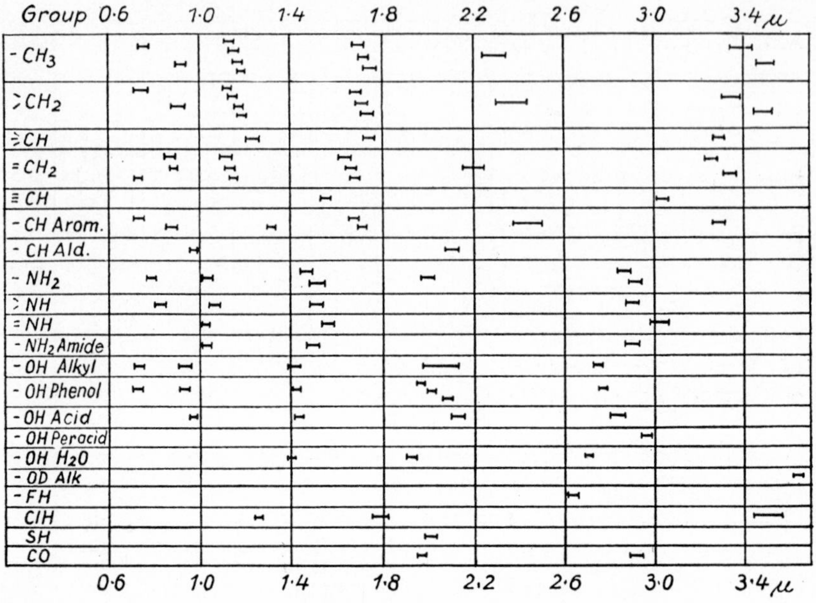

Group	0·6	1·0	1·4	1·8	2·2	2·6	3·0	3·4 μ
-CH₃								
>CH₂								
≩CH								
=CH₂								
≡CH								
-CH Arom.								
-CH Ald.								
-NH₂								
>NH								
=NH								
-NH₂ Amide								
-OH Alkyl								
-OH Phenol								
-OH Acid								
-OH Peracid								
-OH H₂O								
-OD Alk								
-FH								
C/H								
SH								
CO								

FIG. 110. *Table of assignments for the near infrared region* [according to Kaye (375)].

modes. This is because for the latter, as a result of the multiplicity of possible types of vibration which differ but little in frequency, it is often not possible to make an unambiguous assignment of the bands observed to the theoretical vibration modes.

2. Procedure for the Determination of Chemical Structure

A summary of the characteristic frequencies of atomic groupings is clearly of great value if it enables the chemical structure of a substance to be determined from its infrared spectrum. A condition for such an analysis is that the substance to be investigated which must be in the form of a single compound, highly purified (which may have been carried out by fractionation, etc.).

254

The Interpretation of Chemical Constitution

The first step is to determine the spectrum of the substance at a small layer thickness or at equivalent dilution. The strongest bands are then selected and using the spectral tables the possible structural elements from which they could originate are found. This can seldom be carried out unambiguously using only the spectrum because, as can be seen from the above summary, the expected regions of the different characteristic frequencies overlap each other markedly. Therefore other information has to be used for an unambiguous decision. This can be spectroscopic in nature, e.g. information concerning the intensities to be expected or the intensity relationships between definite bands. It could also be data of another type, e.g. the chemical properties of the substance investigated may exclude or confirm certain structures which are possible on the spectroscopic data. To obtain the most reliable picture of the constitution of the substance in question, it is necessary to combine all the available spectroscopic and non-spectroscopic data. If, in this way, the origin of the strongest bands in the spectrum is more or less unambiguously determined, an attempt should then be made to assign the bands of moderate and weak intensity. This is much more difficult and is often not possible to carry out completely even if the substance under investigation is completely pure, i.e. possesses one single structure. This is because bands of moderate or weak intensity can result from overtones or combination vibrations and these have not been included in the summaries given because of their complicated nature. The final step (when possible) is to carry out a complete vibrational analysis. This is based on the constitution decided, and takes into consideration all the circumstances including the symmetry and spectral activity properties, in attempts to clarify the positions of the fundamentals, overtones, and combination vibrations expected. Using this result, the weaker bands can usually be satisfactorily interpreted. However, it must be noted that the carrying out of such a vibrational analysis is difficult and laborious, and it is frequently impossible or not completely possible for compounds of complicated structure. The final test of the whole investigation is the determination of the spectrum of the synthetic compound corresponding to the structure derived. In some cases it is only at this stage that a final decision can be made between several possible structures.

It should be clear that it is hardly possible to give a general recipe for the determination of structure by means of infrared spectroscopy. Nevertheless, it will be shown by means of two examples how this is usually carried out.

First example: In the high boiling point fraction from the product of a synthesis which was expected to give only hydrocarbons, a substance was found of which the spectrum is given in Fig. 111. Because of the carbonyl

The Methods of Practical Infrared Spectroscopy

band at 5·9 μ, this was to be considered as containing oxygen. The substance was a colourless liquid of boiling point 214°C and molecular formula $C_9H_{10}O$, according to the elementary analysis and molecular weight. Without any knowledge of the chemical properties the following deductions can be made from the spectrum: (1) It must be an aromatic compound (sharp, moderately intense bands at 6·2 and 6·7 μ). (2) The position of the carbonyl band at 5·9 μ and the approximately equal intensity of the aromatic doublet at 6·2 and 6·3 μ indicate a benzoyl structure. (3) The benzene ring is di-substituted in the ortho-positions (intense, dominating band at

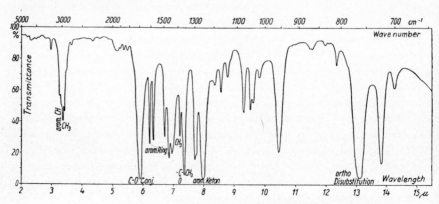

FIG. 111. *Infrared spectrum of o-methylacetophenone as an example of the determination of structure by spectroscopic means.*

13·1 μ). (4) The bands at 3·37, 3·42 and 3·5 μ (the last present only as a shoulder) indicate the presence of CH_3 groups. Combined with the molecular formula, the compound is unambiguously determined as:

ortho-methylacetophenone.

The spectroscopic result was confirmed by means of a classical chemical structure determination and by the comparison of the spectrum with that of authentic *ortho*-methylacetophenone.

Second example: White crystals in the form of plates, melting at 100°C and with molecular formula $C_{13}H_{10}O$, were obtained as the product of a reaction. The spectrum of the crystals in a KBr disc is shown in Fig. 112. The aromatic nature of the substance follows from the molecular formula

256

and the other properties and it is confirmed by the spectrum (=CH—band at $3 \cdot 3 \mu$, aromatic ring bands at $6 \cdot 2$, $6 \cdot 3$, and $6 \cdot 7 \mu$). The ring is di-substituted in the *ortho* position (band at $13 \cdot 3 \mu$). At least one carbon atom is in the form of a saturated methylene group (bands at $3 \cdot 42/3 \cdot 5$ and $6 \cdot 85 \mu$); however, these bands are too weak compared to the aromatic bands to indicate more than one non-aromatic group. Accordingly, two benzene rings connected by a methylene bridge must occur and, because there is no indication that a mono-substituted ring is present, a second bridge must

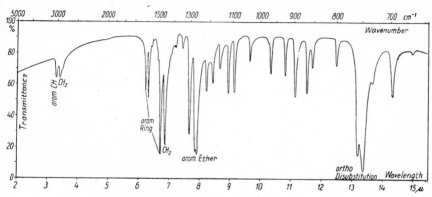

FIG. 112. *Infrared spectrum of xanthene as an example of the determination of structure by spectroscopic means.*

occur between the rings in the *ortho* positions. This must contain the oxygen because the spectrum gives no indication of an OH or C=O group but does show the occurrence of an unsaturated ether linkage (band at $7 \cdot 9 \mu$). Accordingly the structure derived from the spectrum is

xanthene.

This was confirmed by comparing the spectrum with that of an authentic specimen.

Both the examples are concerned with rather simple structures. Obviously, the analysis becomes more difficult with increasing complication of the structure. Examples of this, which would take up too much space to be discussed here, can be found, e.g. in (P).

The determination of chemical structure is much less promising if the substance in question is a mixture of different compounds. Obviously, it is still possible to assign the strongest bands to definite atomic groupings, but no definite deductions as to the structure can be made from these assignments. If an analysis is to be made with success in such a case, it is necessary to separate the components of the mixture or at least to enrich one compo-

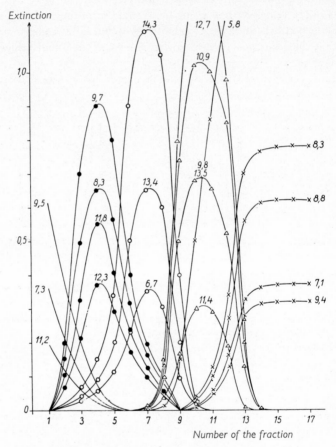

FIG. 113. *Determination of structure by the method of intensity variations by fractionation.*

nent to such an extent that the remainder can be considered as minor impurities corresponding to bands of low intensity only.

If one substance cannot be obtained pure, or if too great an effort would be needed for the complete separation, then the following process is often successful, provided that sufficient amounts of the material in question are available. The methods of infrared structure analysis are applicable if the positions (and possibly the intensities) of the set of bands or at least the main characteristic bands of a molecule are known. In order to find them, the mixture is fractionated in some way, e.g. by distillation, so that the content of the mixture is altered a little from fraction to fraction. The infrared spectrum is obtained for each fraction. Using some characteristic parameter of the fraction, e.g. the fraction number, as abscissa, the extinctions are plotted for every band that can be measured. Provided that not too small a number of fractions is used, it can soon be seen that, as the composition of the mixture changes, certain bands regularly increase, while others decrease, and yet others may retain a constant intensity. All the bands which exhibit the same regularity in the alteration of intensity belong to one set and in general to one and the same molecule. Thus the set of bands required for the analysis is obtained without having prepared the substance even approximately pure. Obviously, this process needs a certain experience in reading the spectra and in recognizing and interpreting special effects. Such an evaluation of a distillation system is shown in Fig. 113; for clarity, many of the weaker bands in each characteristic set have been omitted. Without going into great detail, it can be seen that, if it is assumed that the abscissa represents a continuously changing parameter (which of course is not strictly true), definite compounds are contained in large amounts in fractions 3, 7, 10, and 11, and from 14 onwards, although other substances are still present in these fractions. Fractions 14 onwards result in the spectrum of one definite substance, which was to be expected because these were the fore-run of a distillation. The recognition of the structures of the single substances to which the prominent sets of bands belong now follows according to the treatment described above.

3. Intensity Spectroscopy

Although the knowledge of the group frequencies gained in the preceding treatment is very important, it remains unsatisfactory from a more fundamental point of view. For this treatment indicates only the position of the group frequencies in the spectrum (the interpretation of which is all too often made difficult by the numerous overlaps), i.e. it is based on only *one* numerical parameter of the spectrum. The second possible parameter, i.e. the strength of the bands or, as it is often briefly expressed, the intensity, has up to now been neglected. (In this respect, the conclusions occasionally quoted about the intensity to be expected of individual bands may only be considered to be an expedient). However, it is becoming more and more apparent that '*intensity spectroscopy*' is going to be of equal or greater importance than the '*frequency spectroscopy*' used almost exclusively up to now, and that it makes better use of the fundamental possibilities of the spectrum.

There are both theoretical and instrumental grounds for the initial predominance of frequency spectroscopy. All the prerequisites for the conception 'group frequency' are contained in the theoretical treatment known as 'vibrational analysis'. In addition, the present state of instrumental development allows the transference and comparison of the measured positions of bands. The position with respect to band intensities is different. The measurement of intensities (as has already been mentioned and as will be discussed in detail presently) still depends so much on non-standardized or non-standardizable influences of the apparatus, that transferring measurements from one laboratory to another can only be done with great caution. The position on the theoretical side is just as unsatisfactory. The transition moment or the alteration of the dipole moment (i.e. the alteration of the electron density of a molecule) is responsible for the intensity of an absorption band (for a given number of molecules in a definite vibrational state), as has been shown. Knowledge of the influence of the remainder of the molecule on the distribution of charge in a particular bond is required to comprehend the theory of the alteration of the dipole moment in a way corresponding to the calculation of the spectral positions. Many expressions have been proposed but they have not yet led to a comprehensive theory. As long as this is not attained, because of the difficulties of visualization and mathematics, intensity spectroscopy will remain empirical.

Two quantities are frequently used as measures of the intensity of bands in practical spectroscopy. One of these is the maximum extinction of a band, which is obtainable from the present instruments directly or by means of a simple calculation. If it could be assumed that the instrument were without influence on this quantity (which is unfortunately not the case), then the maximum extinction would unambiguously characterize the band intensity, and would only depend upon the substance under investigation. Together with the position of the band and the natural band width, the maximum extinction would completely characterize the band in question (Fig. 114). As often emphasized, this is not the case and the form of the band is modified by influences of the apparatus as expressed in the so-called slit-function (see page 123). The maximum extinction observed is smaller than the true value, while the band width observed is larger. Only the band position, expressed as the frequency (or wave number, or wavelength) of the maximum extinction remains the same (Fig. 114) (and this only provided no stray light is present). Accordingly, the maximum extinction is a poor measure of the intensity of a band. The integral absorption or extinction, i.e. the integral taken over the whole band observed, is a better measure and, as has already been mentioned, is related to fundamental molecular constants. It has become apparent that the value of this integral is very much less dependent on instrumental effects than the maximum extinction. The band area remains almost constant, although the maximum extinction decreases and the band width increases for increasing slit width. Thus, the integral absorption is the more reliable or even the only correct measure of the intensity of a band. The way in which it is determined, and the difficulties which unfortunately arise thereby, will be discussed later.

Mecke and Thompson and their schools together with a few other workers [Barrow (*44*); Cross and Rolfe (*152*); Francis (*221, 222*); Jones *et al.* (*363*); Lippert and Vogel (*421*); Luther and Czerwony (*437*); Richards and Burton

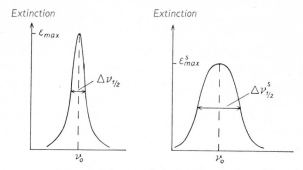

FIG. 114. *Form and characteristic parameters of an absorption band.* (a) *true form determined by the substance alone;* (b) *observed form modified by the influences of the apparatus.*

(*591*); Sensi and Gallo (*640*); Vampiri (*721*)] commenced the study of problems of spectroscopic intensities at an early date. As a result of such studies, Lippert and Mecke (*420*) proposed that the summaries of group frequencies should be replaced by diagrams in which each band is represented by a point in a frequency-intensity plane. This proposal takes into account the two-dimensional character of a spectrum. Undoubtedly, much of the ambiguity of the group frequency tables is thus removed. The importance and value of such a representation for the determination of structure is very high, because it is hardly likely that bands of the same frequency would also be of equal intensity. Obviously, there are further advantages, e.g. in respect of quantitative work or for the development of a suitable theoretical treatment, but this requires an initial accumulation of numerous data. Table 16 gives a few values, from the literature and from measurements by Thompson (*702*). Instruments which will give directly the integrated absorption during the spectroscopic investigation may be expected to become commercially available; this would give the study of intensity spectroscopy further impetus.

TABLE 16. *Absolute Absorption Intensities in* $cm^2\ mol^{-1}\ sec^{-1}$
for several Group Vibrations according to (702)

Group	Position of bands cm^{-1}	Integral absorption ($\times 10^7$)
C=O	*ca.* 1720	
Esters		13
Alkylketones		8
Ketosteroids		9–20
C—O	*ca.* 1200	
Esters		15
Alkylketones		2·5
Acetoxy-Steroids		21
N—H	*ca.* 3400	
Dialkylamines		0·05
Alkylanilines		1·7
Diarylamines		2·2
Pyrroles, Carbazoles		5·4
Indoles		6·6
C≡N	*ca.* 2250	
Alkylcyanides		0·25
CH$_3$	*ca.* 2900	
Hydrocarbons		4·4
—COCH$_3$		0·74
—COOCH$_3$		2·54
CH$_3$	*ca.* 1460	
Hydrocarbons		0·54
—COCH$_3$		1·3
—COOCH$_3$		2·0
CH$_2$	*ca.* 2900	
Hydrocarbons		3·8
—COCH$_2$—		0·5
CH$_2$	*ca.* 1460	
Hydrocarbons		0·23
—COCH$_2$		1·17
—COOCH$_2$—		0·65

A diagram of the type proposed by Mecke is shown in Fig. 115. For external reasons, the primary investigations were carried out on the third overtone of the CH-stretching vibration. The integrated absorption (for the scale, see the discussion in Chapter 3 of this part) is plotted against a linear wavelength scale and in cm^{-1}. Individual substances are shown by means of points and groups of related substances by means of a rectangle with an indication of the class in question. It can be seen that bands, the frequencies of which lie close together, spread out into groups which can be differentiated by the intensity.

If the true integrated absorption of a band for the liquid state is determined by one of the methods given below, the question arises of how far this value also applies to the gas phase. In fact, the energy absorbed depends on

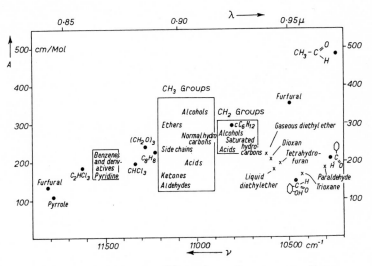

FIG. 115. *Table of infrared intensity assignments according to Mecke using as an example the third overtone of the CH stretching vibration [Lippert and Mecke (318)].*

the strength of the electric fields acting on the molecule; these alter on going from the gaseous state to the liquid state, including solutions. According to the Debye theory of dielectrics, the relationship between the integrated absorption in the dissolved and gaseous states should be given by the equation:

$$\frac{A_{\text{sol.}}}{A_{\text{gas}}} = \frac{1}{n}\left(\frac{n^2+2}{3}\right)^2 \qquad [\text{III}, 1.1]$$

where n is the refractive index of the solvent. On the other hand Mecke (*461*) has given the equation

$$\frac{A_{\text{sol.}}}{A_{\text{gas}}} = \frac{1}{n}\left(\frac{3n^2}{2n^2+1}\right)^2 \qquad [\text{III}, 1.2]$$

263

based on Onsager's theory.† It still appears to be undecided which expression reproduces more exactly the experimental findings. For a transfer from the gas to a liquid (of $n \approx 1 \cdot 5$), the Debye formula [III, 1.1] predicts an increase of the intensity of about 30 per cent, while according to [III, 1.2] the ratio $A_{sol.}/A_{gas}$ is but little different from 1. Measurements by Russell and Thompson (*619*) on the 1200 cm^{-1} band of alkyl esters point to the Mecke expression, while Polo and Wilson (*556*) considered that they had found indications for the correctness of the Debye expression. The decision must be deferred until more experimental data are available [see also Jaffe and Kimel (*778*)].

Apart from the more practical considerations of integrated absorption measurements for quantitative and structural analysis, they are of especial interest because of their relationship to the polar properties of bonds. In simple cases a relationship holds between the integrated absorption A and the rate of change of the electric dipole moment μ with the normal co-ordinates Q_i. Under certain conditions (a quadratic potential function, small amplitude of vibration, normal vibrations independent of each other, no interaction between vibrational and rotational states, and no electric anharmonicity), the relationship is

$$A = \frac{\pi N}{3c} g_i \left(\frac{\delta \mu}{\delta Q_i} \right)^2 \qquad \text{[III, 1.3]}$$

where c is the velocity of light, N is the number of molecules in 1 c.c., and g_i is the degree of degeneracy of the vibration (see page 63). Accordingly, $\delta \mu / \delta Q_i$, the effective vibrating charge, can be deduced from measurements of the integrated absorption. A summary of the most important results to date of such determinations can be found in the work of Thompson (*702*), Hornig and McKean (*333*) and Spedding and Whiffen (*787*).

† However, Polo and Wilson (*556*) have shown that the Onsager theory with reasonable assumptions also leads to the Debye expression [III, 1.1].

2. Qualitative Analysis

1. Qualitative Analysis

Many practical applications of infrared spectroscopy do not concern the problem of the determination or confirmation of the chemical structure of a substance. Instead, a qualitative test is required for the presence of one substance in another, or the purity of a given substance of known constitution has to be investigated. The basis for these uses is the fact that the spectrum of a substance is just as unambiguous as any other characteristic property, e.g. the density. Compared with the specific constants that have usually been used up to now, the spectrum has the advantage of greater specificity because of its two-dimensional character. This applies to any spectrum but especially to the infrared spectrum. Because the vibrations of characteristic structural groups are shown as bands of definite position in the infrared spectral region, the infrared spectrum takes on much the same function as that of a fingerprint in criminal records. A given substance is unambiguously determined by its infrared spectrum. Compounds of similar structure and composition do possess similar spectra in general; however, correspondence down to the last detail is improbable unless the substances investigated are identical. For qualitative analysis by means of infrared spectroscopy, it should be noted that the spectrum of a mixture of substances is generally composed of a simple additive overlapping of the individual spectra. There are some definite exceptions to this. They concern cases of intermolecular interaction between the components of a mixture.

A qualitative analysis consists of confirming the nature of a substance by comparing its spectrum with that of the *pure* substance in question. Here also, the correspondence of the strongest bands should first be tested, then the moderately strong ones, and finally the weakest bands. It is advantageous if the different spectra are obtained with the same apparatus, or at least with instruments of a similar degree of accuracy and resolution; if this is not so, the instrumental influences which always modify the spectrum must be taken into account. This condition often cannot be fulfilled, especially for rare substances or those that are difficult to prepare in a pure state. Experience and an exact knowledge of the subject are then necessary to avoid incorrect conclusions. The appearance of one *single* extra band in the spectrum of one substance compared with that of another proves their

265

qualitative difference or the presence of an impurity. The substance corresponding to the spectrum with fewer bands must obviously be considered to be the more pure.

The qualitative analysis of mixtures is also based on a knowledge of the spectra of the components in the pure state. Provided it is concerned with substances of completely different chemical character, the definite bands characteristic of the classes of substances suffice for their identification and a knowledge of the other, and especially the weaker, bands is hardly necessary. However, if chemically similar substances are present in the mixture, so that the spectral characteristics are almost the same for all, the subsidiary bands must also be used in the analysis. Because of the manifold possibilities, it is recommended that modern methods of documentation such as punched cards should be used for the investigation of systems about which little or nothing is known (see discussion later). Baker *et al.* (*32*) have given directions and examples for this.

2. Determination of Purity

The determination of purity is a special case of qualitative analysis. Whether and to what extent infrared spectra should be used depends on the degree of accuracy required. The determination of the purity of a substance necessitates, as does any qualitative analysis, a knowledge of the spectrum of the substance to be tested in the pure state, but it does not often require the spectra of the impurities. However, if the investigation is extended to the determination of the presence or absence of definite compounds, a knowledge of the spectra, or at the least the main bands, of these is naturally indispensable. Very frequently it is necessary to test whether or not a definite impurity exceeds a certain defined limit, because the further use of the material depends on this decision. This problem may be solved by applying one of the quantitative processes (to be discussed shortly) to one or several of the bands of the impurity. A condition is that the requirements lie in the concentration range applicable to infrared spectroscopy, i.e. the limit of detection is not too low; naturally this limit varies from substance to substance and must be determined empirically in each case.

Such a determination of purity within definite limits may be carried out very simply and elegantly by the use of a double-beam spectrometer and a cell of variable thickness. The process will be described taking the xylenes as an example. It will be assumed that it is required to investigate a sample of o-xylene with respect to its content of the two other isomers, and where certain small concentrations of *p*- and *m*-xylene would be permissible, approximately 2·5 per cent each. The spectra of the three xylene isomers are given in Fig. 116. The problem may be solved by determining whether

certain preselected bands of the minor constituents, e.g. that at $12 \cdot 6 \, \mu$ for
p-xylene and that at $13 \, \mu$ for m-xylene, exceed a certain intensity in the
spectrum of the substance under investigation. If an equivalent mass of the
pure substance to be tested is placed in a cell of variable thickness in the
comparison optical path of a double-beam instrument, the spectrum is automatically compensated by the apparatus because it appears in both the

optical paths. Only the spectrum of the
impurities now appears and this is usually
much easier and simpler to interpret than
the system as a whole. The two lowest
spectra in Fig. 116 show very clearly the
superiority of the compensation method.
Naturally, a mixture of the substance in
question with just those concentrations of
impurities permissible could be placed in
the comparison beam. Then bands observed in the spectrum would at once
indicate that the permitted limit had been
exceeded. If the analytical sample were to
contain less impurity than that permissible, so-called over-compensation would
occur, i.e. the substance analysed would
show a transmittance of more than 100 per
cent at the positions of the bands of the impurities because the comparison substance absorbs more strongly here.

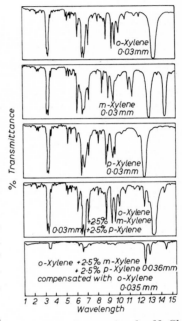

FIG. 116. *Spectra in the NaCl
region of the three xylene isomers,
a mixture, and a compensated
mixture for purity tests (after the
original of the Perkin-Elmer
Corp.).*

The following procedure according to
Washburn (*828*) is recommended for the
compensation: (1) A pair of matched cells
is filled with a solution containing a few
per cent of the *pure* principal components
in a suitable solvent. A spectrum is then
recorded under the usual conditions with one of the cells in the measuring
beam, and all the bands with absorption of more than 93 per cent are noted
as being unsuitable for compensation purposes. (2) Now the second cell is
placed in the comparison beam, without removing the first cell. After
suitable setting of the radiation source, slit programme and amplification,
another spectrum is *slowly* recorded above that already obtained; it is then
obvious how good the compensation is. (3) The measuring cell is now filled
with a solution, of equal strength and in the same solvent, of the sample to
be investigated, and a third spectrum is slowly recorded above the two

already obtained. The interpretation of this final spectrum can usually be done without difficulty.

Frequently the substance needed (principal component of the mixture) for the compensation of the background absorption is not available at the necessary level of purity; because of this all the analyses are fundamentally in error. Here, the use of a substitute substance which is obtainable pure is sometimes helpful [Washburn and Mahoney (*829*)]. Naturally, this must possess an absorption spectrum very close to that of the real substance in the spectral region used for the analysis.

In certain cases a very sensitive differential analysis can be carried out by use of the phenomenon of over-compensation [Robinson (*601*)]. This procedure is specially valuable if it is desired to determine a minute concentration of one substance in another and when unfortunately the single band useful for this purpose lies on the steep slope of a principal band of the main component of the mixture. The quantitative determination of cyclohexane (band at $3 \cdot 5 \, \mu$) in amounts under $0 \cdot 1$ per cent in benzene (band at $3 \cdot 3 \, \mu$) can be carried out by this technique. Usually the sample to be analysed is placed in the measuring beam and the pure substance in the comparison beam, but the essential point of differential analysis lies in the use of two such measurements on one and the same sheet of paper according to the scheme:

	Measurement beam	*Comparison beam*
First measurement:	cell A	cell B
	analytical sample	pure substance
Second measurement:	cell A	cell B
	pure substance	analytical sample

Obviously the first measurement gives only the bands of the substance investigated in the usual manner. The second measurement gives the same bands but in the opposite sense as a result of over-compensation. Therefore the distance between the two curves obtained is double that in the usual process between the band and the 100 per cent transmittance line. If, as is frequently the case, the two cells A and B used are of somewhat different thickness, this difference is eliminated by using initially the analytical sample in cell A against the pure substance in cell B and then the analytical sample in Cell B against the pure sample in cell A. Hammer and Roe (*294*) have given further examples of differential analysis. Reference is again made to the precautions which are necessary with differential measurements and which have been discussed on page 224 ff.

A special type of differential analysis has been used by Elder and Benesch (*190*) in the instrument built by them and called the 'absorption spectrum sorter'. The principle can be seen from Fig. 117. A full cell 1 in the measuring beam against an empty cell 2 in the comparison beam of an ordinary double-beam spectrometer (*a*) gives the usual absorption bands (*b*). The introduction of a constant absorption, e.g. an optical-wedge attenuator, in the measuring beam (*c*) displaces the absorption null point and simultaneously lowers the absorption bands previously measured (*d*). A cell 3 is now introduced into both the optical paths containing a substance *S* absorbing at another wavelength (*e*). This would cause no reading in a normal double-beam spectrometer, but, because of the inequality of the beams induced by the optical wedge, an absorption band of this new substance appears, with the bands pointing in the opposite direction to the original spectrum (*f*). According to the spectral positions of the two bands and the relationship of their intensities, the result can be very different. Bands lying close together and of the same intensity give a recording of type (*g*). Superimposed bands give (*h*) if the substance *S* absorbs more intensely and (*i*) if *S* absorbs more weakly. Obviously this process can be directly applied to qualitative analysis, the determination of purity, etc., pro-vided it is possible to investigate mixtures. For this the spectrum of

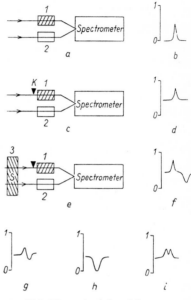

FIG. 117. *The principle of the spectrum sorter of Elder and Benesch (190).*

the mixture is initially recorded in the usual way. Then this is repeated with, for example, the component 1 in the cell 1 and the component 2 in the cell 3. The assignment of the bands of the spectrum initially obtained to the individual components is then simplified because each component shows only bands on one side of the null line determined by the fixed absorption in one of the optical paths. As the authors have shown (*190*), an increase in resolution, which is often valuable, is also obtained with this process. Fundamentally, it is similar to the negative filtering of non-dispersive spectroscopy.

3. Methods of Cataloguing and Documentation

The successful completion of a qualitative infrared spectroscopic analysis, whatever the type, depends very much on the availability of *comparison*

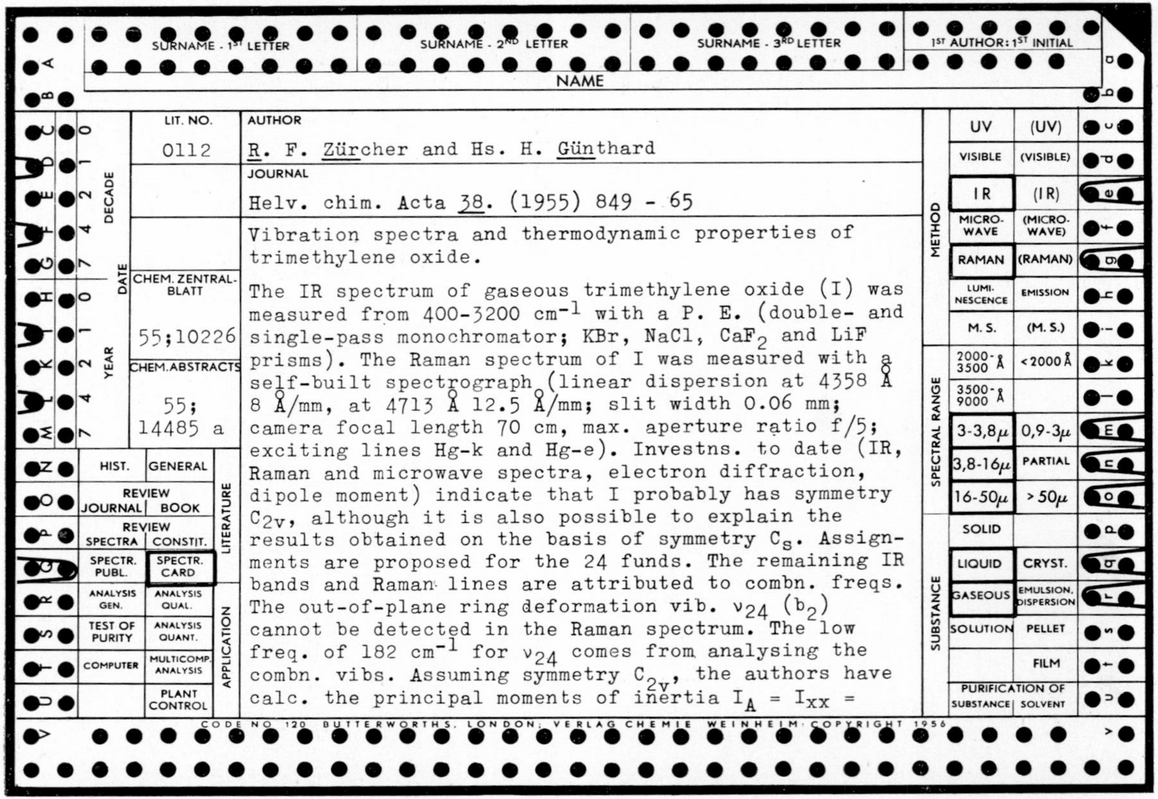

NAME

SURNAME · 1ST LETTER | SURNAME · 2ND LETTER | SURNAME · 3RD LETTER | 1ST AUTHOR: 1ST INITIAL

LIT. NO.
0112

CHEM. ZENTRAL-BLATT
55;10226

CHEM. ABSTRACTS
55;
14485 a

AUTHOR
R. F. Zürcher and Hs. H. Günthard

JOURNAL
Helv. chim. Acta 38. (1955) 849 – 65

Vibration spectra and thermodynamic properties of trimethylene oxide.

The IR spectrum of gaseous trimethylene oxide (I) was measured from 400-3200 cm-1 with a P. E. (double- and single-pass monochromator; KBr, NaCl, CaF_2 and LiF prisms). The Raman spectrum of I was measured with a self-built spectrograph (linear dispersion at 4358 Å 8 Å/mm, at 4713 Å 12.5 Å/mm; slit width 0.06 mm; camera focal length 70 cm, max. aperture ratio f/5; exciting lines Hg-k and Hg-e). Investns. to date (IR, Raman and microwave spectra, electron diffraction, dipole moment) indicate that I probably has symmetry C2v, although it is also possible to explain the results obtained on the basis of symmetry Cs. Assignments are proposed for the 24 funds. The remaining IR bands and Raman lines are attributed to combn. freqs. The out-of-plane ring deformation vib. ν_{24} (b_2) cannot be detected in the Raman spectrum. The low freq. of 182 cm-1 for ν_{24} comes from analysing the combn. vibs. Assuming symmetry C_{2v}, the authors have calc. the principal moments of inertia $I_A = I_{xx} =$

METHOD
UV	(UV) VISIBLE
VISIBLE	
IR	(IR)
MICRO-WAVE	RAMAN (RAMAN)
LUMI-NESCENCE EMISSION	
M.S.	(M.S.)

SPECTRAL RANGE
2000·3500 Å	<2000 Å
3500· 9000 Å	
3-3,8μ	0,9-3μ
3,8-16μ	PARTIAL
16-50μ	>50μ

SUBSTANCE
SOLID	
LIQUID	CRYST.
GASEOUS	EMULSION, DISPERSION
SOLUTION	PELLET
	FILM
PURIFICATION OF SUBSTANCE	SOLVENT

LITERATURE / GENERAL / APPLICATION
HIST.	
REVIEW JOURNAL BOOK	
REVIEW SPECTRA CONSTIT.	
SPECTR. PUBL.	SPECTR. CARD
ANALYSIS GEN.	ANALYSIS QUAL.
TEST OF PURITY	ANALYSIS QUANT.
COMPUTER	MULTICOMP ANALYSIS
PLANT CONTROL	

DECADE
YEAR
DATE

FIG. 118a.

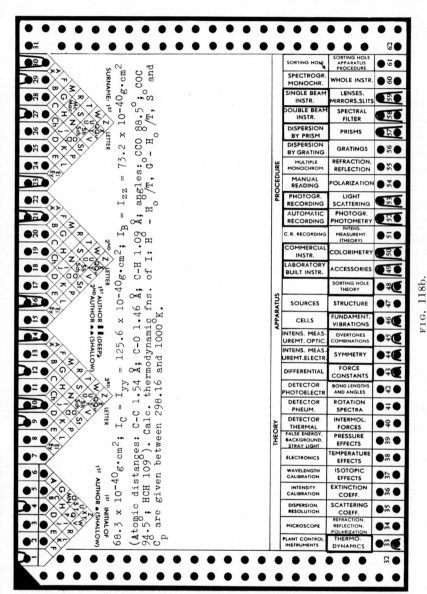

FIG. 118. *Literature card of the DMS.* (a) *front;* (b) *back.*

FIG. 118b.

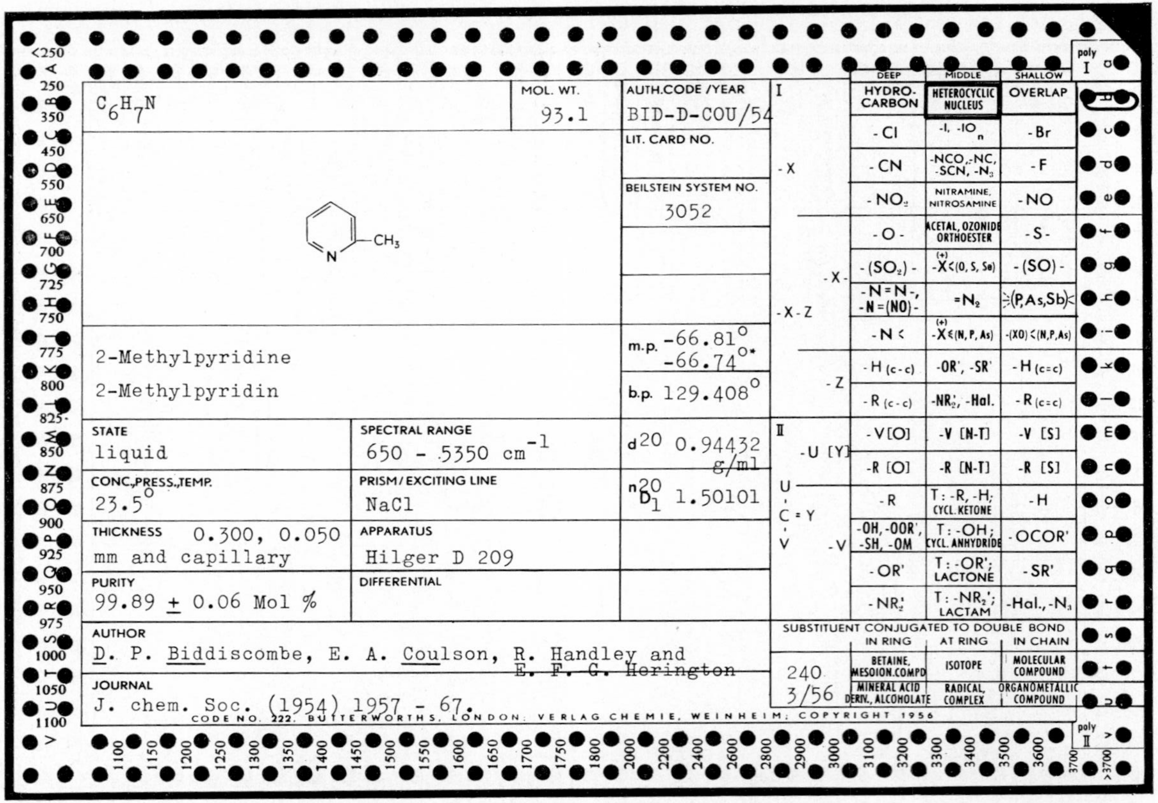

FIG. 119a.

The card contains the following data fields:

C_6H_7N

MOL. WT. 93.1

AUTH.CODE /YEAR BID-D-COU/54

LIT. CARD NO.

BEILSTEIN SYSTEM NO. 3052

2-Methylpyridine
2-Methylpyridin

m.p. $-66.81°$
$-66.74°$

b.p. $129.408°$

d_4^{20} 0.94432 g/ml
n_D^{20} 1.50101

SPECTRAL RANGE
650 – 5350 cm^{-1}

STATE
liquid

PRISM/EXCITING LINE
NaCl

APPARATUS
Hilger D 209

CONC.,PRESS.,TEMP.
23.5°

THICKNESS 0.300, 0.050
mm and capillary

DIFFERENTIAL

PURITY
99.89 ± 0.06 Mol %

AUTHOR
D. P. Biddiscombe, E. A. Coulson, R. Handley and
E. F. G. Herington

JOURNAL
J. chem. Soc. (1954) 1957—67.
CODE NO. 272 BUTTERWORTHS, LONDON. VERLAG CHEMIE, WEINHEIM. COPYRIGHT 1956

240
3/56

Column headers:
HYDRO-CARBON (DEEP) | HETEROCYCLIC NUCLEUS (MIDDLE) | OVERLAP (SHALLOW)

-Cl | -I, -IO$_n$ | -Br
-CN | -NCO, -NC, -SCN, -N$_3$ | -F
-NO$_2$ | NITRAMINE, NITROSAMINE | -NO
-O- | KETAL, OZONIDE, ORTHOESTER | -S-
-(SO$_2$)- | -X<(0, S, Se) | -(SO)-
-N=N-, -N=(NO)- | =N$_2$ | ≥(P,As,Sb)<
-N< | -X<(N,P,As) | -(XO)<(P,As)
-H(c-c) | -OR', -SR' | -H(c-c)
-R(c-c) | -NR'$_2$, Hal. | -R(c-c)
-V(O) | -V (N-T) | -V [S]
-R(O) | -R (N-T) | -R [S]
-R | T: -R, -H; CYCL.KETONE | -H
-OH, -OOR, -SH, -OM | T: -OR'; CYCL. ANHYDRIDE | -OCOR'
-OR' | T: -OR'; LACTONE | -SR'
-NR'$_2$ | T: -NR'$_2$; LACTAM | -Hal., -N$_3$

SUBSTITUENT CONJUGATED TO DOUBLE BOND
IN RING | AT RING | IN CHAIN
BETAINE, MESDION,COMPD | ISOTOPE | MOLECULAR COMPOUND
MINERAL ACID DERIV., ALCOHOLATE | RADICAL COMPLEX | ORGANOMETALLIC COMPOUND

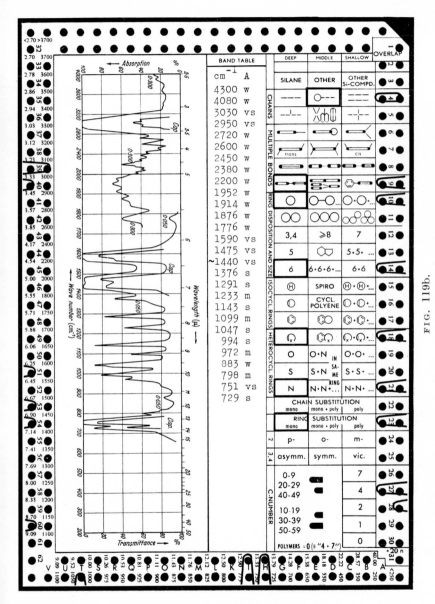

FIG. 119b.

FIG. 119. *Substance card of the DMS:* (a) *front;* (b) *back.*

spectra of the substance in question in the pure state, as may be seen from the preceding discussion. In the course of any problem it cannot be assumed that the necessary pure substances can be obtained at once. Therefore it is important that the relevant literature should be followed and recorded to a considerable extent. At the same time, a catalogue of spectra should be collected. This is no easy task considering the amount of the material already available and increasing every year, especially if the critical evaluation necessary is taken into account. For laboratories commencing infrared spectroscopy, there are several ways in which this task can be facilitated.

With regards to the current literature appearing, the well known international abstracting journals give information relevant to their scope. For many years, several hundred papers have been published annually in the field of infrared spectroscopy. Summaries with detailed literature references can be found in the American journal, *Analytical Chemistry* [Barnes and Gore *(39)*, Gore *(267* to *270)*, Gore *(804)*]. These extraordinarily valuable summaries, written by an outstanding infrared spectroscopist, and widely recognized, appeared annually up to 1952 but since then only every two years. However, the summary for 1955 [Gore *(271)*] made it clear that, because of the quantity of material appearing, the summaries would have to be ended, as no one individual could collect the material, let alone summarize it. For the years 1945–57, 16,000 literature references from the most important journals and arranged according to substances have been summarized in the book, *Infrared Absorption Spectra, Index for 1945–57*, by Hershenson *(807)*. Since the beginning of 1956 the *Documentation of Molecular Spectroscopy* (DMS)† guarantees a specialist recording and interpretation of the whole current literature in the field of molecular spectroscopy. This commercial scheme grew from the intensive collaboration of German and English spectroscopists. Initially it will deal only with infrared and Raman spectra, but the inclusion of UV and other types of spectra is only a question of time and finance [Bergmann and Kresze *(58)*; Thompson *(703)*].

The publication committee works in close collaboration with the editors of *Chemisches Zentralblatt*. Every paper concerned with molecular spectroscopy in every available journal is communicated to all subscribers in the form of a punched literature-card (Fig. 118) according to a predetermined key and with a short summary. Further, all the spectra recorded in a paper are cited in detail in the form of compound cards (Fig. 119), provided they appear for the first time in the DMS or they give more information than one

† Obtainable from Verlag Chemie, Weinheim (Bergstrasse), Germany, or Butterworths Ltd., London.

that has already appeared. The literature card contains details of the technique used and the purpose of the investigation abstracted and its connection with larger related issues. The substance card contains a key to the spectrum in the punched holes (the spectrum is also drawn or given as a table of bands) and characterizes the substance in question according to its chemical structure. For this purpose, definite holes are assigned to certain structural features according to a prearranged key. By means of these cards, a desired compound can be quickly sorted out together with its spectrum. In addition definite spectral features can be correlated with structural features or vice versa. The use of cards punched at the edge instead of the Hollerith cards of an American system of documentation, to be discussed presently, allows the system to be used without complicated and expensive sorting machines.† Probably, about 2000 cards will be issued annually at a subscription price of $750 \cdot 0$ DM or £63. This system includes all the new literature and it is to be hoped that earlier years will be made up. In addition, it provides a collection of spectra which is continually growing, especially as a group of workers will be concerned with the issue of the spectra of pure substances.

In the U.S.A. this problem, which is just as acute there, has been attacked in another way. Instead of the systematic coverage of the current literature, the effort is made to take available spectral material and selected spectra from the literature and make them available in a handy and easily interpretable form. This work is done by the 'American Society for Testing Materials'. The basis of the system is one that was used by the Wyandotte Chemical Corp. [Kuentzel (*396*)]. Using the appliances of the highly developed modern office and organizational methods, the essential content of the spectra and other relevant information, together with the chemical structure of a compound, are punched on to cards of the Hollerith type which can be interpreted and sorted by machines (Fig. 120). At the present, about 15,000 cards exist for infrared spectroscopy and thousands more for Raman, UV, X-ray, and mass spectroscopy. In contrast to the cards of the European DMS system, the Hollerith cards contain almost no text and also no diagram of the spectrum; they are thus almost valueless without a sorting machine. Their capacity for storing information is larger, as a result of the larger number of holes (12×80 against 2×104 for the DMS card). The ease of manipulation does not set a limit to the number of the cards.‡ With respect

---

† However, it is probable that in the future cards punched over their whole surface will have to be used because of their superior capacity for storing information.

‡ A normal 'slow' Relais sorting machine can sort 24,000 cards in an hour; however, only those questions can be dealt with simultaneously that concern information in one and the same column.

to the coverage, the essential difference is that the ASTM system is principally based on spectral material already available (at least at present, although this may be altered some time in the future), essentially on the API collection and the Sadtler atlas. The ASTM system is much less expensive,† provided the cost of the sorting machine is not taken into account.

A further American documentation project is that known in specialist circles as the Creitz system. This consists of marginal punched cards‡ issued by the Committee on Spectral Absorption Data of the National Research Council in collaboration with the National Bureau of Standards

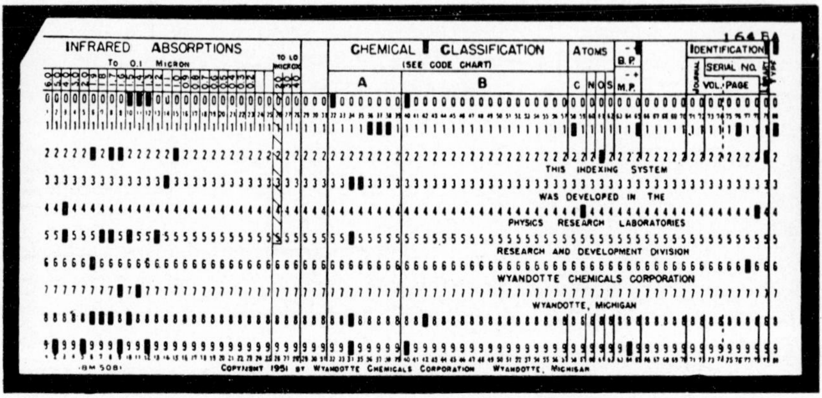

FIG. 120. *Hollerith card of the ASTM for infrared spectroscopic and chemical purposes punched according to the code of the Wyandotte Chemical Corp.*

in Washington. This system is similar to the European DMS system; however, the project was apparently not founded on sufficient basis. Accordingly, it no longer has any importance compared with the ASTM system, and is to be incorporated in this. Other systems of documentation, which have been suggested or described in the literature [Clark (*118*), Shreve and Heether (*652*), Baker *et al.* (*32*)], are only of historical or local importance.

All the systems of documentation (unless they include a sufficiently satisfactory diagram of the spectrum treated) do not satisfy the condition for a comprehensive collection of spectra available to everyone. Every practical spectroscopist needs such a collection as the basis of his daily work. If possible the type of representation of the spectra should not differ too much from that to which he is accustomed, i.e. the spectra should have been

† The ASTM set of Hollerith infrared cards consists at present of almost 13,000 cards at a price of $160·0; for UV spectra, there are about 7000 cards for $90·0.

‡ More than 700 cards at a price of $0·10 each.

attained with one of the usual modern double-beam instruments. In the course of time, a valuable stock of such spectra will be collected in every spectroscopic laboratory, but the possibility of solving a problem is frequently dependent on the availability of definite comparison spectra in not too small a number; usually it is not possible to obtain the necessary spectra quickly oneself.

Before dealing with the large commercial collections of spectra which are available, some smaller collections will be mentioned which have been published in books and journals:

Landolt-Börnstein	Band 1, Teil 2	458 spectra of all sorts
Barnes–Gore–Liddel– Williams	*Infrared Spectroscopy*	363 spectra of all sorts
Randall–Fuson–Fowler– Dangl	*Infrared Determination of Organic Structures*	354 spectra connected with the penicillin problem
Dobriner–Katzen– ellenbogen–Jones	*Infrared Absorption Spectra of Steroids*	306 spectra of steroids
Cannon–Sutherland	*Spectrochim. Acta* **4**, 373 (1951)	104 spectra of aromatic hydrocarbons
Kendall–Hampton– Hausdorff–Pristera	*Appl. Spectr.* **7**, 179 (1953)	79 spectra of plasticizers
Hausdorff	Perkin–Elmer Corp.	32 spectra of plastics
Whistler–House	*Anal. Chem.* **25**, 1463 (1953)	35 spectra of mono-saccharides
Hunt–Wisherd–Bonham	*Anal. Chem.* **22**, 1478 (1950)	68 spectra of inorganic compounds
Miller–Wilkins	*Anal. Chem.* **24**, 1253 (1952)	160 spectra of inorganic compounds
Hummel	*Identifizierung von Kunststoff und Lackrohstoffen*	ca. 500 spectra of plastics, plasticizers, etc.

These collections contain spectra which are in part superseded and are often of variable value. In addition, there are two large commercial collections of infrared spectra. The first is the collection of the American Petroleum Institute Research Project 44 (API 44) containing at present about 1800 spectra mostly of hydrocarbons.† The spectra are issued for the spectral region 2 to 15, and frequently to 25 μ in a largely standardized representation and in part with additional table of bands. The second is the very comprehensive collection, 'Catalogue of Infrared Standard Spectra',‡ which consists

† Issued by the Carnegie Institute of Technology, Pittsburg (Dr Rossini) in collaboration with the National Bureau of Standards. Price per sheet, $0·10.

‡ Issued by the firm J. P. Sadtler and Sons, Philadelphia, Pa. Price for bulk purchase and on subscription $0·40 per spectrum up to 1956, and then after $0·25; for 3 spectra on a sheet, the sheet costs $0·10.

at present of about 11,000 spectra and is increased by 150 every month. Spectra recognized as unsatisfactory are exchanged for better ones without charge. These spectra comprise the spectral region of 2 to $15 \cdot 5\,\mu$. They are issued either in loose-leaf form of approximate size 45×17 cm (recently about 32×13 cm), or reduced in size so that three spectra appear on a normal sheet, or in the form of micro films. In addition, there is a further collection, 'Sadtler Commercial Spectra', consisting of approximately 1000 spectra at the moment, which deals with the infrared spectra of industrial products the composition of which has not been exactly given.

Very recently collections of spectra for the near infrared region have become available. One such collection is being produced by the firm Sadtler and Sons, Philadelphia, already mentioned, and the other by Anderson Physical Laboratory, Champaign, Ill., U.S.A.

Collections of spectra of such size are understandably not easy to handle. The danger exists that they lose their value because it is not possible to select at required moments the particular spectrum needed or desired. In this respect it is noteworthy that both the API and the Sadtler collection are integrated components of the ASTM documentation system mentioned above and their spectra can be incorporated into the ASTM system retaining the same number. Therefore, the Hollerith cards of the ASTM system are also the quickest and simplest means of selecting any desired spectrum from these comprehensive collections.

All the present documentation systems are based on spectra which were recorded and published for other purposes; the spectral characteristics are later converted into documental records. It is perhaps a fruitful thought for the future that the spectra should be recorded in a manner suitable for documentation simultaneously with their measurement, as has been proposed by King *et al.* (*385*). They recommend recording the spectra in numerical form on Hollerith cards and have indicated suitable accessories for this to the usual spectrometers. They suggest that a spectrum of an average number of bands should be divided into about 8000 wavelength or wave number divisions and that the transmittance or a corresponding quantity for each should be punched on Hollerith cards according to a binary scale. For a single spectrum, 20 punched cards would be needed in this way; they would be punched simultaneously as the curve were recorded. Obviously such a process would save the large amount of time now needed for a subsequent documentation treatment. It would also have several other advantages, spectra could be directly compared with each other by using Hollerith machines, fully automatic analysis could be carried out, and similar procedures could be mechanized. However, the realization of this scheme would need to overcome considerable difficulties including that of expense.

278

3. Quantitative Analysis

1. The Law of Absorption

To extend the qualitative to a quantitative analysis, a numerical relationship is needed between the spectral characteristics observed and the number of molecules in which they originate. Therefore, the alterations in the radiation used to measure the substance investigated must be recorded and described with respect to their quantity as well as their quality. The interaction of radiation with matter can lead to three types of changes in the energy. (1) *Reflection*, in which a part of the incident radiation energy is deflected back. The reflected radiation as a fraction of the incident is the power of reflection of the substance. (2) *Scattering*, a part of the radiation energy incident at a definite direction is scattered in all directions. This is caused by interaction depending on the relationship of the radiation wavelength to the particle size of the body investigated and certain optical flaws such as inhomogeneities, i.e. differences in refractive index, etc. (3) *Absorption*, by which is designated the conversion of radiation energy into heat, i.e. energy of non-ordered molecular motion. The portion of the energy transformed, expressed as a fraction of the incident radiation energy, is known as the power of absorption.

These three effects depend in a characteristic manner on the wavelength of the radiation and also on the specific properties of the substance investigated. For quantitative investigations in the infrared regions, with few exceptions, only the last effect of absorption is important. Therefore this will be discussed exclusively. Reflection can be eliminated in the theoretical discussion by considering the radiation which passes into the body instead of that which is incident to the body; the difference between these quantities is obviously the reflected part. Scattering can be avoided by a suitable form and preparation of the irradiated body and by its complete optical homogeneity. Further possible alterations of the radiation in a geometrical respect, i.e. alterations in direction by the refraction on entry into the body, can be eliminated by assuming *perpendicular* incidence on a *planar* surface of the body investigated. In addition, the incident radiation will initially be assumed to be *parallel*. In practice, the two last conditions are only of importance for the investigation of solid and liquid substances at large layer thicknesses.

The Methods of Practical Infrared Spectroscopy

The radiant energy of a definite monochromatic wavelength λ or wave number ν passing into the body is denoted by J_0 and the same quantity after passing through a layer of the substance limited by two parallel planes and of thickness d is denoted as J. The relationship between these quantities then follows from theoretical considerations and from experiment [Sippel (*668*), Strong (*689*)]:

$$J = J_0 \cdot e^{-\alpha \cdot d} \qquad \text{[III, 3.1]}$$

or if the e function, which is somewhat disadvantageous for practical calculations, is avoided:

$$J = J_0 \cdot 10^{-\alpha' \cdot d} \qquad \text{[III, 3.1a]}$$

This relationship is known as the absorption law and is usually named after Lambert-Bouguer. The index found on the right-hand side αd or $\alpha' d$ is called the *optical density* or the absorbance and the quantity α or α' is called the (Bunsen's) *extinction coefficient*. The layer thickness d is generally measured in centimetres. The optical density is an index of a power and must be dimensionless, thus the extinction coefficient is expressed in cm^{-1}. However, when comparing data of different origin it is advisable to note the units used, because in practical spectroscopy layer thicknesses are often quoted in millimetres. The extinction coefficient can be visualized as the reciprocal value of that layer thickness which the radiation energy must pass through to be reduced to the e-th part, i.e. *ca.* 37 per cent, or to one-tenth if the equation [III, 3.1a] is used.

The extinction coefficient is a characteristic quantity of the substance investigated and is independent of the dimensions of the sample. On the other hand, the optical density and therefore the radiant energy J transmitted are dependent on the thickness used. However, all three are functions of the wavelength of the radiation. A graphical representation of the optical density or the radiation energy transmitted J, both for a definite layer thickness d, or the extinction coefficient α as a function of the wavelength give the spectrum of the substance investigated. However, other quantities are usually used in the representation of spectra, namely the *transmittance D* or the *absorbance A*, both given as a percentage of the incident radiation. The first is defined as the ratio of the radiation energy transmitted to that passing in:

$$D = \frac{J}{J_0} = e^{-\alpha \cdot d} \qquad \text{[III, 3.2]}$$

The absorbance A is defined as

$$A = 1 - D = 1 - e^{-\alpha \cdot d} \qquad \text{[III, 3.3]}$$

280

Accordingly the extinction† is

$$E = -\ln D = \ln \frac{1}{D} = \ln \frac{J_0}{J} \qquad \text{[III, 3.4]}$$

The extinction coefficient is given by

$$\alpha = \frac{E}{d} = \frac{1}{d}\ln\frac{1}{D} = \frac{1}{d}\ln\frac{J_0}{J} \qquad \text{[III, 3.5]}$$

For special purposes, but seldom in the field of experimental spectroscopy, the absorption coefficient k or the absorption index $\varkappa$ are used instead of the extinction coefficient. These quantities are derived from the extinction coefficient by the relationships

$$\alpha = \frac{4\pi}{\lambda}.k = \frac{4\pi}{\lambda}.n\varkappa \qquad \text{[III, 3.6]}$$

where n is the refractive index of the substance at wavelength λ.‡§

The quantities which have appeared in the preceding discussion are true parameters of the substance investigated alone and are independent of the influences of apparatus and the technique of measurement. The experimentally determined quantities do not coincide with them because the necessary conditions are not fulfilled. For example, in almost all instruments the incident radiation energy is measured, instead of that which actually passes into the substance, and is used as a basis for calculating the transmittance, etc. The mathematical relationships naturally remain the same, and observed quantities are obtained instead of the true ones. Instead of J_0 and J, the radiation energies experimentally measured are denoted by T_0 and T, which are dependent on the instrument and the conditions of measurement. The most important of these influences originate from the slit width used, from reflection and scattering losses, from stray light; there are a number of other less important effects.

---

† From now on only natural logarithms will be used, i.e. the discussion is based on the form [III, 3.1] of the absorption law. Transforming to logarithms to base 10, i.e. based on the form [III, 3.1a], which are often advantageous in practice for calculations, can be carried out using the well known relationship $\ln x = 2\cdot303 \log x$. The transmittance is not altered thereby, but the optical density and extinction coefficient are.

‡ The assignment of these names to the different units often varies from author to author. When reading and comparing data in the literature the convention being used should always be noted. See also Hughes (*337*).

§ Stevens (*684*) has proposed that the power of absorption of a substance should be given in the decibel units used in the fields of high frequencies and acoustics, instead of as percentage absorption or optical density. This is only an alteration of the scale for the optical density, for by definition $1\,db = 20.\log(1/D) = 8\cdot69.E$.

The Methods of Practical Infrared Spectroscopy

The influence of the slit width has already been discussed (see page 123) and will be further discussed below. The reflection losses J_r/J_e (where J_e denotes the incident radiation) can be calculated from the refractive indices n_1 of the surroundings (cell window) and n of the substance as

$$\frac{J_r}{J_e} = 2\left(\frac{n-n_1}{n+n_1}\right)^2 \qquad [\text{III, 3.7}]$$

provided only one reflection at each surface of the layer investigated (which usually suffices) is taken into account. The value J_r/J_e is usually less than 8 per cent. For the investigation of pure liquids, this is best compensated by means of a single plate of the material of the cell windows in the comparison beam. On the other hand, for solutions, compensation is best by means of a cell full of the solvent and for gases by an empty cell or one filled with N_2.

However, especially for very thin layers, multiple reflections have often to be taken into account. Yasumi (761) has recently drawn attention to this, and to the extent of the resulting errors, using as an example carbon tetrachloride in the region of its fundamental vibrations. Taking into consideration multiple reflections and provided the cell windows do not absorb, it can be shown that the transmittance observed is

$$T = D.\frac{1+R_{12}^4-2R_{12}^2\cos 2\gamma}{1+R_{12}^4.e^{-(8\pi kd/\lambda)}-2R_{12}^2.e^{-(4\pi kd/\lambda)}.\cos\left(\frac{4\pi kd}{\lambda}+2\gamma\right)} \qquad [\text{III, 3.8}]$$

where

$$R_{12}^2 = \frac{(n-n_1)^2+k^2}{(n+n_1)^2+k^2}, \qquad \tan\gamma = \frac{2n_1 k}{(n+n_1)^2+k^2} \qquad [\text{III, 3.9}]$$

The deviation between D and T is more than 30 per cent at the maximum for a layer thickness of 10^{-4} cm in the example given by Yasumi (761).

When highly concentrated solutions are investigated, the difference in absorption for the solvent caused by replacing the solvent by the solute in the solution must be considered. Neglecting reflection and diffraction losses, the relationship between the optical density of the solvent in the solution and in the pure solvent is

$$\frac{E_{\text{solution}}}{E_{\text{solvent}}} = \frac{c_2}{c_1+c_2} = \frac{\dfrac{1000\rho}{M}}{c_1+\dfrac{1000\rho}{M}} \qquad [\text{III, 3.10}]$$

where ρ is the density of the solvent (and for simplification also that of the solution), M is its molecular weight, c_1 is the concentration of the solution and c_2 that of the solvent. For the usual solvents, the ratio deviates by 1 per cent from the value of 1 at concentrations of $0 \cdot 1$ molar, and the deviation increases rapidly with increasing concentration. In spite of this, this effect can usually be neglected except when unusually high concentrations are used or in regions of strong solvent absorption.

The radiation losses by scattering, which are measured as a decreased transmittance and therefore simulate an absorption, are less easily calculated, but they are also less important. Their magnitude depends on the diameter of the scattering particles compared with the wavelength of the radiation. If it is possible to keep the diameter smaller than 1μ, the scattering is small and negligible for the wavelengths of interest here. In fact, a certain degree of scattering for short-wave radiation, e.g. as a result of

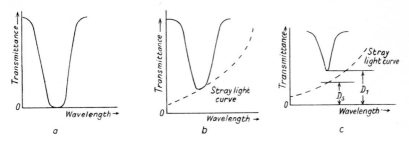

FIG. 121. *Stray light:* (a) *absorption without stray light, and* (b) *absorption with stray light;* (c) *elimination of the stray light for quantitative purposes.*

dulling of the cell windows from attack of moisture, is quite desirable for the infrared region above about 8μ. This is because the so-called stray light, i.e. radiation of another, in fact a shorter, wavelength than that selected at the moment, is thus decreased; some instruments deliberately make use of this phenomenon to eliminate short-wave stray light.

The stray light influences the measurements in two ways: firstly in a displacement of the band centre and secondly in an increase of the transmittance which simulates a decrease in the absorption. The qualitative test for stray light is simple. The transmittance is measured for a substance which is known to absorb completely for the wavelength in question (easily obtainable by increasing the layer thickness) but is not completely non-transparent to infrared radiation. This is done after the transmittance null line has been exactly determined by means of a metallic stop. Instead of the expected curve of Fig. 121a for no stray light, the curve of Fig. 121b is obtained, characterizing a stray light curve which generally rises with increasing wavelength. Obviously, the band centre is displaced to smaller wavelength, as a result, and the more so the more intense the band. If it is not possible to work with an optical density sufficiently small so that the stray light effect can be neglected, the exact position of the band can be determined by plotting the recorded wavelength as a function of the concentration (or thickness) and extrapolating to zero. To eliminate the stray light for quantitative investigations, the stray light curve is initially determined by means of measurements with completely absorbing samples (metals should not be used). If the transmittance for the band in question is D_1 without taking into account the stray light, i.e. with respect to the null point of the instrument (Fig. 121c), and the amount of stray light at this wavelength is D_s, then the true transmittance is

$$D = \frac{D_1 - D_s}{100 - D_1} \qquad \text{[III, 3.11]}$$

neglecting reflection and scattering losses. This is naturally conditional on the stray light curve having been measured under the same conditions as those for the D_1 measurement. Possible alterations in D_s must be taken into account, especially if the substance investigated itself absorbs strongly in the short-wave region where the stray light originates.

The convergence of the beam of radiation also influences the transmittance observed. This must be considered especially if an extinction coefficient, etc., is to be determined from the transmittance measured. Obviously, the convergence increases the active layer thickness, but this effect is only of importance when a micro-illuminator is used (see page 204), for the layer thicknesses normally used in infrared spectroscopy.

Clearly, in any given case all or several of the effects mentioned must be taken into consideration. The various correction factors can be calculated independently.

2. Beer's Law

The law of absorption [III, 3.1] gives a first quantitative unit for spectral measurements. Obviously, the absorption of a body is larger the greater the thickness d irradiated, and the larger its extinction coefficient α for the wavelength in question. In the last resort, however, the number of molecules absorbing in the volume of the body irradiated is responsible for the absorption. Accordingly, the extinction coefficient suffices completely for the description of the absorption power of a substance provided the number of molecules in the volume under measurement remains constant. This applies to solid and liquid pure substances but not to mixtures and solutions which are by nature of variable composition. Also, it does not apply to gases for which the number of molecules per unit volume is dependent on the

284

pressure applied to the gas. Yet, just these cases are the most important in experimental spectroscopy. Therefore a relationship must be derived between the *concentration* of a substance and the *absorption* it causes.

Because the number of molecules responsible for the absorption depends directly on the concentration of the substance (for gases on the pressure) it is logical to write the absorbance as a linear function of the concentration or of the pressure c:

$$E = \epsilon.c.d \qquad [\text{III}, 3.11a]$$

This law is named after its discoverer Beer. Usually it is written together with the Lambert-Bouguer law of absorption [III, 3.1] by substituting [III, 3.11a] in [III, 3.1] and the combined relationship is known as the Lambert-Beer law.† The proportionality factor ϵ is another characteristic of the substance dependent on the wavelength or wave number; it is called the extinction coefficient as was α above. The dimensions of ϵ depend upon the unit selected for the concentration c. It is usual, but not absolutely necessary, to give c in mol/litre; ϵ, now called the molar extinction coefficient, has the dimensions litre. cm^{-1}. mol^{-1}.

The equation [III, 3.11a] is the basis of the whole of infrared spectral analysis, and it is necessary to discuss the conditions for which it holds. The linear relationship of optical density with the concentration is logical and corresponds to the predominating effect found in natural science that such relationships are as simple as possible. Indeed, it can be shown empirically that it is justified for certain regions of concentrations, but certainly not for all cases. If the optical density, according to [III, 3.11a], is plotted as a function of the concentration for a given cell thickness and substance, then a straight line is obtained. Deviations from Beer's law cause deviations from this straight line. It is found that Beer's law is always obeyed for sufficiently small pressures or concentrations, but that for larger concentrations or pressures deviations appear, for some substances earlier than for others.

There are two reasons for the appearance of such deviations from Beer's law. The first is due to the substance investigated and appears when the absorption measured is no longer simply the sum of equal absorption for all the individual molecules, i.e. is no longer proportional to their number. This is due to the formation of molecular aggregates, regardless of whether these aggregates are formed by the association of the molecules of the substance with each other or by interactions of a definite type with molecules of the environment, e.g. the solvent. For sufficiently small concentrations or pressures such aggregates do not appear and therefore Beer's law is obeyed. The

† See also Hughes (*337*) for naming the absorption laws and for other questions of nomenclature; however, not all of this author's recommendations are accepted.

aggregates only appear with an increase in concentration or an increase in pressure and then the deviations from Beer's law mentioned can be observed. Such deviations are a means of studying intermolecular interactions.

The other reason for such deviations is the fundamentally unsatisfactory nature of the instruments used. Beer's law holds rigorously only for truly monochromatic radiation. This condition is not realizable because of the finite slit width of the monochromators, which is necessary for instrumental reasons, and because stray light can never be completely eliminated. It follows that the effect measured is dependent on the slit width used, measured as its ratio to the width of the band investigated. The magnitude of the deviation also depends on the position in the band at which Beer's law is tested, i.e. whether at maximum absorption or somewhere on the flank of the band [Brodersen (87)]. This is illustrated for an example in Fig. 122 [Robinson (600)]. The positions on the band used for measurement are shown in a. Straight lines corresponding to agreement with Beer's law for several ratios of the slit width to the band width are shown in b, c, and d, together with actual measurements at the positions of the band characterized as 0, 1, 2, and 3. It can be seen from the figure that the deviations are very slight for small slit widths and can then only be noticed at the steepest point (i.e. point No. 1) on the side of the band. For increasing slit width the deviations become more noticeable; they are always the largest for the steepest place on the side of the band, while they are still very small for the band wings.

Obviously, experimental deviations from Beer's law, which are frequently found, arise from both sources. Thus the use of Beer's law is only possible if certain precautions are taken. However, nothing prevents an *empirical* determination of the actual relationship between the optical density and the concentration for any particular case, keeping the instrumental factors constant. Such a standardization curve, empirically obtained, is a sufficient basis for quantitative infrared spectral analysis, provided the conditions of analysis once selected are kept constant. In general, the validity or not of Beer's law for any definite case can only be decided by experiment, by ascertaining how far the optical density can be represented as a linear function of the concentration.

3. Integrated Absorption

The extinction coefficient ϵ with respect to the selected unit of concentration introduced in [III, 3.11a] is dependent on the wavelength or wave number, just as is α. In the following discussion, only the dependence on the wave number will be referred to explicitly. This is done for convenience and always implies a corresponding dependence on the wavelength. The corresponding relationships can be obtained directly using the fundamental connection

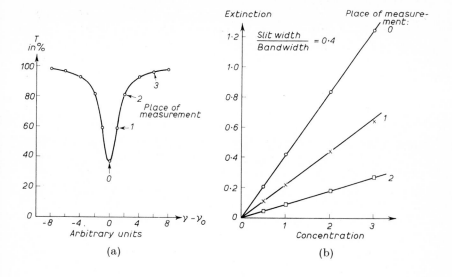

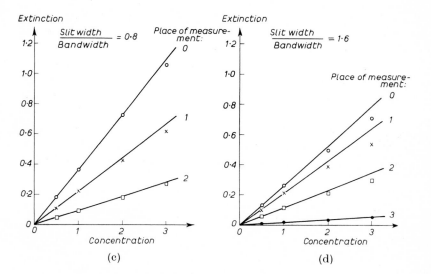

FIG. 122. *Test of Beer's law:* (a) *form of the band;* (b), (c), *and* (d) *results for different values of the ratio slit width to band width, and for measurements at different places on the band* (*from Robinson (600)*].

between wavelength and wave number.† Every absorption vibrational band occupies a certain wave number *region*. This is the result of the fundamental defects of available instruments and also of other effects including the so-called natural width, the broadening by the Doppler effect, and the interaction with neighbouring molecules. For convenience, it will be assumed that a single, and therefore isolated, band is being dealt with in a spectrum which is otherwise free of absorption. Then the extinction coefficient will commence to differ from zero at a definite ν value, it will then gradually increase with the wave number, pass through a maximum value, decrease again, and finally become zero once more. The whole region in which ν differs from zero must be considered the region of the band. Provided there are no special circumstances, the band is symmetrical about the position of the maximum, but initially nothing further will be said about the form of the ϵ curve. Obviously, the essential features of the ϵ function representing the band are described by giving the extinction coefficient (or the optical density itself) at the *maximum* (say at position ν_0, this position defines the spectral position of the band) and also one of the wave number values $\nu' = \nu_0 \pm \frac{1}{2}\Delta\nu$ which lie symmetrically on either side of ν_0 for which $\epsilon_\nu = \frac{1}{2}\epsilon_{\max}$. The interval $\Delta\nu_{1/2}$ thus defined is called the *half-band width*; if necessary a tenth band width, etc. can be defined correspondingly.

It is not possible to describe the ϵ-function mathematically in a concise simple form which completely coincides with reality. Only approximations to this are possible, the validity of which depend on the basic treatment and simplifications. However, details of the various mathematical expressions, which have been discussed for the form of the band (usually those for liquids and solids but under special conditions also those of gases), will not be given here. A reference to the relevant literature must suffice: Fox and Martin (*218*); Ramsay (*568*); Pirlot (*540*); Philpotts *et al.* (*533*); Willis (*745*); Richards and Burton (*591*); Eberhardt (*182*); Crawford and Dinsmore (*148*). Provided that only the non-instrumental influences on the form of the band mentioned above are taken into account (and of these especially the last one, because the others have little influence), it has been found advantageous to represent the optical density as a *resonance curve*. This is frequently referred to in the literature as a Lorentz curve:

$$\ln\frac{J_0}{J} = \frac{a}{(\nu-\nu_0)^2+b^2} \qquad [\text{III, } 3.12]$$

† Nevertheless, the use of the wave number is indispensable where the integrated absorption is calculated from the band height and the band width (i.e. not by a planimeter) because the bands are completely symmetrical only when wave numbers are used. Accessories for the evaluation of spectra linear in wavelength for the purpose of obtaining integrated absorption values have been described by Oswald (*514*).

Here, a and b are constants which determine the maximum optical density and the half band width:

$$\ln\left(\frac{J_0}{J}\right)_{\nu_0} = \frac{a}{b^2}, \qquad \Delta\nu_{1/2} = 2b \qquad [\text{III}, 3.13]$$

where ν_0 denotes the band centre and $\Delta\nu_{1/2}$ the half band width. This treatment applies to the bands of liquids and solids but not directly to those of gases. For gases, as has been seen, the shape of a vibrational band is modified markedly by the rotational fine structure. Only when this rotational fine structure has been eliminated by suitable means can the above treatment be applied. For example, this can be done by pressure broadening. The gas under investigation is measured either at high pressures (with simultaneous decrease of the cell thickness in order to retain an optical density which is readily measurable) or with the addition of a non-absorbing gas (H_2, N_2, O_2, or one of the rare gases); here the partial pressure of the gas to be measured remains constant but the total pressure is raised. Luck (*431*) has discussed the theoretical treatment of this problem, especially the validity of the Lambert-Beer law. Examples can be found in the work of Coggeshall and Saier (*127*), and Penner and Weber (*524*). If the rotational influence on a vibrational band is eliminated in such a manner, the treatment mentioned for characterizing the shape of the bands can also be applied to gases. In addition, the total absorption, which is discussed below, can be introduced and measured. Under certain conditions [see Matossi (*451*), Matossi and Rauscher (*452*)] this quantity should theoretically vary with $p^{1/2}$, while experiments indicate a variation with $p^{1/4}$. The last word has not been written on this matter.

The instrumental influences, essentially those of the finite slit width, are principally a result of the fact that the quantities J and J_0 appearing in [III, 3.8] apply only to completely monochromatic radiation. This could only be obtained with infinitely small slits, i.e. in practice it can never be obtained. The radiation emitted from the exit slit of a monochromator set for a definite wave number is therefore not monochromatic. It shows a characteristic distribution of the radiation energy as a function of the wave number. This distribution is known as the *slit function* and is here denoted as $\rho\,(\nu, \nu')$ for brevity. The slit function defines the amount of radiation of wave number ν that occurs when the monochromator is set for the wave number ν', or in other words the influence of the limited resolving power of the spectrometer used. If such an instrument is applied to a band of the true form [III, 3.12], the true transmittance J/J_0 is not measured. Instead, a transmittance modified by the influence of the slit width is measured which will be denoted here as the apparent transmittance. Using the ratio of the

U

apparent radiation energies T and T_0 with and without the absorbing substance, and eliminating any influence of reflection, this is found to be

$$\frac{T}{T_0} = \frac{\int \rho(\nu, \nu') \cdot e^{-a/[(\nu - \nu_0)^2 + b^2]}\, d\nu}{\int \rho(\nu, \nu')\, d\nu} \qquad [III, 3.14]$$

A further condition is that the slit width should be the same for the entry and for the exit slit, and that both the energy of the incident radiation and the degree of resolution within the width of the slit remain constant from a spectral point of view. Under these conditions, which are easily fulfilled, the slit function can be considered to be an isosceles triangle of definite height and a base equal to double the slit width s in wave numbers (Fig. 31; see also the explanation there).

If such a slit function is used as the basis of the treatment, then equation [III, 3.14] can be numerically evaluated by using different values of a, b, and s. Nevertheless, no concise mathematical expression is found for the apparent shape of the band thus obtained. Therefore it appears to be convenient to relate the results [i.e. the band height and the band width expressed as the maximum *apparent* optical density $\ln (T_0/T)_{\nu_0}$ and the *apparent* half band width $\Delta \nu_{1/2}^s$] to the ideal relationships, i.e. to the maximum *true* optical density $\ln (J_0/J)_{\nu_0}$ and the *true* half band width $\Delta \nu_{1/2}$. $\ln (T_0/T)_{\nu_0}$ and $s/\Delta \nu_{1/2}^s$ are selected for tabulation. The results of such calculations by Ramsay (*568*) are given in Tables 17 and 18. The values of the quantity $\ln (T_0/T)_{\nu_0}$ tabulated correspond to easily measurable conditions and those of $s/\Delta \nu_{1/2}^s$ are selected so that the highest value corresponds approximately to

TABLE 17. *The Ratio of the True to the Apparent Maximum Optical Density for Various Apparent Maximum Optical Densities and Slit Widths of a Band in the Form of a Lorentz Curve* [Ramsay (*568*)]

$\ln (T_0/T)_{\nu_0}$	$s/\Delta \nu_{1/2}^s$													
	0·00	0·05	0·10	0·15	0·20	0·25	0·30	0·35	0·40	0·45	0·50	0·55	0·60	0·65
2·0	1·00	1·00	1·01	1·02	1·03	1·04	1·06	1·09	1·13	1.18	1·24	1·32	1·44	1·61
1·8	1·00	1·01	1·01	1·02	1·03	1·04	1·06	1·09	1·13	1·18	1·24	1·32	1·44	1·59
1·6	1·00	1·01	1·01	1·02	1·03	1·04	1·06	1·09	1·13	1·18	1·24	1·32	1·43	1·58
1·4	1·00	1·01	1·01	1·02	1·03	1·04	1·06	1·09	1·13	1·18	1·24	1·32	1·43	1·57
1·2	1·00	1·01	1·01	1·02	1·03	1·04	1·06	1·09	1·13	1·17	1·24	1·31	1·42	1·56
1·0	1·00	1·01	1·01	1·02	1·03	1·04	1·06	1·09	1·13	1·17	1·24	1·31	1·42	1·55
0·8	1·00	1·01	1·01	1·02	1·03	1·04	1·06	1·09	1·13	1·17	1·24	1·31	1·41	1·54
0·6	1·00	1·01	1·01	1·02	1·03	1·04	1·06	1·09	1·13	1·17	1·24	1·31	1·41	1·53
0·4	1·00	1·01	1·01	1·02	1·03	1·04	1·06	1·09	1·13	1·17	1·23	1·30	1·40	1·53
0·2	1·00	1·01	1·01	1·02	1·03	1·04	1·06	1·09	1·13	1·17	1·23	1·30	1·40	1·52

TABLE 18. *Ratio of the Apparent to the True Half Band Width for Various Apparent Maximum Optical Densities and Slit Widths of a Band in the Form of a Lorentz Curve* [Ramsay (568)]

$\ln (T_0/T)_{\nu_0}$	$s/\Delta\nu^s_{1/2}$													
	0·00	0·05	0·10	0·15	0·20	0·25	0·30	0·35	0·40	0·45	0·50	0·55	0·60	0·65
2·0	1·00	1·00	1·01	1·02	1·03	1·04	1·06	1·09	1·13	1·17	1·22	1·28	1·36	1·51
1·8	1·00	1·00	1·01	1·02	1·03	1·05	1·07	1·10	1·14	1·18	1·23	1·29	1·38	1·53
1·6	1·00	1·00	1·01	1·02	1·03	1·05	1·07	1·10	1·14	1·18	1·24	1·31	1·40	1·55
1·4	1·00	1·00	1·01	1·02	1·03	1·05	1·08	1·11	1·15	1·19	1·25	1·33	1·43	1·58
1·2	1·00	1·00	1·01	1·02	1·03	1·05	1·08	1·11	1·15	1·20	1·27	1·35	1·45	1·60
1·0	1·00	1·00	1·01	1·02	1·03	1·05	1·08	1·12	1·16	1·21	1·28	1·37	1·48	1·63
0·8	1·00	1·00	1·01	1·02	1·03	1·06	1·09	1·12	1·16	1·22	1·30	1·39	1·51	1·66
0·6	1·00	1·00	1·01	1·02	1·04	1·06	1·09	1·13	1·17	1·23	1·31	1·41	1·54	1·70
0·4	1·00	1·00	1·01	1·02	1·04	1·06	1·09	1·13	1·18	1·24	1·32	1·43	1·57	1·75
0·2	1·00	1·00	1·01	1·02	1·04	1·07	1·10	1·14	1·19	1·25	1·34	1·46	1·60	1·80

the condition $s = \Delta\nu_{1/2}$. As can be seen, the effect of the true half band width is much greater than that of the maximum optical density.

It follows, from the whole of this discussion, that a band is certainly not completely and unambiguously defined by the experimental determination of the maximum extinction coefficient $\epsilon_{\max}$ at the position $\nu = \nu_0$. For complete definition the whole course of the ϵ curve must be given, at least to such wave numbers on each side of the maximum that the remaining values of ϵ are negligibly small. For example, for a slit width equal to half the true half band width of a band, the true and observed maximum extinction coefficients differ by about 20 per cent. It is a logical step to use the area lying between the abscissa and the ϵ curve as characteristic of and as a means of measurement of the band intensity, instead of using the extinction coefficient at one or more wave numbers. In other words, the curve of the extinction coefficient is integrated and transition is made from monochromatic absorption to an *integral* or *total absorption* of the band. Mathematically this is expressed by

$$A = \int \epsilon_\nu \, d\nu = \frac{1}{c.d} \int \ln \frac{J_0}{J} \, d\nu \qquad \text{[III, 3.15]}$$

In principle $-\infty$ and $+\infty$ should be taken as the limits of the integral; accepting an error that can be made as small as is desired, finite limits can also be used for the integral, e.g. $\nu_0 \pm 5\Delta\nu_{1/2}$ or something similar. Experience has shown that this integrated absorption is little influenced by the slit width and thus really is a useful unit of measurement.

If [III, 3.12] is taken as the true form of the band, [III, 3.15] can be directly evaluated as

$$A = \frac{1}{c.d} \int\limits_{-\infty}^{+\infty} \frac{a}{(\nu - \nu_0)^2 + b^2} d\nu$$

$$= \frac{1}{c.d} \cdot \frac{a}{b} \arctan \left[\frac{\nu - \nu_0}{b} \right]_{-\infty}^{+\infty} = \frac{1}{c.d} \cdot \frac{\pi a}{b} \qquad [\text{III, 3.16}]$$

Taking [III, 3.13] into account, this can be written as

$$A = \frac{1}{c.d} \cdot \frac{\pi}{2} \ln \left(\frac{J_0}{J} \right)_{\nu_0} . \varDelta \nu_{1/2} \qquad [\text{III, 3.17}]$$

As only experimental quantities are available, it is advantageous to substitute these for the true quantities in the equation:

$$A = \frac{K}{c.d} \ln \left(\frac{T_0}{T} \right)_{\nu_0} . \varDelta \nu_{1/2}^s \qquad [\text{III, 3.18}]$$

Here

$$K = \frac{\pi}{2} \cdot \frac{\ln \left(\dfrac{J_0}{J} \right)_{\nu_0}}{\ln \left(\dfrac{T_0}{T} \right)_{\nu_0}} \cdot \frac{\varDelta \nu_{1/2}}{\varDelta \nu_{1/2}^s} \qquad [\text{III, 3.19}]$$

is a factor which varies with the maximum optical density and with the slit width. Values for it are given in Table 19 in a way analogous to the earlier

TABLE 19. *Values of the Coefficient K of the Equation [III, 3.19] for Various Values of the Apparent Maximum Optical Density and Slit Width of a Band in the Form of a Lorentz Curve [Ramsay (568)]*

$\ln (T_0/T)_{\nu_0}$	$s/\varDelta \nu_{1/2}^s$													
	0·00	0·05	0·10	0·15	0·20	0·25	0·30	0·35	0·40	0·45	0·50	0·55	0·60	0·65
2·0	1·57	1·57	1·57	1·57	1·57	1·57	1·57	1·57	1·58	1·59	1·60	1·62	1·65	1·68
1·8	1·57	1·57	1·57	1·57	1·57	1·57	1·57	1·57	1·57	1·58	1·58	1·60	1·62	1·64
1·6	1·57	1·57	1·57	1·57	1·57	1·57	1·57	1·57	1·57	1·57	1·57	1·58	1·59	1·60
1·4	1·57	1·57	1·57	1·57	1·57	1·56	1·56	1·56	1·56	1·56	1·56	1·56	1·56	1·56
1·2	1·57	1·57	1·57	1·57	1·56	1·56	1·56	1·55	1·55	1·54	1·54	1·54	1·54	1·53
1·0	1·57	1·57	1·57	1·56	1·56	1·56	1·55	1·54	1·54	1·53	1·52	1·52	1·51	1·50
0·8	1·57	1·57	1·57	1·56	1·56	1·55	1·54	1·53	1·52	1·51	1·50	1·49	1·48	1·47
0·6	1·57	1·57	1·56	1·56	1·55	1·55	1·54	1·53	1·51	1·50	1·48	1·46	1·45	1·43
0·4	1·57	1·57	1·56	1·56	1·55	1·54	1·53	1·52	1·50	1·48	1·46	1·44	1·41	1·38
0·2	1·57	1·57	1·56	1·56	1·55	1·54	1·53	1·51	1·49	1·47	1·44	1·41	1·37	1·33

tables of this chapter; it can be seen that its value is generally not very different from $\pi/2$.

The direct calculation of the true total absorption is obviously only possible when the *form* of the band and the relationship between the true and the apparent absorption are known. This is frequently not the case. Using the fact that the observed transmittance and thus the observed integrated absorption of a band can be measured, the true integrated absorption may be determined by extrapolation from such measurements at different optical densities, obtained by altering the concentration or the cell thickness. The methods for this originated from Bourgin (*78*) and Wilson and Wells (*747*).

It will be assumed that the integrated absorption of a band has been determined for various concentrations or cell thicknesses. Then the true integrated absorption can be obtained by extrapolation to zero concentration or thickness. Conditions for this are that the incident radiation energy should be constant over the slit width and that the degree of resolution should be constant over the band width. In addition the extrapolation depends on the band contour and the degree of resolution used. For carrying out this process, it is advantageous to plot the apparent integrated absorption $\int \ln (T_0/T)d\nu$ against the apparent maximum optical density $\ln (T_0/T)_{\nu_0}$ for different concentrations or cell thicknesses. Assuming a band contour corresponding to [III, 3.12], a curve is obtained the extrapolation of which to the value $\ln (T_0/T)_{\nu_0} = 0$ gives the true integrated absorption as the intersect on the ordinate. This plot may be represented as a straight line with a slight negative gradient $A \tan \vartheta$. The value of $\tan \vartheta$ depends on the ratio of the spectral slit width used s to the true half band width $\Delta\nu_{1/2}$ of the band. It is given in Table 20 for several values of this ratio according to

TABLE 20. *Values for* $\tan \vartheta$ *of Equation* [*III, 3.20*] *for Various Slit Widths for a Band Contour of the Lorentz Type* [*Ramsay (568)*]

$s/\Delta\nu_{1/2}$	0·1	0·2	0·3	0·4	0·5	0·6	0·7	0·8	0·9	1·0
$\tan \vartheta$	$-0{\cdot}002$	$-0{\cdot}004$	$-0{\cdot}008$	$-0{\cdot}013$	$-0{\cdot}020$	$-0{\cdot}028$	$-0{\cdot}036$	$-0{\cdot}046$	$-0{\cdot}057$	$-0{\cdot}070$

Ramsay (*568*). Therefore the relationship between the apparent integrated absorption B and the true absorption A can be represented mathematically as

$$B = A + A \cdot \tan \vartheta \cdot \ln \left(\frac{T_0}{T}\right)_{\nu_0} \qquad \text{[III, 3.20]}$$

The Methods of Practical Infrared Spectroscopy

As a result of the small magnitude of tan ϑ, the deviation between B and A is small and usually lies within the accuracy of 2 to 3 per cent attainable for such measurements. The correction for the wings of bands is much more important. Normally, the integration of the band measured is carried out only in a limited wave number region on both sides of the maximum. The wings of the band, which theoretically extend to infinity, are thus not included. Assuming the validity of equation [III, 3.12] and an integrated wave number interval of $\nu - \nu_0$ on both sides of the maximum, the ratio of the area integrated to the total area under the band is arctan $\left[\dfrac{\nu - \nu_0}{b}\right]$ to $\pi/2$. Accordingly, the percentage correction to be applied to the integrated surface is $100\left(\dfrac{\pi}{2}\arctan^{-1}\left[\dfrac{\nu - \nu_0}{b}\right] - 1\right)$. These values are given in Table 21 for the range $4 < \dfrac{\nu - \nu_0}{b} < 20$. It usually suffices in practice to carry out the integration itself to such wave numbers that the correction is less than 10 per cent.

TABLE 21. *Wing Correction for Integrated Absorption Curves* [*Ramsay* (568)]

$\dfrac{\nu - \nu_0}{b}$	%	$\dfrac{\nu - \nu_0}{b}$	%	$\dfrac{\nu - \nu_0}{b}$	%	$\dfrac{\nu - \nu_0}{b}$	%
4·0	18·5	6·2	11·3	8·4	8·1	11·5	5·8
4·2	17·5	6·4	10·9	8·6	7·9	12·0	5·6
4·4	16·6	6·6	10·6	8·8	7·8	12·5	5·3
4·6	15·8	6·8	10·3	9·0	7·6	13·0	5·1
4·8	15·1	7·0	9·9	9·2	7·4	13·5	4·9
5·0	14·4	7·2	9·6	9·4	7·2	14·0	4·7
5·2	13·8	7·4	9·4	9·6	7·0	14·5	4·6
5·4	13·2	7·6	9·1	9·8	6·9	15·0	4·4
5·6	12·7	7·8	8·8	10·0	6·8	16·0	4·1
5·8	12·2	8·0	8·6	10·5	6·4	18·0	3·7
6·0	11·8	8·2	8·4	11·0	6·1	20·0	3·1

Another method for determining the true integrated absorption from experimental values is to use the absorption defined by equation [III, 3.3] (monochromatic) [Bourgin (78)] the integration of which gives

$$A' = \int \left(1 - \frac{J}{J_0}\right) d\nu \qquad \text{[III, 3.21]}$$

Substitution of the T for the J values is permissible because the area under absorption curve is independent of the resolving power and the band contour (slit width) [Dennison (*167*); Nielsen *et al.* (*499*)]. The Lambert Bouguer law [III, 3.1] is now introduced, taking into consideration Beer's law [III, 3.7], and the e function is expanded. Then, after taking the factor $c \times d$ out of the integral and placing it on the left-hand side, the following equation is obtained

$$\frac{A'}{c.d} = \int \left[\epsilon_\nu - \frac{\epsilon \nu^2}{2!} . cd + \frac{\epsilon \nu^3}{3!} . (cd)^2 - \dots \right] d\nu \qquad \text{[III, 3.22]}$$

If an extrapolation is made to zero optical density, the result is

$$\lim_{cd \to 0} \frac{A'}{c.d} = \int \epsilon_\nu \, d\nu = A \qquad \text{[III, 3.23]}$$

i.e. the true integrated absorption. If the band form is taken as that of [III, 3.12], then the following relationship is found

$$\frac{A'}{c.d} = A \left\{ 1 - \frac{1}{2!} . \frac{1}{2} \ln \left(\frac{J_0}{J} \right)_{\nu_0} \right.$$

$$\left. + \frac{1}{3!} . \frac{3}{8} \left[\ln \left(\frac{J_0}{J} \right)_{\nu_0} \right]^2 - \frac{1}{4!} . \frac{5}{16} \left[\ln \left(\frac{J_0}{J} \right)_{\nu_0} \right]^3 + \dots \right\}$$

$$\text{[III, 3.24]}$$

In order to obtain A, A'/cd must be plotted against $\ln (J_0/J)_{\nu_0}$ and extrapolated to $\ln (J_0/J)_{\nu_0} = 0$. Closer investigation of the expression $\frac{1}{A} . \frac{A'}{cd}$ and $A . \frac{cd}{A'}$ shows that the first is markedly curved so that extrapolation is difficult whereas the latter is almost a straight line of slope $0 \cdot 25$. For evaluation it is advantageous to write

$$A = \phi . \frac{A'}{c.d} \qquad \text{[III, 3.25]}$$

with

$$\phi = \frac{1}{1 - \frac{1}{2!} . \frac{1}{2} \ln \left(\frac{J_0}{J} \right)_{\nu_0} + \frac{1}{3!} . \frac{3}{8} \left[\ln \left(\frac{J_0}{J} \right)_{\nu_0} \right]^2 - \dots} \qquad \text{[III, 3.26]}$$

Values of ϕ are given in Table 22 for various values of $\ln (J_0/J)_{\nu_0}$. This method also needs a correction for the band wings and this correction is given in Table 23 as a percentage.

A comparison of the three methods discussed for the determination of the true integrated absorption of a band from experimental data [Jones *et al.* (*363*)] shows that they give results which lie within about 5 to 10 per cent of

The Methods of Practical Infrared Spectroscopy

TABLE 22. *Values of the Factor ϕ of the Equation [III, 3.25] for Various Values of the True Maximum Optical Density for a Band Corresponding to a Lorentz Curve [Ramsay (568)]*

$\ln\left(\dfrac{J_0}{J}\right)_{\nu_0}$	ϕ	$\ln\left(\dfrac{J_0}{J}\right)_{\nu_0}$	ϕ	$\ln\left(\dfrac{J_0}{J}\right)_{\nu_0}$	ϕ	$\ln\left(\dfrac{J_0}{J}\right)_{\nu_0}$	ϕ
0·0	1·000	0·5	1·125	1·0	1·248	1·5	1·368
0·1	1·025	0·6	1·149	1·1	1·272	1·6	1·391
0·2	1·050	0·7	1·174	1·2	1·296	1·7	1·415
0·3	1·075	0·8	1·199	1·3	1·320	1·8	1·438
0·4	1·100	0·9	1·223	1·4	1·344	1·9	1·462
						2·0	1·484

TABLE 23. *Wing Corrections for Integrated Absorption Curves [Ramsay (568)]*

$\ln\left(\dfrac{J_0}{J}\right)_{\nu_0}$	$\dfrac{\nu-\nu_0}{b}$											
	4	5	6	7	8	9	10	11	12	13	14	15
2·0	29·9	22·9	18·6	15·6	13·4	11·8	10·6	9·4	8·6	7·9	7·3	6·8
1·8	28·7	22·0	17·9	15·0	13·0	11·4	10·2	9·1	8·3	7·6	7·0	6·5
1·6	27·5	21·1	17·2	14·4	12·5	11·0	9·8	8·8	8·0	7·4	6·8	6·3
1·4	26·3	20·2	16·5	13·8	12·0	10·6	9·4	8·5	7·7	7·1	6·6	6·1
1·2	25·0	19·3	15·7	13·2	11·5	10·1	9·0	8·1	7·4	6·8	6·3	5·8
1·0	23·8	18·4	15·0	12·6	11·0	9·7	8·6	7·8	7·1	6·5	6·0	5·6
0·8	22·6	17·5	14·3	12·0	10·5	9·3	8·3	7·5	6·8	6·3	5·8	5·4
0·6	21·4	16·6	13·6	11·5	10·0	8·8	7·9	7·2	6·5	6·0	5·5	5·2
0·4	20·3	15·8	12·9	10·9	9·5	8·4	7·5	6·8	6·2	5·7	5·3	5·0
0·2	19·3	15·0	12·3	10·5	9·1	8·1	7·2	6·5	5·9	5·4	5·0	4·7

each other. Indeed, the last two methods give results which coincide still more closely. The importance of such calculations is that they alter fundamentally the nature of infrared spectroscopy. Up to now the principal aims of infrared spectroscopic investigations have been to determine as precisely as possible the absorption bands within the spectrum and to assign them to definite vibrations or bonds of atomic groups; considerations of intensity were only of lesser importance. With the development of precise intensity measurements and their application to theoretical discussions of structure, a transition is being made from positional infrared spectroscopy to the still more valuable intensity spectroscopy. Doubtless this change will not

296

only give an impulse to structural determination but will also benefit infra-red spectral analysis even though simpler methods are used at present.

4. Problems and Methods of Evaluation

Subsequent to this more fundamental and theoretical treatment, the prac-tical problems of quantitative infrared spectral analysis will be discussed. The multitude and variety of problems which occur means that it is cer-tainly not possible to indicate a method of solution for every case. The discussion will be limited to characteristic groups of problems and an indi-cation of methods of evaluation. It must be left to the skill and experience of the individual spectroscopist to apply this general information to any given problem.

Initially only those cases will be considered in which it is possible to find, for each component to be estimated in a mixture, an isolated band for analysis, i.e. one which is not influenced by the other components. Whether such bands exist must be found by a preliminary qualitative investigation. If several bands are found to satisfy this condition, the one which would appear to offer the greatest accuracy is used for the analysis itself and the others for supplementary estimations to confirm the results. It is clear that the requirement of such a band for analysis limits the application of the method to such mixtures which, because of their small number of com-ponents, and because of the simplicity of their spectra, are expected to fulfil this condition.

The simplest method of analysis is to measure directly, at the selected wavelength, the transmitted radiation energy for a series of *standard mix-tures* with known amounts of the substance to be investigated. The content of the analytical sample is then found by comparison of its spectrum with these standard spectra, sometimes with interpolation. A requirement for the use of this method is that the intensity of the incident radiation J_0 and all the other conditions should be held constant throughout the record-ing of all the necessary curves. Fig. 123a shows the use of this method in the determination of an aldehyde in an alcohol by means of the carbonyl band at $5 \cdot 8 \, \mu$ using low resolution. The upper curve is that recorded for the pure alcohol; the bands appearing in it originate from the atmospheric water vapour. The other curves were obtained for increasing percentages of aldehyde in the alcohol; for clarity, some of these have been omitted.

If the available spectral instruments give the percentage transmittance directly, as do many modern instruments, or if this can be calculated from the curves of Fig. 123a, then this can be plotted as a function of the concen-tration of the substance investigated (Fig. 123b). Using this plot and the measured transmittance of the analytical sample, the concentration to be determined can be found. Instead of the transmittance, the optical density, i.e. the logarithm of the reciprocal of the transmittance, can equally well be plotted (Fig. 123c). If Beer's law is obeyed, then a straight line results (as in this example up to about 5 per cent) and interpolation or extrapolation is especially easy.

A further possibility is to use Beer's law directly (after testing its validity) for carrying out the analysis. This is done in the following way. The extinction coefficient is determined for the component to be investigated in a standard sample of known composition and at the wavelength selected. Using this value, the content of the analytical sample of this substance is determined from the experimentally determined optical density using equation [III, 3.11]. This method, which appears very simple, is usually not applicable for two reasons. Firstly, the exact determination of the extinction coefficient necessitates an exact measurement of transmittance and also an

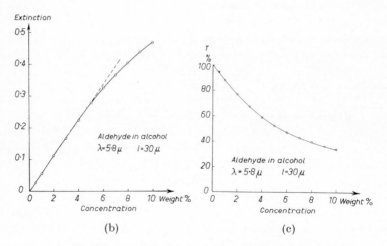

FIG. 123. *Quantitative analysis:* (a) *by recording energy curves for standardized mixtures;* (b) *from the measured percentage transmittance;* (c) *from the measured optical density.*

298

exact knowledge of the thickness of the substance irradiated. In the region of the fundamental vibrations, the useful thicknesses are between about 10 and 100 μ for liquids and solids. Especially for the smaller values, precise measurements, accurate to about 1 per cent, are hardly possible because of the necessity for planarity and parallelism of the cell windows. Secondly, the difference between the measured transmittance and 100 per cent is not due only to the true absorption as is assumed in the equation [III, 3.3]. Reflection and scattering influences also play a part, the size of which is unknown and only estimated with difficulty. In addition to these, there is the unsatisfactory nature of the instrument with respect to the monochromatic character of the radiation used. It has to be taken into account that some radiation of wavelengths other than that desired always occurs. However, this will be relatively unimportant for instruments of higher quality. Such stray radiation, especially that from regions of intense emission in places of small emission, obviously distorts the transmittance to an extent which often cannot be estimated easily. Accordingly to use Beer's law directly necessitates well-considered precautions, which the previously described comparative methods do not need, as the disturbing influences can be kept constant for the duration of the measurements.

As already mentioned, the preceding processes are applicable, when several components have to be determined, provided an *isolated* band for analysis can be found for each. As an example, the simultaneous determination of normal- and iso-butane in a mixture with N_2 will be discussed. Fig. 124 shows the spectra (*a*) of the two substances obtained with a recording double beam spectrophotometer of the older type. The band at $8 \cdot 6 \, \mu$ for iso-butane and that at $10 \cdot 3 \, \mu$ for n-butane are suitable as analysis bands. The third spectrum shown (*b*) is that of the analytical sample of unknown composition. Fig. 124c shows the standardization curves obtained with samples of known mixtures of iso- or n-butane in N_2. The spectrum of the analytical sample shows an absorption of $40 \cdot 6$ per cent at $8 \cdot 6 \, \mu$ and $36 \cdot 4$ per cent at $10 \cdot 3 \, \mu$. Using the standardization curves this shows that the content of n-butane is $40 \cdot 5$ per cent and that of iso-butane is 39 per cent.

The condition assumed up to now, that the other components do not absorb at the position of the analysis band selected, is not fulfilled in the majority of cases. Even if no specific band of a component occurs at this position, a certain so-called *background absorption* has still to be taken into account. This originates from the wings of intense neighbouring bands quite apart from scattering and reflection effects. There are numerous ways of eliminating such background absorption. For example, the intensity of the selected analysis band A can be measured as a ratio of a neighbouring band B of the basic substance and the ratios A/B, $(B-A)/A$ or $(B-A)/B$

can be plotted as a function of the concentration of the component being investigated. For this to be applicable, the band of the main substance must not be distorted by influences in the analytical sample which do not appear in the standardization.

The use of so-called *isosbestic points* is especially advantageous for the elimination of the effect of background absorption, provided such points occur in the spectrum investigated. An isosbestic point is defined as a wavelength or wave number and an absorption or optical density at which all

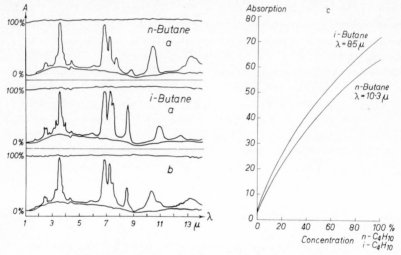

FIG. 124. *Simultaneous quantitative estimation of two components (normal- and iso-butane): (a) spectra of the pure substances; (b) spectrum of the analytical sample; (c) standardization curve [from Lehrer (412)].*

the spectral curves of a series of concentrations of a mixture coincide; i.e. the absorption is the same for all concentration ratios. Fig. 125 shows such points for the spectrum of the binary system normal butanol-isobutanol. Obviously, isosbestic points can really only occur in binary systems. They occur between two bands of which one belongs to the first and the other to the second component; i.e. the absorption of one band decreases through the series of concentrations and that of the other band increases. By making all measurements with reference to these isosbestic points, the background absorption is eliminated. However, the more the system investigated deviates from a binary system, the more this characteristic property is lost.

The so-called *base line* method has found very wide application [Wright (758); Heigl et al. (316)]. It was initially developed for spectrometers which record energy curves successively with and without the sample instead of the percentage transmittance directly. In such cases, the percentage transmittance would have to be calculated point by point, but this method does not require the determination of the true energy curve without the absorbing substance. With obvious modifications, the method is also applicable to percentage spectra. Essentially, the method consists of replacing the unknown course of the curve without sample absorption near the analysis band (in percentage spectra the line 100 per cent transmittance) by a

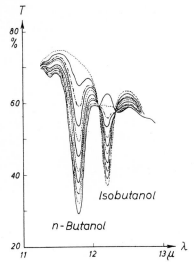

FIG. 125. *Part of the spectra of a series of different concentrations of the binary mixture normal-butanol and iso-butanol with isosbestic points.*

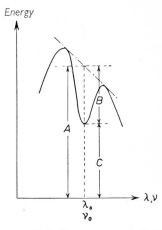

FIG. 126. *Base-line method of eliminating the background absorption by means of a tangent connecting the maxima flanking the analytical band [from Wright (758)].*

straight line. In principle this line could be drawn anywhere but in practice it is placed according to definite rules. For example, if the band selected for analysis is flanked by two marked transmittance maxima (Fig. 126), then the line connecting these tangentially is taken as the base line. Using this line, the logarithm of the ratio A/C (corresponding to the usual optical density) is taken from the recorded spectra. This quantity is plotted as a function of the concentration and the process repeated using the base line initially defined. Sometimes the base line cannot be drawn in this way because the recorded curve is too steeply inclined or the flanking maxima are not sufficiently marked. In these cases Pirlot's (*538*) method can be used. Two further wavelengths λ_1 and λ_2 are selected at certain, not necessarily

301

The Methods of Practical Infrared Spectroscopy

equal, distances from the analytical wavelength λ_0. The line connecting the points where the ordinates of these wavelengths coincide with the recorded curve is taken as the base line (Fig. 127).

The base line method has been extensively and critically examined by Pirlot (*538*). He was able to show that in general the 'optical density' obtained was not proportional to the concentration of the substance causing the band. Nor was it the sum of the 'optical densities' of several substances present, even when Beer's law was obeyed. The above relationships are only fulfilled when the extinction coefficients at the two additional wavelengths λ_1 and λ_2 are equal; this is the condition for the application of optical density calculations without standardization curves. In the case of a binary system of which one component has to be estimated, this requirement is satisfied if the arbitrary base line is drawn parallel to the true energy curve without the absorbing substance. The method for this case has also been given by Wright. If several substances absorb, the condition can hardly be satisfied. Yet this does not affect the utility of the procedure provided special standardization curves are obtained, which is usually worth while.

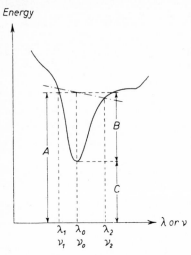

FIG. 127. *Base-line method according to Pirlot* (*538*).

Pirlot (*538*) has recommended the use of a quantity described by Seyfried and Hastings (*641*) as '*compensated optical density*' instead of the base-line technique. The compensated optical density is quite generally an expression of the form

$$Q = \sum k_i E_{\lambda_i} \qquad [\text{III, 3.27}]$$

i.e. a linear function of the true optical densities measured at the wavelengths λ_i. The form of the function, i.e. the coefficients k_i, should be selected so that the advantages of the base-line method (no need for the energy curve without the absorbing substance) remain. If Beer's law is obeyed, then

$$\sum k_i E_{\lambda_i} = c.d \sum k_i \epsilon_{\lambda_i} \qquad [\text{III, 3.28}]$$

where $\sum k_i \epsilon_{\lambda_i}$ can be taken as a 'compensated extinction coefficient'. Independent of whether Beer's law is obeyed, it can be shown that the com-

302

pensated optical density of a mixture is equal to the sum of the compensated optical densities of the components. In the simplest case [III, 3.27] can be verified by the expression

$$Q = E_{\lambda_0} - E_{\lambda_1} \qquad \text{[III, 3.29]}$$

In order to obtain this compensated optical density, the true optical densities should accordingly be measured at the two wavelengths λ_0 and λ_1. It now follows from the definition of optical density that

$$Q = \ln \frac{J_0^{(\lambda_0)}}{J^{(\lambda_0)}} - \ln \frac{J_0^{(\lambda_1)}}{J^{(\lambda_1)}} = \ln \frac{J_0^{(\lambda_0)}}{J_0^{(\lambda_1)}} - \ln \frac{J^{(\lambda_0)}}{J^{(\lambda_1)}} \qquad \text{[III, 3.30]}$$

The first term on the right can be determined once and for all. It involves the emission intensities of the radiation source used at the wavelengths λ_0 and λ_1 for the conditions applying at the time. However, even if these alter, the constancy of the intensity ratio can be assumed, provided the two wavelengths are not too far apart and that the alteration in the emission is not too large. The second term contains the ratio of the radiation intensities at the selected wavelength for the energy curve of the absorbing substance. Thus only this last curve has to be determined (and indeed only at the wavelength selected) just as for the base-line procedure.

The relationships are more complicated when the other components absorb at the specific wavelengths selected for the first component [Brattain *et al.* (*82*); Fry *et al.* (*235*); Kent and Beach (*380*); Mecke and Mutter (*464*); Rasmussen (*579*)]. The more components present or to be determined, and the more complicated the spectra of the components, the greater are the difficulties expected. For the further treatment of the problem, the assumption will be made (which must be tested empirically) that, for every band selected for analysis, the measured total optical density is equal to the sum of the optical densities originating from the individual components. It will be assumed that four components are to be determined, designated respectively 1, 2, 3, 4. The necessary four analytical wavelengths are $\lambda_1, \lambda_2, \lambda_3, \lambda_4$. According to the preceding assumption

$$E_{\lambda_i} = E_{\lambda_i}^{(1)} + E_{\lambda_i}^{(2)} + E_{\lambda_i}^{(3)} + E_{\lambda_i}^{(4)} \qquad \text{[III, 3.31]}$$

applies to each of them, where the lower index applies to the absorption wavelength and the upper to the component in question. Further

$$E_{\lambda_i}^{(k)} = \epsilon_{\lambda_i}^{(k)} \cdot c^{(k)} \cdot d \qquad \text{[III, 3.32]}$$

where $\epsilon_{\lambda_i}^{(k)}$ is the extinction coefficient of the k-th component at position λ_i, $c^{(k)}$ is the concentration of the k-th component, and d is the cell thickness used. Thus, [III, 3.31] represents a system of linear equations, in the present case four equations for the four unknowns $c^{(1)}$, $c^{(2)}$, $c^{(3)}$, $c^{(4)}$. The coefficients of this system are the 16 products $\epsilon_{\lambda_i}^{(k)} . d$ and the four quantities measured E_{λ_i}. The 16 products must be determined once by recording standardization curves of the pure substances for every wavelength. The extension of the system of equations [III, 3.32] to more than four components should be clear. Its solution for the concentrations required is easy in principle, although it may involve difficulties of calculation. These difficulties may be overcome by known algebraic methods: for many component systems it is advantageous to use calculating machines if many measurements are to be made [Adcock (5); Berry and Pemperton (62); Comrie (137); Frost and Tamres (234); Hardy and Deuch (298); Hughes and Wilson (338)]. In addition to calculating machines, punched cards have been shown to be valuable for the evaluation of such systems of equations (King and Priestley (383); Opler (510)]. It is clear that these difficulties are greater, the larger the number of components to be determined. Therefore all available facilitating circumstances should be used to the full. An important aid is that one of the selected absorption wavelengths is frequently found to depend on only one of the required components; the corresponding equation of the system [III, 3.31] then reduces to one term on the right hand side. It also frequently happens that each analytical band is essentially affected by only one component while the others contribute but little to its absorption. Initially, these contributions can be neglected and an approximate value for the concentration of the corresponding substance be determined from each analytical band. Commencing with these approximate values, exact values can be obtained by considering stepwise the contributions of the other components. This calculation is simpler than that involving direct treatment of the whole system of equations. Nevertheless, all these are but palliatives. Because of the great effort needed for solving the system of equations for more than three components, and considering the fact that even small errors of measurement can cause very large errors in such calculations, it should be ascertained in every case whether the problem could not be solved in another way. This might involve further chemical treatment of the analytical material. Such efforts would not be wasted, as without them one advantage of infrared spectral analysis, i.e. its speed, is lost because of the time needed for the interpretation.

In addition to the optical density itself, Pirlot's compensated optical density can also be used for the analysis for many-component systems. As an example, the simultaneous determination of the three isomers in a xylene mixture will be discussed [Duyckaerts and Pirlot (181)]. Fig. 128 shows the spectra of the pure substances. The analytical bands selected are at $13 \cdot 48 \, \mu$

for o-xylene, at $13 \cdot 0\,\mu$ for m-xylene, and at $12 \cdot 59\,\mu$ for p-xylene. To obtain compensated optical densities, the following further wavelengths are used: for o-xylene $13 \cdot 46\,\mu$, for m-xylene $12 \cdot 85$ and $13 \cdot 16\,\mu$, for p-xylene $12 \cdot 44$ and $12 \cdot 74\,\mu$. Fig. 128 shows the contribution to the optical density given by each component at the analytical wavelengths. Suitable concentrations in CS_2 (expressed here in $g/100\,cm^3$) are measured to obtain them. To compare the individual curves, the assignment to the corresponding

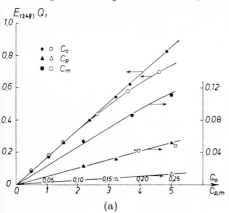

(a)

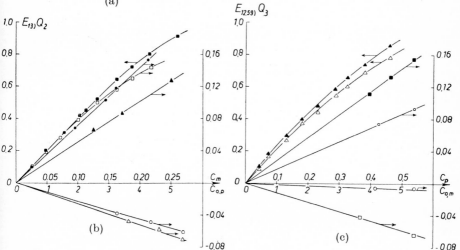

(b) (c)

FIG. 128. *Quantitative analysis of the xylenes, as an example of a many component system, by means of the optical density and the compensated optical density methods according to Pirlot [from Duyckaerts and Pirlot (181)].*

co-ordinate axis should be noted. The following expressions are used for the compensated optical densities:

for o-xylene: $Q_1 = E_{13,48} - E_{13,64}$,
for m-xylene: $Q_2 = E_{13,0} - E_{12,85} - 0 \cdot 5 . (E_{13,16} - E_{12,85})$,
for p-xylene: $Q_3 = E_{12,59} - E_{12,44} - 0 \cdot 5 . (E_{12,74} - E_{12,44})$.

X

They are given in Fig. 128 for the standard mixtures. The results for the analysis of several mixtures of unknown composition are given in Table 24 for both the methods. Both methods give almost the same values with much the same accuracy. Nevertheless, the smaller time needed for the method of

TABLE 24. *Infrared Spectroscopic Analysis of Xylene Mixtures (Concentrations in g/100 cm³ CS₂) [from Duyckaerts and Pirlot (181)]*

			Found					
Weighed out			Normal optical density			Compensated optical density		
o	m	p	o	m	p	o	m	p
1·666	0·027	0·027	1·664	0·028	0·028	1·668	0·026	0·027
1·666	0·165	0·027	1·669	0·159	0·030	1·675	0·159	0·028
1·666	0·027	0·161	1·662	0·032	0·160	1·668	0·028	0·158
1·666	0·164	0·161	1·664	0·165	0·162	1·675	0·157	0·159
0·018	2·429	0·039	0·024	2·413	0·049	0·017	2·431	0·038
0·015	2·429	0·039	0·114	2·411	0·048	0·105	2·430	0·038
0·018	2·429	0·173	0·026	2·412	0·182	0·017	2·432	0·168
0·105	2·429	0·173	0·113	2·416	0·178	0·100	2·417	0·166
0·017	0·028	1·289	0·019	0·026	1·289	0·017	0·029	1·288
0·017	0·165	2·577	0·017	0·158	2·584	0·017	0·168	2·574
0·104	0·028	2·577	0·107	0·024	2·578	0·098	0·032	2·579
0·104	0·165	2·577	0·108	0·156	2·581	0·098	0·165	2·582
0·208	0·329	0·322	0·200	0·318	0·315	0·200	0·320	0·314
0·156	0·493	0·241	0·153	0·480	0·240	0·150	0·498	0·238
0·156	0·246	0·483	0·164	0·244	0·466	0·157	0·240	0·468

compensated optical density, when the standardization has been carried out, is definitely advantageous for a long series of measurements.

5. Errors

After having treated the most important cases and methods of quantitative infrared spectroscopic analysis, it is relevant to discuss the *accuracy* attainable. Such considerations give information about the expected errors of infrared determinations and also information about the experimental *conditions* which lead to the smallest errors and which should therefore be kept to as nearly as possible. In general, this discussion will be limited to the optical density or the extinction coefficient. It will be left to the reader to estimate the effect of an error in one of these quantities on the result of a quantitative investigation, i.e. on the concentration found. It will be assumed that Beer's law is obeyed and the optical density E and the extinction coefficient ϵ will be defined by the equation

$$E = \epsilon.c.d = \ln J_0 - \ln J \qquad \text{[III, 3.33]}$$

According to the particular conditions and the object in view, the method and results of the error treatment vary. In view of the increasing importance of infrared spectroscopy in the last few years, it is not surprising that a whole series of very diverse investigations on this subject (sometimes with inconsistent results) can be found in the literature. Starting with the requirement that a given error in the measurement of transmittance should have a minimum influence on the concentration to be found, Barnes, Gore, Liddel and Williams (*A*) have indicated that an optical density of 1 or a transmittance of about 37 per cent is optimal. This requirement determines the optimum cell thickness to be used, which should be chosen accordingly. This simple result applies only to the case of a single component in a medium which is otherwise free of absorption. If several absorbing components are present, the optimum value of transmittance is lower. Nevertheless, the effects can be shown to be so small that it is almost immaterial which value within the range 25 to 50 per cent transmittance is taken. Cole (*129*) and Robinson (*600*) have come to similar conclusions.

Martin (*447*) has made a more general treatment of the problem. He distinguishes between the determination of an extinction coefficient of a substance for a definite wavelength and the real aim of quantitative analysis. The latter naturally includes several measurements, e.g. of a solution and of a pure solvent. The former is influenced principally by the errors of measurement in J and J_0, while the concentration c and the cell thickness d can be taken as almost free of errors. It is different in the second case, as measurements of J and J_0 are required for both the sample and the comparison substance (pure substance, solvent, etc.). Each of these measurements can, of course, be erroneous, but they can be assumed to have been made under comparable conditions. In addition, there is a series of further possible errors. For every optical density measurement, the following principal sources of error must be taken into account. (1) Erroneous determination of J and J_0 as a result of unavoidable variations in the emission of the radiation source and in the measuring system. (2) Errors resulting from stray light of other than the wavelength being investigated. (3) Errors resulting from inequalities in the cells used. (4) Errors resulting from recording the measuring system non-linearly.

It is clearly disadvantageous to carry out spectroscopic measurements for transmittances near 0 per cent or 100 per cent; in the first case, even very

307

small amounts of stray light cause a large error in the optical density. In the second case, a very small error of measurement in the transmittance of the analytical sample or the comparison substance has a very large effect on the optical density. To determine an optimum region of transmittance, it is advantageous to assume definite magnitudes (corresponding to those expected under experimentally) for the errors enumerated. The following will be taken: (1) The final instrumental reading can be measured to 1 per cent. (2) The stray light contribution is 1 per cent of the radiation energy transmitted by the empty cell or that filled with pure solvent. (3) The inequality in the cells is 1 per cent. (4) The non-linearity of the recording system is such that a reading of 50 per cent of the final amplitude is caused by a value of 51 per cent of the final amplitude. The errors enumerated as (2), (3), and (4) can be taken to be constant over long periods. The error (1) is naturally variable. It must still be differentiated whether this error is caused solely by the variation of the emission of the radiation source [case (a)] or from the general uncertainty in measurement as a result of the noise level [case (b)], which will be taken to be the uncertainty inherent in every reading. In case (a) the error in J_0 or J can accordingly always be taken to be $\pm 0 \cdot 01 . J_0$ or $\pm 0 \cdot 01 . J$. However, in case (b) the error in the measurement for J is $J \pm 0 \cdot 01 . J_0$. Cases (a) and (b) are identical if $J = J_0$. The former is important for low amplification, i.e. broad absorption bands and wide slits. The latter is important for high amplification, i.e. narrow bands and slits.

The individual calculation of the various influences of the errors will not be given here; it can be found in the work of Martin (*447*). The results are given in Table 25 for the range of transmittance 10 to 90 per cent. This table confirms again the well known rule that, when possible, working with

TABLE 25. *Percentage Error in the Optical Density for Individual Errors of 1 per cent Deviation [Martin (447)]*

Type of error	Percentage transmittance measured								
	10	20	30	40	50	60	70	80	90
Measurement of J_0 (case a)	0·43	0·62	0·83	1·09	1·44	1·96	2·80	4·48	9·49
Measurement of J (case a)	0·43	0·62	0·83	1·09	1·44	1·96	2·80	4·48	9·49
Measurement of J_0 (case b)	0·43	0·62	0·83	1·09	1·44	1·96	2·80	4·48	9·49
Measurement of J (case b)	4·34	3·11	2·77	2·73	2·89	3·26	4·01	5·60	10·55
Stray light	4·14	2·56	1·98	1·67	1·47	1·33	1·22	1·13	1·07
Cell inequality	0·43	0·62	0·83	1·09	1·44	1·96	2·80	4·48	9·49
Non-linearity	1·56	1·99	2·33	2·62	2·88	3·14	3·36	3·58	3·80
Maximum total error (case a)	6·99	6·41	6·80	7·56	8·67	10·4	13·0	18·2	33·3
Maximum total error (case b)	10·9	8·90	8·74	9·20	10·1	11·7	14·2	19·3	34·4

transmittances below 20 per cent and above 60 per cent should be avoided. If this is done, there is no individual error above $3 \cdot 3$ per cent of the optical density obtained, while the possible maximum error (the sum of the individual errors) does not rise above 12 per cent. Generally, the individual errors will, at least in part, mutually compensate each other. To determine the total error expected from the laws of probability, the two cases previously mentioned, i.e. determination of an extinction coefficient or of a concentration, must be distinguished. In the first case (neglecting the error due to non-linearity), it must be distinguished whether the errors due to

TABLE 26. *Percentage Error in Optical Density for Various Interactions of Individual Errors according to Martin (447)*

	Measured transmittance (%)								
	10	20	30	40	50	60	70	80	90
I. Probable error in the measurement of J_0 and J (case *a*)	0·61	0·88	1·17	1·54	2·04	2·77	3·96	6·34	13·4
II. Probable error in the measurement of J_0 and J (case *b*)	4·36	3·17	2·89	2·94	3·23	3·80	4·89	7·17	14·2
III. Sum of the errors as a result of stray light and cell inequality	4·57	3·18	2·81	2·76	2·91	3·29	4·02	5·61	10·6
IV. Difference of the errors as a result of stray light and cell inequality	3·71	1·94	1·15	0·58	0·03	0·63	1·58	3·35	8·42
I.+III.	5·18	4·06	3·98	4·30	4·95	6·06	7·98	12·0	24·0
I.+IV.	4·32	2·82	2·32	2·12	2·07	3·40	5·54	9·69	21·8
II.+III.	8·93	6·35	5·70	5·70	6·14	7·09	8·91	12·8	24·8
II.+IV.	8·07	5·11	4·04	3·52	3·26	4·43	6·47	10·5	22·66

stray light and to the inequality between the cells reinforce or oppose each other. The latter, which is the advantageous case, occurs if the unequal cells are so used that the more transparent serves as the empty or comparison cell. Taking the probable error in the measurement of J and J_0, Table 26 is obtained. According to the size of the individual errors, somewhat different final values are obtained. However, in the region of 20 to 60 per cent transmittance they are hardly more than 6 per cent. In every case the probable total error shows a quite broad minimum in the region of 30 to 50 per cent transmittance. A considerable compensation of the errors denoted above by (2), (3), and (4) is obtained in measurements on unknown analytical and known comparison samples for the determination of concentrations, if it ensured that both measurements are carried out at equal transmittance. Table 27 also shows Martin's results for this. The usual procedure for obtaining a standardization curve (optical density as a function of the concentration) with optimum fitting of the curve to the points measured, reduces the

TABLE 27. *Percentage Errors in the Spectroscopic Determination of Concentrations using Equi-transparent Comparison Samples according to Martin (447). Compensation of the Errors resulting from Stray Light and Cell Inequality is assumed*

	Measured transmittance (%)								
	10	20	30	40	50	60	70	80	90
Probable error in case (a)	0·87	1·24	1·66	2·18	2·88	3·92	5·60	8·96	19·0
Probable error in case (b)	6·17	4·48	4·09	4·16	4·57	5·37	6·91	10·1	20·1

chance errors to below the values given in the table. In case (a), the use of a small transmittance is to be recommended. Yet in practice instruments belong neither to case (a) nor (b); on the contrary they lie between these extremes. Thus the transmittance region about 30 per cent is also suitable for measurements with the aim of the determination of concentration.

It should be noted that the results mentioned also apply to the modern automatically compensated instruments which record the percentage transmittance. For these, an error in the 100 per cent or in the 0 per cent absorption lines must be considered in addition to the errors of the measurement of transmittance itself. Relevant details can be found in the work of Robinson (*600*) already cited. A summary of the attainable accuracy for routine analysis has been given by Childers and Struthers (*115*) and Schnurmann and Kendrick (*635*); the latter consider especially difference techniques.

4. Reflection, Polarization and Microspectroscopic Measurements

1. Possibilities and Limits of Reflection Spectroscopy

Formerly, infrared spectroscopic reflection measurements had a much greater importance for the investigation of the internal structure of substances and for characterization purposes than they have today. For certain cases, and at a certain period (about the turn of the century), they even surpassed the absorption measurements which today are used almost exclusively. Many results of infrared reflection measurements have long since become a part of general physics, e.g. the concept connected with the term 'residual rays'. Many important advances in the knowledge of the lattice structures of simple, and also complex, crystals were due to reflection measurements. This was done long before the technique of absorption measurements had been developed so that they could afford equal or more refined information. Yet this also indicates the reason for the decrease in importance of reflection spectroscopy to almost nil today. In principle, reflection gives characteristic information for substances. Nevertheless, although the procedure is often technically simple, the method of interpretation of the reflection spectra is extraordinarily complicated, time consuming, and in its precision is not at all comparable with that of absorption spectra. Only in those cases where insurmountable difficulties are encountered in the preparation of the samples needed for absorption measurements are reflection measurements available as a makeshift. This is the case where especially strong absorption necessitates the preparation of extremely thin layers; a requirement which today can nevertheless almost always be fulfilled. Obviously, it is only certain solids, in crystalline and amorphous forms, and possibly a few liquids, of especially strong absorption in the fundamental region, which come into consideration. However, if the reflection method is used, it yields, in principle, the refractive indices as well as the absorption constants of the substance investigated. Thus together with the spectrum, the dispersion curve is also obtained, an advantage which sometimes makes the difficult interpretation worth while.

311

The Methods of Practical Infrared Spectroscopy

The results of that earlier golden age of the reflection method can be found in the older works of infrared spectroscopy. However, in the more recent works, less place is allocated to these results on account of the preponderance of the absorption method. Therefore for a detailed treatment the older works must be consulted. Reflection measurements can serve the same purpose as absorption measurements, namely the clarification of molecular structure and the characterization of substances. In addition, there are technical measurements which are more in the nature of documentation and which are comparable with the colour measurements in the visible spectral regions. They will not be further discussed here, because this technique is of little importance. For the purposes of this book they are only worth mentioning because of the way in which they are carried out (see page 196 ff.).

2. The Fundamental Equations of Reflection Spectroscopy

The *reflective power* takes the place in reflection spectroscopy of the spectral transmittance in absorption spectroscopy. It is defined as the ratio of the radiation energy reflected to that incident at any wavelength. It is usually given as a percentage just as is the absorption or transmittance. As well as on the wavelength, it is also dependent on the angle of incidence and on the state of polarization of the radiation; the latter is comparable to the dependence of the transmittance for dichroic crystals. The direction of vibration of the radiation is taken with respect to the plane of incidence at the object investigated. The component of the radiation vibrating parallel to this plane is denoted by the index π, and that vibrating perpendicularly by the index σ. The reflective power for each component is given as a function of the angle of incidence ϕ and refracting angle χ by the Fresnel formulae

$$r_\sigma = \frac{\sin^2(\phi-\chi)}{\sin^2(\phi+\chi)}, \qquad r_\pi = \frac{\tan^2(\phi-\chi)}{\tan^2(\phi+\chi)} \qquad \text{[III, 4.1]}$$

The angles which occur therein are connected by the Snell's law of refraction

$$\sin\phi = n'\sin\chi \qquad \text{[III, 4.2]}$$

where n' is the refractive index of the substance investigated at the wavelength in question. If non-polarized, so-called natural radiation is used, then the power of reflection is

$$r = \tfrac{1}{2}(r_\sigma + r_\pi) \qquad \text{[III, 4.3]}$$

For absorbing media, the refractive index is expressed as the complex quantity

$$n' = n(1-i\kappa) \qquad \text{[III, 4.4]}$$

312

Here n is the real part of the refractive index, κ is the absorption index, and $k = n\kappa$ is the absorption coefficient. The latter quantity is connected with the extinction coefficient of the Lambert absorption law by the equation [III, 3.6]. From the complex refractive index, a complex angle of refraction follows from [III, 4.2], which is given by the expression

$$n' \cos \chi = (n'^2 - \sin^2 \phi)^{1/2} = a - ib \qquad [\text{III, 4.5}]$$

Now the components of the reflective power can be written as functions of a and b. The complex calculations will be omitted and the results given directly

$$r_\sigma = \frac{a^2 + b^2 - 2a \cos \phi + \cos^2 \phi}{a^2 + b^2 + 2a \cos \phi + \cos^2 \phi} \qquad [\text{III, 4.6}]$$

$$r_\pi = r_\sigma \frac{a^2 + b^2 - 2a \sin \phi \tan \phi + \sin^2 \phi \tan^2 \phi}{a^2 + b^2 + 2a \sin \phi \tan \phi + \sin^2 \phi \tan^2 \phi} \qquad [\text{III, 4.7}]$$

$$r = r_\sigma \frac{a^2 + b^2 + \sin^2 \phi \tan^2 \phi}{a^2 + b^2 + 2a \sin \phi \tan \phi + \sin^2 \phi \tan^2 \phi} \qquad [\text{III, 4.8}]$$

In addition to the functions of the angles, only a and the combination $a^2 + b^2$ occur and these can themselves be given as functions of n, κ, and ϕ from [III, 4.5] and [III, 4.4] as:

$$a^2 + b^2 = \tfrac{1}{2}[n^2(1 - \kappa^2) - \sin^2 \phi] + \tfrac{1}{2}[\{n^2(1 - \kappa^2) - \sin^2 \phi\}^2 + 4n^4 \kappa^2]^{1/2}$$
$$+ \frac{2n^4 \kappa^2}{n^2(1 - \kappa^2) - \sin^2 \phi + [\{n^2(1 - \kappa^2) - \sin^2 \phi\}^2 + 4n^4 \kappa^2]^{1/2}}$$
$$[\text{III, 4.9}]$$

$$2a = [\tfrac{1}{2}\{n^2(1 - \kappa^2) - \sin^2 \phi\} + \tfrac{1}{2} < \{n^2(1 - \kappa^2) - \sin^2 \phi\}^2 + 4n^4 \kappa^2 > ^{1/2}]^{1/2}$$
$$[\text{III, 4.10}]$$

Using these complicated expressions, r_σ and r_π also become functions of n, κ, and ϕ. However, these very complicated formulae will not be given here. Inversely, the characteristic quantities n and κ of the substance can be calculated for known r_σ and r_π values and the given angle of incidence ϕ. A reflection measurement under known geometrical conditions (measurements at two different angles of incidence are necessary) is thus capable of giving the absorption properties of the compound investigated in addition to its dispersion. Yet, a glance at the complicated formula shows that the direct calculation in every individual case is a very wearisome and laborious process, if not completely hopeless. It appears advantageous to calculate once and for all the three quantities r_σ, r_π, and r, which characterize the reflection of a substance, as functions of n and κ for certain values of the angle of incidence suited to practical conditions. These values could then be placed in tables with provision for sufficient possibilities of interpolation. The effort needed for this, both in time and in instrumentation (calculating machines are the least requirement), is only worth while if the reflection method is desired to be used, or must be used, often. Because the tables are also inconvenient, it is recommended that they should be recorded graphically and on this basis the practical interpretation of the results can be carried out. Such a representation has been published by Šimon (*661*), where further instructions for the experimental reflection analysis can be found, with examples. Oswald and Schade (*516*) and Neuroth (*496*) have discussed the accuracy attainable.

For the methods discussed, two measurements are always needed at angles not too close to one another, e.g. 20° and 70°. This doubles the time of measurements compared with the absorption technique. The reason for this is the simultaneous appearance of n and κ in the reflective power. n and κ can only be determined together and coupled with each other from reflection measurements using these procedures. Robinson (*602*) has recently pointed out that measurements using only one angle of incidence suffice if this is 0° and a frequency range of sufficient width is measured. As angles of 0° are not easily obtained technically, one near 0° is used, with a correction for the angle of refraction which occurs at small angles of incidence. The Fresnel formulae can then be written

$$r^\circ = \frac{(n'-1)^2}{(n'+1)^2}$$
[III, 4.11]

with n' from [III, 4.4]. Robinson's method is based on considerations and calculations similar to those used in the theory of electrical conductors, especially the knowledge that damping and phase displacement through an electrical conductor are not independent of each other. Thus the complex refractive index n' or n and κ are both calculable from the measured r°. Robinson used the graphical methods of conductor theory, especially the well known Smith conductor diagram, for the numerical interpretation of the measurements, which is no less complicated than that of the earlier methods. The power of this method has been clearly shown by investigations on polytetrafluoroethylene and urea crystals [Robinson and Price (*603*)].

3. Polarization Measurements

The need for, and the use of, experiments with polarized radiation has been referred to several times already, and the instruments for this discussed.

The nature of such investigations is as follows. As emphasized earlier, every absorption band results from an alteration in the dipole moment of the molecule during the corresponding vibration. For substances with the molecules in non-ordered positions, i.e. gases, liquids, and amorphous solids, the direction of this alteration is of no importance with respect to the excitation of the vibration. For unpolarized radiation, the vibrational direction alters continuously about the direction of propagation (but remains, of course, perpendicular to this), so that in a very short period all the possible directions of a dipole moment alteration will be passed through. Accordingly, the excitation of all the vibrations possible is equally probable. This applies also to the application of polarized radiation to the investigation of disordered substances. This is because, for constant direction of vibration of this radiation, the same probability of excitation is ensured by the statistical distribution of the positions of the molecules. The situation is different for substances with a regular orientation of the molecules, i.e. for crystals. Now probability of excitation, i.e. observed intensity of absorption of a band, depends on the angle between the direction of the alteration of the dipole moment and the direction of the vibration of the polarized radiation. If they lie at right angles to one another, then no excitation takes place; if they lie parallel to one another maximum excitation takes place; between these two cases there is a gradual transition.

Experimentally, investigations with polarized radiation are usually absorption measurements of the usual sort. Naturally reflection measurements are also possible and the discussion of the preceding sections applies to these. Two spectra of the substance investigated are obtained, one for each of the two directions of vibration of the polarized radiation with reference to a preferred or reference direction. This is defined according to the special circumstances of the investigation, e.g. a definite direction of elongation of the material, a crystal axis, or the direction of the slits. The two spectra are differentiated by a more or less variable intensity of the same bands. This phenomenon is called *dichroism*. More precisely, σ-dichroism is said to occur if the absorption is stronger for the component vibrating perpendicular to the direction of propagation, and π-dichroism if it is stronger for the parallel-vibrating component of the polarized radiation. The *dichroic ratio* serves as a quantitative description of dichroism. It is defined as the ratio of the optical density of the σ to the π spectrum:

$$\vartheta = \frac{E_\pi}{E_\sigma} = \frac{\ln \dfrac{1}{D_\pi}}{\ln \dfrac{1}{D_\sigma}} \qquad [\text{III, 4.12}]$$

Thus, σ-dichroism occurs for $\vartheta < 1$ and π-dichroism for $\vartheta > 1$ [Ambrose *et al.* (*16*); Glatt and Ellis (*256*)].

The importance of the dichroic ratio lies in its connection with the internal order of the sample investigated. If this is a crystal, the order can be assumed to be ideal. There is an unambiguous relationship between the dichroic ratio and the position of the crystallographic axes, the orientation during the investigation, and the nature of the molecular vibrations in question. The rather complicated formulae will not be given here. They can be found in the work of Ambrose *et al.* (*17*) for various crystal systems. Vice versa, conclusions as to the type of vibration can be drawn from the dichroic behaviour if the crystal arrangement is known.

The dichroic ratio gives information even in cases where instead of an ideally ordered crystal only a more or less good approximation to this occurs. Such examples are plastics which were originally crystalline or have been crystallized by some external influence. It will be assumed that the molecule in question possesses an axis to which the vibration, i.e. the dipole moment alteration, can be referred. This applies, for example, to the rubber molecule. The case of a rubber sample oriented by sufficient extension will be taken. This means that the molecules in the sample lie more or less parallel to the direction of stretching. In reality, the order obtained will certainly not be ideal; the degree of order attained is characterized by the mean angle which the molecular axes make with the direction of extension. The smaller the angle is, the better is the order enforced from without. The dichroic ratio ϑ is connected with the angle characterizing the degree of orientation ϵ by the expression

$$\vartheta = 2 \cdot \cot^2 \epsilon \qquad \text{[III, 4.13]}$$

if the vibration investigated is connected with an alteration of the dipole moment parallel to the molecular axes. On the other hand, if the alteration is perpendicular, the formula is

$$\vartheta = \frac{2 \sin^2 \epsilon}{2 - \sin^2 \epsilon} \qquad \text{[III, 4.14]}$$

By determining the dichroic ratio for vibrations of known symmetry, the mean orientation of the molecules with respect to the desired direction of extension can be simply determined [Ambrose *et al.* (*16*)].

In the preceding discussion, ϑ naturally signifies the true dichroic ratio given solely by the properties of the material investigated. The transmittance measurements for different azimuths of the polarizer can give considerably deviated values as a result of imperfections in the apparatus. The main source of errors is the so-called apparatus polarization which is unavoidable. It is caused essentially by the reflections at an angle at the surfaces of the prism and by the incomplete polarization caused by the usual polarizers. For the experimental determination of the apparatus polarization a measurement is made (without absorbing substance) first with radiation vibrating parallel to the slit and then with radiation of the same intensity vibrating perpendicular to the slit, i.e. plane of incidence of the polarizer respectively parallel and perpendicular to the slit. The energies emitted from the apparatus are denoted by J_π and J_σ. The ratio

$$K = \frac{J_\pi}{J_\sigma} \qquad\qquad [\text{III, 4.15}]$$

or the corresponding degree of polarization, often given as a percentage

$$P_\% = 2\frac{J_\pi - J_\sigma}{J_\pi + J_\sigma} = 2\frac{K+1}{K-1} \qquad\qquad [\text{III, 4.16}]$$

serves as a measure of the apparatus polarization, sometimes also called depolarization. The influence of the apparatus polarization can accordingly be described as the alteration of the σ-component of the radiation by a factor K compared with its original equality with the π-component. This can be written

$$J_\sigma = \frac{J_\pi}{K} \qquad\qquad [\text{III, 4.17}]$$

For two reflections at prism surfaces, depolarization between $1 \cdot 2$ and $1 \cdot 6$, or a degree of polarization of 18 to 46 per cent, are usually found. These values naturally increase with the number of reflections at prisms or similar surfaces and thus especially in multiple monochromators.

Elliot *et al.* (*195*) have recommended measurements with radiation which vibrates at 45 per cent to the base of the prism as a means of eliminating apparatus polarization in dichroic measurements. However, Charney (*114*) has pointed out that the influences of incomplete action of the polarizer still remain. The measured intensities yield the dichroic ratio from the equations

$$\vartheta^{\text{pol}} = \frac{\ln[(t_x + t_y t_m)/(t_x D_\pi + t_y t_m D_\sigma)]}{\ln[(t_y + t_x t_m)/(t_y D_\sigma + t_x t_m D_\pi)]} \qquad\qquad [\text{III, 4.18}]$$

or

$$\vartheta^{\text{sample}} = \frac{\ln[(t_y + t_y t_m)/(t_y D_\pi + t_x t_m D_\sigma)]}{\ln[(t_y + t_x t_m)/(t_y D_\sigma + t_x t_m D_\pi)]} \qquad\qquad [\text{III, 4.19}]$$

according to whether the polarizer azimuth or that of the sample is altered. In these equations, x, y, z, define a fixed system of co-ordinates, the z-axis of which is identical with the direction of the radiation and the x- and y-axes respectively are horizontal and vertical. Further, π and σ are two mutually perpendicular axes defining the orientation of the sample; t_x and t_y are the coefficients of transmittance of the prism for radiation vibrating parallel to the x- and y-axes respectively; and t_m is the coefficient of transmittance of the polarizer for the unwanted component (that which vibrated perpendicularly to the plane of incidence of the polarizer, see page 199). Indeed, both equations reduce to the expression [III, 4.12] for an ideal polarizer ($t_m = 0$) and no prism polarization ($t_x = t_y$). The first condition can be fulfilled in principle by the use of a suitable reflection polarizer (with a sufficient number of plates), but the second condition cannot be fulfilled. If the procedure recommended by Elliot *et al.* (*195*) is used, then the dichroic ratio is

$$\vartheta^{45°} = \frac{\ln[(1+t_m)/(D_\pi+t_m D_\sigma)]}{\ln[(1+t_m)/(D_\sigma+t_m D_\pi)]} \qquad [\text{III, 4.20}]$$

Here also $\vartheta^{45°} \rightarrow \vartheta$, when $t_m \rightarrow 0$. However, with the usual six plates of the AgCl and also the Se polarizer, t_m still differs notably from zero and its value decreases but slowly with increasing number of plates as can be seen from Table 11 (page 200). In order to obtain the true value of the dichroic ratio from such measurements, which is needed to draw structural conclusions, a series of important precautions must be taken.

In addition to the purposes already described, investigations with polarized radiation are valuable for assigning the vibrational bands of a crystalline substance to the individual vibrational species [Pimentel and McClellan (*536*); Mecke (*462*)]. Vibrations belonging to the same species are characterized by the same dichroic behaviour. Therefore if the optical densities of the bands in question are determined for various polarizer azimuths, then the vibrations belonging to each species show the same dependence on the azimuth. Because several species could by coincidence show the same dependence, it is advantageous to alter the orientation of the sample in the optical path. This can be done by inclining it, for which a goniometer table is advantageous; the course of variation of the optical density is then determined again. In general, bands belonging to different species, which previously were not differentiated, can now be distinguished. An example of this is shown in Fig. 129. It can be seen that in this example (naphthalene crystal) the bands which belong to the species B_{1u} and B_{3u} cannot be distinguished by the dichroic measurements at various azimuths alone, but can be by altering the angle of incidence.

Finally it is noted that measurements of the radiation reflected from a substance can serve to determine its optical constants, i.e. refractive index and absorption coefficient. This method is advantageous and frequently the only one applicable in cases of extremely strongly absorbing media, e.g. metals. The reflected radiation is elliptically polarized; this complicates the interpretation and for this reference must be made to the literature [Conn and Eaton (*139, 140, 141*)].

A further example of the application of measurements with polarized radiation is the determination of the power of optical rotation, especially of organic compounds in the infrared [Heidiger and Günthard (*315*)].

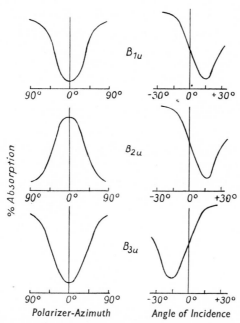

FIG. 129. *The dichroic behaviour of different species of vibration using naphthalene as an example [from Pimentel and McClellan (536)].*

4. Microspectroscopic Measurements

The use of a microscope (micro-illuminator) for infrared spectroscopic investigations necessitates the consideration of many points which are not found in usual spectroscopy. They are naturally related closely to the small dimensions of the sample to be investigated and the setting of the instrument suited to such problems.

Firstly, the strong convergence of the beam of radiation must be considered, because otherwise only weakly convergent radiation is normally used. The geometric relationships are shown in Fig. 130. It is assumed that the beam of radiation of the microscope, of angle of aperture α, is focused on

319

the upper surface of a sample, limited by parallel planes, of thickness s, refractive index n, and surrounded by air. The angle of convergence of the beam is reduced to a value β in the substance which can be calculated by the refraction law. The thickness passed through by the edge ray is $s/\cos\beta$; the mean thickness is therefore smaller than this value and larger than s. Accordingly, a larger absorption is measured than by using parallel radiation at perpendicular incidence. For the calculation of the extinction coefficient, a certain portion of the measured absorption must be subtracted. The size of this correction depends on the amount of the absorption and the

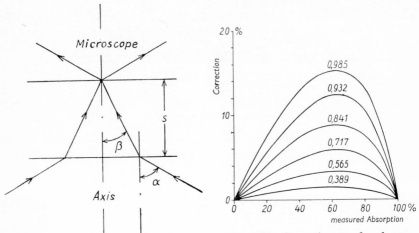

FIG. 130. *Transit of convergent radiation through a thick sample.*

FIG. 131. *Corrections to the absorption for convergent radiation according to Blout et al. (72).*

index of refraction and the convergence of the radiation. Blout *et al.* (72) have investigated the relationships in detail. Their results are shown in Fig. 131; the numbers written on the curves indicate the ratio of the numerical aperture of the microscope to the refractive index of the substance investigated. It can be seen that the errors can easily reach 10 per cent. For samples of variable thickness, such as fibres, the corresponding treatment is still more complicated.

The consideration of the attenuator relationships is extremely complicated. The unavoidable energy losses require the use of a high degree of amplification in the electronic apparatus. In order to obtain a useful signal-noise ratio, the monochromator slits must be opened wide. As a result, the slit or its image at the position of the object can then be larger than the object. In this case, so-called 'spectral dilution' occurs, i.e. a part of the radiation falling on the detector has not passed through the sample at all. The result of this is a displacement of the null point, similarly to that which occurs for the presence of stray light; but now, even for 100 per cent absorption of the sample, stray radiation of the selected wavelength still acts on the detector. It is possible for this effect to occur during the recording of the spectrum only when the slit is opened wider for the longer wavelengths. To avoid spectral dilution, a variable diaphragm is provided in the microscope which is closed to such an extent that its opening is somewhat less than the dimensions of the sample. Obviously the action of this stop must be related to the monochromator slits, the resolution, and the signal-noise ratio. For automatically adjusted slits, the results for different bands of a

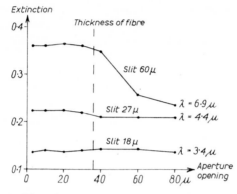

FIG. 132. *The microspectroscopy of samples of limited size.*

spectrum can vary according to the relationships of the three quantities: width of sample, slit width, and opening of the diaphragm. Fig. 132 shows, as an example, three bands of a polyacrylonitrile fibre 36 μ thick.† The measured maximum optical density of the bands at $3 \cdot 4 \mu$ (CH stretching vibration), $4 \cdot 4 \mu$ (CN stretching vibration) and $6 \cdot 9 \mu$ (CH scissor vibration) are plotted against the diameter of the stop. The slit width (projected in the plane of the sample) is given for each band. It can be seen from the figure that no spectral dilution occurs for the two bands of shorter wavelengths because the slit width is smaller than the breadth of the sample and therefore the breadth of the stop is without influence (except for extremely small values). The situation is different for the $6 \cdot 9 \mu$ band. The slit width of 60 μ is here considerably greater than the breadth of the sample. As long as the diameter of the stop is less than that of the sample no difficulties occur.

† These measurements were kindly supplied by Dr C. R. Bohn of E.I. du Pont de Nemours and Co., Wilmington, Delaware, U.S.A.

Y

Yet as soon as the stop becomes broader than the sample, spectral dilution occurs to an increasing degree, resulting in a decrease in the optical density observed.

The resolution is also influenced by the size of the sample. As long as samples of practically unlimited width are used as in normal infrared spectroscopy, the resolution is most easily increased by transition through a resolving substance of high dispersion. This no longer generally applies to the samples of limited size of microspectroscopy. For higher dispersion, an increase in slit width is always required to retain the signal-noise ratio and this can lead to spectral dilution if the slit width becomes greater than the size of the sample, so that it becomes necessary to reduce the opening of the stop. Thus the optimum resolution is dependent on the dimensions of the sample. Instead of an increase in dispersion for an increase in resolution, sometimes it is necessary to increase the intensity of the source emission in microspectroscopy, because this can be accompanied by a reduction in the slit width.

PART IV

Results and Applications;
Bibliography

1. The Infrared Spectra of some Important Classes of Substances

This chapter will be concerned with the spectra of certain important *classes* of compounds. It will deal with characteristics common to whole groups and not with peculiarities of individual molecules. Differences between the spectra for the various aggregation states will be neglected. The following discussion is a sequel and extension of the previous treatment of the characteristic frequencies.

1. Paraffins and Cycloparaffins

Hardly any other class of compounds has been the subject of so many infrared spectroscopic investigations as the saturated hydrocarbons. The most extensive collection of their spectra is the catalogue of the American Petroleum Institute Research Project 44, published by the Carnegie Institute of Technology, Pittsburgh, in collaboration with the National Bureau of Standards, Washington. The subject has been reviewed by McMurry and Thornton (*460*), and by Sheppard and Simpson (*649, 650*).

As a consequence of the simple chemical structure of the open chain paraffins, only the following vibrations are to be expected: (1) Stretching vibrations of the CH linkage. The various possibilities, i.e. $-\overset{\backslash}{C}H$, $\overset{\backslash}{C}H_2$, and $-CH_3$ groups can be distinguished and give information about the degree of branching of the chain. (2) Bending vibrations of the HCH and HCC angles; again the possibility exists of differentiation between the various groups. (3) CC Stretching vibrations. (4) CCC Bending vibrations. (5) Torsional vibrations about the CC bond.

Vibrations of the type last mentioned cause absorption below about 200 cm^{-1} and are thus usually not observed in the infrared spectrum. All the other vibrational types are represented once or more in the spectrum. The stretching vibrations of the CH bond lead to bands in the spectral region near 3000 cm^{-1} or about $3 \cdot 3$ μ. The resolution attained by NaCl

Results and Applications

prisms of the ordinary size usually gives a broad absorption band with little or no structure. Using higher resolution (LiF or CaF_2 prisms or a diffraction grating), the bands corresponding to the possibilities enumerated above are obtained. This can be then used for structural investigation and for analysis, but overlapping still occurs and must always be taken into account. Fig. 133 gives the positions for such resolved CH stretching vibrations (Fox and Martin (*218*); Barchewitz and Chabbal (*35*); Saier and Coggeshall (*621*); Hastings *et al.* (*308*)]. Selective deuteration can be used in this case, as in other cases involving hydrogen, as a help to unambiguous assignments of the bands to definite vibrations or atomic groups. The substitution of hydrogen by deuterium, which is twice as heavy, displaces the stretching vibration to the 2000 cm^{-1} region. The deformation vibrations of the CH_2

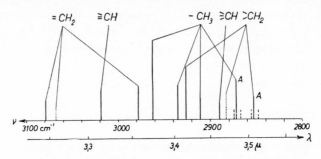

FIG. 133. *The position of paraffinic and olefinic CH stretching vibration bands according to Fox and Martin (218). The bands marked A are sometimes split; those dotted do not always occur.*

and CH_3 groups occur at about 1460 and 1380 cm^{-1} or 6·8 and 7·2 μ. Here also, the mutual overlap of the different vibrations is considerable and a large degree of resolution is necessary for their unambiguous isolation. The symmetrical vibration generally corresponds to the smaller wave number.

The types of vibration so far discussed cause absorption at spectral positions relatively little influenced by the size of the molecule. This does not apply to the CC stretching and bending vibrations, or to the remaining HCH and HCC bending vibrations. The more CC bonds that are present, the larger is the mutual interaction of the members of the chain. This is important for the practical application of infrared spectroscopy to long-chain paraffins and many important investigations have recently been devoted to its theoretical clarification. Whitcomb *et al.* (*734*) first noted that two types of vibrations are to be expected for long chains; these were denoted as *end vibrations* and *chain vibrations*. This concept is shown schematically in Fig. 134. It was later extended and refined by other workers. In spite of all the

326

efforts, only rather general non-specific information is available about the spectral positions of the corresponding bands. The limits of the region of the

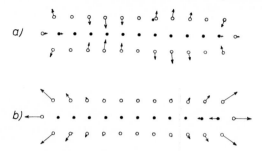

FIG. 134. *Schematic diagram of chain vibrations* (a) *and end vibrations* (b) *of long-chain molecules [from Whitcomb et al. (734)].*

chain-stretching vibrations are at about 800 and 1100 cm^{-1} for unbranched paraffins and at somewhat lower values for branched paraffins. For mole-

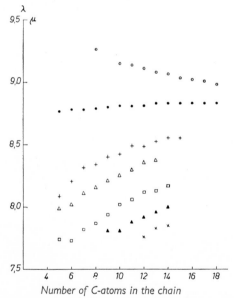

FIG. 135. *Influence of the chain length on the spectral position of some bands of paraffins from McMurry and Thornton (460).*

cules with more than *ca.* 10 to 14 carbon atoms in the chain, the spectra become almost identical and specific conclusions can no longer be drawn from them (Fig. 135). Finally, the bending vibrations of the carbon chain

Results and Applications

all lie below 500 cm^{-1} and are thus reached in the infrared spectrum only by the use of a KBr or CsBr prism; the most suitable technique for their investigation is Raman spectroscopy.

An absorption band occurs at about 730 cm^{-1} for open chain paraffins and derivatives with more than three CH$_2$ groups in the chain. This band has now been recognized as a CH$_2$ rocking vibration, on theoretical grounds, and from experimental results using deuterated paraffins (Sheppard and Sutherland [648]). The remaining wagging and rocking vibrations of the CH$_2$ and CH$_3$ groups are still not completely clarified [Brown et al. (95, 96)]. However, it has been possible to determine the influence of the chain length and other effects on some of the vibrations by the systematic investigation of certain bands in the spectra of homologous series. Fig. 135 gives examples

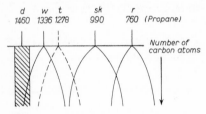

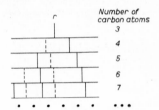

FIG. 136. Displacements of the positions of some of the vibrations of paraffins with the length of the chain [from Brown et al. (96)].

FIG. 137. Splitting of the rocking vibration of paraffins [from Brown et al. (96)].

of the displacement of the frequency for some of these bands with increasing number of carbon atoms in the chain [McMurry and Thornton (460)]. The various types of vibrations of the CH$_2$ groups, as given in Fig. 99, are of especial interest, because these become the dominating structural element of the normal paraffins at long-chain lengths. For example, the wave numbers of the various bending vibrations in n-propane are as follows [Brown et al. (96)]: deformation 1460, wagging 1336, twisting (active in the Raman spectrum only) 1278, rocking 760 cm^{-1}; the CC stretching vibration occurs at 990 cm^{-1}. With increasing chain length, the CH$_2$ deformation band retains the same value but all the others split and gradually extend the region in which they occur at both ends, as shown schematically in Fig. 136. Obviously, the ranges mutually overlap increasingly and assignment becomes increasingly difficult with increase of chain length and the resulting complication of the spectra by the splitting. Assignments are then only possible with the help of special techniques such as a low-temperature spectroscopy. The splitting of the individual types of vibration can be investigated theoretically as indicated schematically in Fig. 137 for the

328

rocking mode. Here, a new vibrational form occurs for every new CH_2 group; these vibrations are alternately infrared inactive (even number of carbon atoms) and active (odd number of carbon atoms) as a result of the symmetry properties of the normal paraffins. The spectra thus predicted are in good agreement with experiments.

With sufficient resolution, the CH stretching vibration region can give information about the important question of whether paraffins are *branched* [Simpson and Sutherland (*664*)]. A NaCl prism provides sufficient resolution for this to be done using the spectral region between 1100 and 1400 cm^{-1} or about $7 \cdot 2$–9μ. The symmetrical deformation vibration of the CH_3 group is used principally; it is split into several components, the ratios of the intensities of which are sometimes characteristic. For example, the isopropyl group is characterized by a triplet at 1340, 1365, and 1380 cm^{-1} with intensity increasing in that order, and by two further bands at 1145 and 1170 cm^{-1}. For tertiary butyl groups, a doublet with components of almost the equal intensity is found at 1360 and 1385 cm^{-1}, together with two further bands at 1205 and 1260 cm^{-1}. Finally, for the group

$$CH_3 \diagdown \diagup$$
$$C$$
$$CH_3 \diagup \diagdown$$

the doublet lies at the same position but the shorter wave component is the weaker and only one further band occurs, at 1195 cm^{-1}.

A whole series of investigations has recently been devoted to the *cycloparaffins* (*naphthenes*) [Bartleson *et al.* (*46*); Condon and Smith (*138*); Plyler *et al.* (*553*); Plyler and Acquista (*547*); Derfer *et al.* (*168*); Josien and Fuson (*365*); Miller and Inskeep (*476*); Marrison (*444*); Roberts and Chambers (*598*); Francis (*223*, *224*); Wiberley and Bunce (*741*); Voetter and Tschamler (*723*); Larnaudie (*405*); Haszeldine (*309*); Slabey (*669*); Roberts and Simmons (*599*); Baker and Lord (*31*)]. The relationships are not very favourable for analytical interpretation. The spectra of cyclic and open-chain paraffins are very similar and this makes the differentiation of alkyl substituted cycloparaffins from ordinary paraffins especially difficult. Characteristic bands at 810, 890, 1010, 3025, and 3095 cm^{-1} have been given for the cyclopropyl group and at 910–930, 1220, and 2980 cm^{-1} for the cyclobutyl group. Circumstances are especially unfavourable for cyclopentane because its key bands are almost all overlapped by those of other hydrocarbons; a band at 900 cm^{-1} can be mentioned, and the CH stretching vibration at 2955 cm^{-1}. For the cyclohexyl group, the following bands have been described: one between 860 and 890 cm^{-1}, a very useful one limited to hydrocarbons at 1260 cm^{-1}, and a CH stretching vibration at 2925 cm^{-1}. Usually, rather high resolution is required for analytical purposes. In addition, the content of other hydrocarbons must frequently be taken into consideration [Francis (*224*)].

Near infrared spectroscopy, i.e. the investigation of the region of the overtones of the CH stretching vibrations, has recently gained importance, especially for the paraffins. The characteristic absorptions of the various groups lie much further apart for the overtones than for the fundamental vibrations, because of differing anharmonicities. This automatically clarifies the spectra. For example, the second overtones of the CH stretching modes for CH_2 group lie at 8235, the CH_3 group at 8360, the cyclopentyl group at 8375, the cyclohexyl group at 8400, and the aromatic CH band at $8710 \, cm^{-1}$ [Hibbard and Cleaves (*325*)]. Because high resolution is easily obtained in this spectral region, investigations could be carried out on long chain hydrocarbon with up to 34 carbon atoms in the chain [Evans *et al.* (*201*)].

2. Olefins and Cyclo-olefins

The occurrence of a $C{=}C$ bond manifests itself in the infrared spectrum in three ways [Rasmussen and Brattain (*580*); Rasmussen *et al.* (*582*); Kletz and Sumner (*393*); Creitz and Smith (*151*); Sheppard and Simpson (*650*); Barnard *et al.* (*38*); Haszeldine (*309*)]: (1) A band of moderate intensity at $3 \cdot 3$ and $3 \cdot 25 \, \mu$ (usually observed only as a shoulder on the paraffin band at $3 \cdot 4 \, \mu$) assigned to the stretching vibration of $={}CH{-}$ and $={}CH_2$ groups. (2) The stretching vibration of the $C{=}C$ group at $1660 \, cm^{-1}$ or *ca.* $6 \, \mu$; the exact position depends somewhat on the substituents (Fig. 138). For identical substitution at both ends, the vibration is inactive on symmetry grounds, and otherwise it is only weakly excited. Thus, the absence of this band in the spectrum does not necesssarily indicate the absence of $C{=}C$ bonds. (3) The most important manifestation of the $C{=}C$ bond in the infrared spectrum are bands of the wagging modes of the $={}CH{-}$ groups. These are found as one or more bands, of much higher intensity than the corresponding paraffinic vibrations, in the region of 10 to $17 \, \mu$. Mono-substituted ethylenes absorb at 910 and $990 \, cm^{-1}$ (terminal double bond); trans-disubstituted ethylenes at 966, cis-disubstituted ethylenes at $700 \, cm^{-1}$ (internal double bonds); unsymmetrically disubstituted ethylenes at $890 \, cm^{-1}$ (terminal branch double bond); tri-substituted ethylenes between 800 and $850 \, cm^{-1}$ (internal branch double bond). Primary, secondary, and tertiary alkyl groups appear to have their own characteristic effect on the position of the band. Besides these principal features, further bands originating from the $C{=}C$ groups occur in the spectrum, but their behaviour is less characteristic. They are overlapped by other bands originating from the paraffinic portion of the molecule, and this especially means that they are usually of little use in structural investigations, although they can be of importance in special cases.

The direct substitution of an atom other than carbon on the double bond has a considerable effect. In spite of sparse data it appears that the conclusion is that the substitution of electronegative groups such as Cl, OR, and O.CO.R, on the double bond displaces the corresponding stretching vibration to smaller wave numbers with simultaneous increase of the intensity [Hatch *et al.* (*311*); Kitson (*388*)].

The effect of conjugation on the bands originating from the double bonds is especially important. On other evidence it is known that the double-bond character is thereby decreased, and the central single bond lying between the conjugated double bonds is made stronger. This is confirmed by the infrared spectrum; the C=C stretching vibration moves towards smaller wave numbers and the C—C stretching vibration towards larger ones. Further, interaction between the conjugated double bonds splits the stretching vibration into two components which appear as bands at 1650 and 1600 cm^{-1}, provided they are both infrared active. In order to assess the size of the displacement

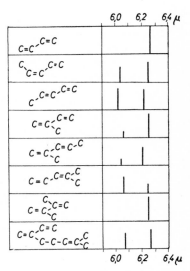

FIG. 138. *Position of the C=C stretching vibration bands for conjugated olefins [from Rasmussen (Q)].*

for comparison with other similar effects, the mean of the spectral positions of the two bands is taken, i.e. 1625 cm^{-1}. Thus, there is a displacement of 35 cm^{-1} compared with the normal position. For 1,3-butadiene only the band at 1600 cm^{-1} occurs. For alkyl substitution of various types (Fig. 138), the band at 1650 cm^{-1} appears in some cases but not in others. The band at 1600 cm^{-1} is always observed and with only small displacements, this is quite characteristic of conjugated C=C bonds. Conjugation with an aromatic ring has somewhat less influence than that with an aliphatic double bond (displacement of about 20 cm^{-1}). Conjugation with a C=O bond leads to a similar displacement of the C=C band of about 35 cm^{-1}, but the band does not split because the two dissimilar double bonds do not interact. Table 28 gives the position of the C=O and also the C=C bands for a series of conjugated aldehydes, ketones, and esters. Together with the displacement, the C=C band is intensified to a degree unusual for

331

Results and Applications

olefins, although the intensity of the C=O band is never attained. Finally, conjugation of a double bond with a triple bond displaces the C=C bond to $6 \cdot 2 \mu$ (e.g. vinyl acetylene). Cumulation of double bonds, as in allene, displaces the wave number of the C=C stretching vibration to the spectral region of the triple bonds. The bands are considerably split as a result of the strong mutual interaction; e.g. for allene itself, the bands occur at 1980

TABLE 28. *Position of the Bands Corresponding to the Double Bonds for conjugation of C=O with C=C or a Benzene Ring [from Rasmussen (Q)]*

Compound	C=O Stretching vibration	C=C Stretching vibration
Aldehydes:		
Acrolein	5·88	6·17
Methacrolein	5·90	~6·1
Crotonaldehyde	5·92	6·08
Benzaldehyde	5·86	
m-Tolualdehyde	5·87	
p-Isopropylbenzaldehyde	5·86	
Ketones:		
Methyl-vinyl-ketone	5·93	6·17
4-Methyl-3-penten-2-one	5·92	6·15
3-Methyl-3-buten-2-one	5·91	6·13
Acetophenone	5·93	
Propiophenone	6·93	
Esters:		
Methyl acrylate	5·80	6·15
Ethyl acrylate	5·77	6·11; 6·17
Methyl methacrylate	5·81	6·12
Ethyl crotonate	5·83	6·06
Methyl benzoate	5·80	
Ethyl benzoate	5·83	

and 1031 cm^{-1} [Herman and Shaffer (320); Miller and Thompson (478); DeHeer (166); Lord and Venkateswarlu (426); Mizushima (483); Overend and Thompson (518); Wotiz (755)].

Comparatively little is yet known about the spectrum of the cyclic olefins [Epstein *et al.* (199); Lord and Walker (429)], especially the terpenes [Frank and Berry (226)]; Table 29 gives some details on the position of the C=C stretching vibration. The ring strain, connection with another unsaturated ring, and the degree of substitution at the double bond are all seen to influence the position of the C=C stretching vibrational frequency. The position of the olefinic wagging vibration also appears to be influenced, but the experimental data do not suffice for the formulation of general rules. It appears that the substitution at an unsaturated ring carbon atom displaces

TABLE 29. *Position of the C=C Stretching Vibration in cm⁻¹ for some Cyclic Olefins [from (429)]*

Compound	$\nu_{C=C}$
Cyclobutene	1566
Cyclopentene	1611
Cyclohexene	1646
Cycloheptene	1651
Cyclooctene	1653
Bicyclo-2,2,1-heptene	1598
Bicyclo-2,2,2-octene	1614
1-Methylcyclopentene	1658
1-Methylcyclohexene	1678
1-Methylcycloheptene	1673

the wagging vibration of the ring double bond, originally at 670 cm⁻¹, to higher values. The infrared spectra of some cyclic olefins have recently been published by Cope and co-workers [literature in (*144*)].

3. Carbonyl Compounds

Bands in the region about $5 \cdot 5$ to $6 \cdot 5 \mu$ originate in the stretching vibrations of C=O groups; the intensity of these bands is generally greater than those of all other double bonds [Hartwell *et al.* (*306*)]. Table 30 summarizes the positions of the bands for some typical simple carbonyl compounds. In general, these positions are little influenced by substituents provided they are either alkyl groups or are not directly connected to the carbonyl atom. The influences sometimes observed (especially for α-substituents) originate from inductive or steric effects.

For saturated *aldehydes* the carbonyl band is generally found at $5 \cdot 7\,9\mu$. Conjugation with a C=C bond or a benzene ring displaces this by about 35 cm⁻¹ to $5 \cdot 90\ \mu$. Confirmation of the assignment of this band with an aldehyde is frequently obtained using the CH stretching vibration of the HCO group at about $3 \cdot 6\ \mu$, provided this is sufficiently well resolved from the other CH bands in the region $3 \cdot 3$ to $3 \cdot 5\ \mu$. [Bonino and Scrocco (*77*); Eggers and Lingren (*186*); Pinchas (*784*)].

Open-chain saturated *ketones* show the C=O band at $5 \cdot 83\ \mu$. Conjugation with another double bond displaces it about 25 cm⁻¹ to $5 \cdot 92\ \mu$; especially strong conjugation sometimes causes further displacement, e.g. for benzophenone [Rasmussen *et al.* (*583*); Gutsche (*280*); Soloway and Friess (*679*);

Results and Applications

Recart (*584*); Hergert and Kurth (*319*); Lecomte (*411*); Mecke and Noack (*467*)]. Ketonic groups in rings of six or more members absorb at the same position, but for small rings the band is successively further displaced to-

TABLE 30. *Position of the C=O Stretching Band in μ for the Liquid State [from Rasmussen (Q)]*

Aldehydes:	
non-conjugated	5·79
conjugated with C=C or benzene ring	5·90
Ketones:	
open chain, non-conjugated	5·83
cyclobutenones, non-conjugated	5·63
cyclopentanones, non-conjugated	5·75
cyclohexanones, non-conjugated	5·83
conjugated with C=C or benzene ring	5·93
α-diketones	5·83
enolized β-diketones	∼6·30
ketenes	4·75
Esters:	
Open chain, non-conjugated	5·75
β-lactones, non-conjugated	5·50
γ-lactones, non-conjugated	5·65
δ-lactones, non-conjugated	5·75
conjugated with C=C or benzene ring	5·81
vinyl or phenyl ester	5·65
enolized β-ketoesters	∼6·00
thiolesters	5·97
Carboxylic acids:	5·65–5·80
Salts of carboxylic acids:	6·20–6·40
Acid chlorides:	
non-conjugated	5·55
conjugated with benzene ring	5·65
Acid anhydrides:	
open chain, non-conjugated	5·50; 5·70
conjugated with C=C	5·60; 5·80
Amides:	
open chain	5·90–6·10
β-lactams	5·75
γ-lactams	5·85
δ-lactams	6·00

wards shorter wavelengths as the number of atoms in the ring decreases. This phenomenon also applies to lactones and lactams and may be correlated with increasing ring strain. The conjugation of carbonyl groups with one another (open chain α-diketones) is without effect on the position of the

334

band [Barnes and Pinkney (*43*)].† However, as a result of simultaneous conjugation and chelate-formation, enolized β-diketones of the form

$$\begin{array}{c} \text{OH}\ldots\text{O} \\ | \quad\quad || \\ -\text{C}{=}\text{C}-\text{C}- \\ | \end{array}$$

show a spectrum which is markedly altered both in the carbonyl and in the hydroxyl regions. The carbonyl band is displaced to about $6\cdot3\,\mu$ and considerably intensified; the hydroxyl band is displaced to $3\cdot7\,\mu$ (in contrast to its normal position at about $2\cdot8\,\mu$) with simultaneous weakening and broadening. The CC bond of a benzene ring has the same effect as that of an aliphatic C=C group. In ketenes, $R_2C{=}C{=}O$, the two double bonds interact markedly and cause absorption at $4\cdot75$ and $9\,\mu$ [Whiffen and Thompson (*733*); Harp and Rasmussen (*301*); Halverson and Williams (*292*); Miller and Koch (*475*); Drayton and Thompson (*177*)].

Acyclic saturated *esters* show the carbonyl band at $5\cdot75\,\mu$ with the exception of formates for which it occurs at $5\cdot80\,\mu$ [Hampton and Newell (*296*); Rasmussen and Brattain (*581*); Sinclair *et al.* (*666, 667*); Stahl and Pessen (*680*); Leonard *et al.* (*416*); Searles *et al.* (*638*); Shreve *et al.* (*653*)]. The displacement compared with ketones originates in the electronegative character of the alkoxyl group compared to the alkyl group itself. In addition to the canonical forms

$$\begin{array}{cc} \begin{array}{c} \text{O} \\ || \\ \text{R}'{-}\text{C}{-}\text{O}{-}\text{R} \\ \text{(I)} \end{array} & \begin{array}{c} \text{O}^- \\ | \\ \text{R}'{-}\text{C}{=}\text{O}^+{-}\text{R} \\ \text{(II)} \end{array} \end{array}$$

there is the further canonical form

$$\begin{array}{c} \text{O}^+ \\ ||| \\ \text{R}'{-}\text{C} \quad \text{O}^-{-}\text{R} \\ \text{(III)} \end{array}$$

which gives the C=O bond some triple-bond character and which in the case of the esters is more important than structure (II). If the —OR group is replaced by a more electropositive group such as —SR or —NR$_2$, then forms of type (II) are again of more importance, so that the C=O bond takes on more of the character of a single bond and accordingly the carbonyl band appears at longer wavelengths. The conjugation of the ester carbonyl group with C=C bonds or benzene rings causes a smaller displacement towards longer wavelengths than that in aldehydes or ketones, i.e. only about $15\,\text{cm}^{-1}$ to $ca.\,5\cdot80\,\mu$. Lactones are influenced by the number of atoms in the ring in the

† This applies only to trans-forms. For cis-forms, splitting occurs and the bands move to higher frequencies with increasing strain [Alder *et al.* (*10*)].

Results and Applications

same way as cyclic ketones, as already mentioned; chelation effects are also similar to those found in ketones. The considerable displacement of the carbonyl band in esters derived from unsaturated alcohols, to $5 \cdot 65 \mu$, is probably a result of a further increase in the effect of the structures described above. In addition to the carbonyl bands, bands occur for unsaturated esters at $8 \cdot 0$ to $8 \cdot 3 \mu$ and $9 \cdot 0$ to $9 \cdot 5 \mu$, which doubtless arise from the C—O vibration and which facilitate the recognition of carbonyl groups belonging to esters. For conjugation of the ester carbonyl group with a double bond (esters derived from unsaturated carboxylic acids) these bands are displaced to *ca.* $7 \cdot 8$ and $8 \cdot 5 \mu$.

For *carboxylic acids* as free molecules (gas or very dilute solution) the hydroxyl band is found at $2 \cdot 8 \mu$ and the carbonyl band at $5 \cdot 65 \mu$. However, as a result of the well-known tendency of these substances to dimerize and form hydrogen bonds, these bands are found at $3 \cdot 4$ and $5 \cdot 90 \mu$ for compounds in the liquid state, and the hydroxyl band is very broad and often extends to about 4μ. As with the other types of carbonyl vibration, the study of the finer correlations between the structure and band position is still being investigated for acids [Flett (*213*); Sinclair *et al.* (*666, 667*); Jones *et al.* (*364*); Freeman (*229, 230*); Hadži and Sheppard (*284*); Shreve *et al.* (*653*); Guertin *et al.* (*279*)]. A band in the $6 \cdot 2$ to $6 \cdot 4 \mu$ region is characteristic for the COO-ion. This relatively high value indicates a resonance effect of the type O^-—CR=O $\leftrightarrow$ O=CR—O^-, which weakens the double-bond character of the C=O group. The two bonds between the oxygens and the carbon atom, which were originally well differentiated, now become identical and the band mentioned is only one component of a pair. The other component appears at about $7 \cdot 0 \mu$ but is usually overlapped by CH deformation vibrations. Investigation of the 7 to 12μ region has recently become important for questions of the linearity or branching of acids [Wenzel *et al.* (*789*)].

Several recent investigations by Fuson and his co-workers (*238, 366, 367, 368*) have dealt with the influence of structure on C=O bands in polycyclic compounds and allow the formulation of certain rules. Other workers have also investigated quinones [Flett (*212*); Hadži and Sheppard (*285*); Otting and Staiger (*517*); Yates *et al.* (*763*) and Flaig and Salfeld (*775*)].

Carbonyl bands for open-chain *anhydrides* are found near $5 \cdot 50$ and $5 \cdot 70$ μ. These positions show that for anhydrides the resonance structure (III), mentioned above for the esters, also gives the double bond some triple-bond character. The fact that two bands occur can be explained as a result of the interaction of the two individual C=O bonds. Conjugation results in the expected displacement to about $5 \cdot 60$ and $5 \cdot 80 \mu$.

Details about the amide group CO.NH are given under 'nitrogen compounds'.

336

4. Other Compounds containing Oxygen

According to the environment of the oxygen atom, bands are obtained which are characteristic for the particular bond type. These bands, together with those originating in the other parts of the molecule, are thus characteristic of the corresponding class of compounds.

The *hydroxyl group* generally causes a band in the region $2 \cdot 6$ to $3 \cdot 0 \mu$ corresponding to its stretching vibration. Liquid water itself, at low concentration, shows two absorption maxima at $2 \cdot 75$ and $2 \cdot 83 \mu$ and the HOH deformation vibration at about $6 \cdot 1 \mu$. At higher concentrations, the two short wavelength bands merge into a single one which is considerably displaced towards 3μ as a result of association. As a result of the well-known tendency of the H_2O molecules to form various types of aggregates, the spectrum is characterized by a general non-specific absorption (Fig. 94). This arises from the various interactions of the individual groups with one another and their smeared-out frequency position. In quantitative work, a small amount of water can be very disadvantageous because this background absorption disturbs the standardization (Fig. 92).

The position and the appearance of the hydroxyl band for *alcohols* shows the existence of strong association [Smith and Creitz (*675*)]. This association is the more pronounced the further the band is displaced towards longer wavelengths. If the degree and nature of the association in the various aggregations is uniform, the band remains more or less sharp. However, if the hydrogen bonding is of varying strength, the band broadens, because a different displacement occurs for every degree of hydrogen bonding and the many non-resolvable bands merge to a single one. The strong band (assigned to the C—O bond) between 8 and 10 μ is also characteristic of the spectra of alcohols [Zeiss and Tsutsui (*768*); Plyler (*545*); Shreve *et al.* (*653*); Stuart and Sutherland (*691*)]. This occurs at 1040 for primary alcohols, at 1110 for secondary alcohols, and at 1160 cm^{-1} for tertiary alcohols. If the carbon atom carrying the OH group is unsaturated, the C—O band lies between 1200 and 1250 cm^{-1}. The OH deformation vibration is found at about 1020 cm^{-1} [Quinan and Wiberley (*565, 566*)]. A further characteristic is a broad absorption band which usually commences at 12 μ and reaches a maximum at 15 μ. Based on the interpretation of the methanol spectrum, this is assigned to the torsional vibration of the OH group about the C—O bond as axis.

The hydroxyl band for *phenols* and phenol derivatives is similar to that of the alcohols. In addition to the OH stretching vibration, phenol itself is characterized by a very strong band at about $8 \cdot 2 \mu$, which corresponds to the bond connecting the OH group to the aromatic ring [Kletz and Price (*392*);

z

Results and Applications

Friedel (*231*); Ingraham *et al.* (*347*); Mecke and Rossmy (*465*)]. Characteristic bands for the phenol ring occur, e.g. at $6 \cdot 25$ and $6 \cdot 7\,\mu$; sometimes the intensity is altered compared with benzene itself (see also Section 6). For polysubstituted phenols of different orientation, characteristic absorption is found especially between 10 and 15 μ as for other benzene derivatives (see also later).

Ethers are characterized by a strong band in the 8 to $9 \cdot 5\,\mu$ region, which originates from the C—O—C bonds. The exact position of these bands depends on the strain in the ether linkage and on its chemical environment. The band occurs near 9 μ for open-chain aliphatic ethers, and near 8 μ for unsaturated and aromatic ethers. The band is split into a doublet at $8 \cdot 9$ and $9 \cdot 2\,\mu$ for acetals [Tschamler and Leutner (*718*)]. For cyclic ethers, there are two characteristic bands, the positions of which depend on the ring size [Barrow and Searles (*45*)]. Epoxides are characterized by a band in the region 11–12 μ (for cis-compounds at $11 \cdot 2\,\mu$, for trans-compounds at 12 μ), and a further band at 8 μ [Field *et al.* (*209*); Shreve *et al.* (*654*); Barrow and Searles (*45*); Zuidema (*770*); Patterson (*523*)]. For increasing ring size, and decreasing ring strain, the C—O—C vibration moves nearer and nearer to the position (9 μ) for the open-chain ethers: a further characteristic ring vibration occurs at 11 μ [Barrow and Searles (*45*)]. Data on the spectra of *peroxides* and *hydroperoxides* has been given by Davison (*163*), Shreve *et al.* (*655*), Milas and Magelli (*474*), Philpotts and Thain (*532*) and Williams and Mosher (*744*). A key band occurs for hydroperoxides at 860 cm^{-1} but is sometimes apparently absent. The peroxide bond itself generally causes absorption at 800 to 1000 cm^{-1}; more exact rules are not yet available.

Finally, the influence of an oxygen atom in the molecule on the position of bands other than those resulting directly from the oxygen bonds, especially the various CH bands, is, according to Pozefsky and Coggeshall (*560*), so small that it can generally be neglected.

5. Compounds containing Nitrogen

NH and NH$_2$ groups manifest themselves by respectively one and two bands near 3 μ. The wavelength of these bands is generally somewhat longer than that of hydroxyl bands. However, the regions of two types overlap because of the marked displacement of the OH band on association. Such displacements are also found for NH bands when hydrogen bonded but not to the same extent [Fuson *et al.* (*239, 240*)]. The characteristic group —N$^+$—H of some ammonium type compounds causes a broad, rather ill-defined, and usually not very strong band near 4 μ, as found in amine hydrochlorides and

338

amino-acids. The HNH deformation mode of the NH_2 group is found at $6 \cdot 15$ to $6 \cdot 25\,\mu$; in the NH group it is displaced to $6 \cdot 5$ with simultaneous decrease in intensity. This allows primary, secondary and tertiary amines to be differentiated.

The possible nitrogen-carbon bonds will now be considered. The C—N single bond gives a band of moderate intensity in the $9\,\mu$ region, the C=N double bond a strong band at $6\,\mu$, and in ring compounds between $6 \cdot 15$ and $6 \cdot 20\,\mu$ [Fabian *et al.* (*202*)]. The C≡N triple bond causes a moderately strong and very sharp band at $4 \cdot 4$ to $4 \cdot 5\,\mu$ which is displaced by about 25 cm^{-1} to longer wavelengths in conjugated compounds [Kitson and Griffith (*389*)]. The isocyanate group absorbs near $4 \cdot 4\,\mu$ [Hoyer (*336*)]. Nitro-groups in aliphatic compounds absorb strongly at $6 \cdot 4$ and $7 \cdot 3\,\mu$, and in aromatic compounds at $6 \cdot 5$ and $7 \cdot 4\,\mu$ [Randle and Whiffen (*570*); Grabiel *et al.* (*274*); Kornblum *et al* (*395*)]. Nitrites are characterized by an intense band at $6 \cdot 1\,\mu$ originating in the N=O bond, by the N—O vibration at $11 \cdot 8\,\mu$, and by the bending mode of the O=N—O— group near 14 to 15 μ. Sometimes cis-trans-isomerism of nitrites causes these bands to split [Tarte (*697, 698*)]. Nitroso-compounds show the N=O stretching vibration at $6 \cdot 2$ to $6 \cdot 7\,\mu$ and the CNO bending mode near 23 μ [Lüttke (*782*)]. Bands characteristic of the N=N group lie at $6 \cdot 35\,\mu$ [Thomas (*700*)], of the azide group at $4 \cdot 7$ and at $7 \cdot 6$ to $8 \cdot 1\,\mu$, and of the tetrazole ring at $8 \cdot 8$ to 10 μ usually with several peaks [Lieber *et al.* (*418*)]. Palm and Werbin (*522*) and Califano and Lüttke (*107, 108*) have worked on oximes.

The *amides* are of great interest because of their structural similarity with the proteins in the linkage between the carbonyl and the NH group. Because of this similarity, information on numerous natural products can be obtained from their spectra. The carbonyl band of the amide group is generally found at *ca.* $6\,\mu$ for sufficiently dilute solutions in which the compounds exist as the free molecules. The exact position is characteristic of the substitution at the nitrogen: for unsubstituted compounds the band is at $5 \cdot 90\,\mu$, for monosubstituted at $5 \cdot 96\,\mu$, and for disubstituted at $6 \cdot 08\,\mu$. For concentrated solutions and solids, all these bands are displaced to the $6 \cdot 08\,\mu$ region. These relatively low values for a carbonyl vibration can be explained by the well-known tendency of nitrogen atoms to donate electrons. This means that resonance structures

$$R—\overset{\overset{\textstyle O^-}{|}}{C}=N^+\Big\langle$$

are important for the amide group, thus the carbonyl group takes on some of the character of a single bond. For amides in which the nitrogen is unsubstituted, or carries only a single substituent, a further band occurs in the

Results and Applications

double-bond region: for unsubstituted amides at $6\cdot10$ to $6\cdot20$ μ and for monosubstituted amides at $6\cdot45$ to $6\cdot65$ μ. This band does not occur for disubstituted amides or lactams; [for the latter, the effect on the carbonyl vibration when the amide group is included in a ring has already been dealt with (Table 30)]. The exact position of this band depends on the state of aggregation: it appears at shorter wavelengths for solids and liquids, i.e. where association can occur, and at longer wavelengths for sufficiently dilute solutions. The assignment of this band has been much disputed and has still not been finally clarified. Richards and Thompson (587) originally assigned it to an HN deformation mode; this was based on the displacement observed from 1560 to 1490 cm^{-1} on deuteration on the nitrogen. This assignment has recently been supported by Cannon (110) and Mecke and Mecke (466). On the other hand, Lenormant (414) has assigned it as a CN stretching vibration of an ionic form $(CO{-}N)^-$, which according to him exists in the amides; Letaw and Gropp (417) agree with this interpretation. Findings of Gierer (253), resulting from very intensive investigations from acetamide and its derivatives, disagree with this view. It appears probable that the band mentioned corresponds with a deformation mode of the

$$-N\big\langle{}^{CH_3}_{H}$$

group. These investigations also give valuable information on the questions whether amides exist in two forms of the amide group (amide-imidol tautomerism) or in only one form (amide) with two different types of hydrogen bonding (stretched or bent).

6. Aromatic and Heteroaromatic Compounds

Much attention has been paid to the infrared spectrum of *benzene* in connection with the clarification of the structure of this compound which is theoretically and practically so important. It is one of the few molecules with a comparatively large number of atoms for which a complete vibrational analysis was possible and in which all the theoretically expected vibrations both in the infrared and in the Raman spectrum could be identified [Pitzer and Scott (541); Ingold and co-workers (324, 345, 346)].†

Using the results obtained from benzene itself, it is possible to interpret the spectra of benzene derivatives, especially substitution products, and to obtain much information of practical use [recent literature with references to the earlier work: Vago *et al.* (720); Cole and Thompson (132); Launer and

---

† References made here only to papers which deal principally with infrared investigations. Purely Raman investigations are of equal or, because of their priority, of even greater importance.

McCaulay (*407*); Garg (*241*); Schlatter and Clark (*634*); Margoshes and Fassel (*443*); Lebas and Josien (*409*); Randle and Whiffen (*571*); Katritzky (*810*)]. The CH stretching vibration of the aromatic ring is the origin of a band at $3 \cdot 27 \mu$, i.e. at the same frequency as that from the olefinic double bond. This band is of especial interest; its intensity varies with the number and orientation of the substituents as a result of the alteration of the symmetry and activity relationship.

The bands which are derived from the CC vibrations of the ring are of great importance. One result of the equivalence of all the CC bonds in the aromatic ring is an extremely marked interaction between the vibrations of the individual bonds which leads to a series of bands in the region of 6 to 10 μ. Four of these bands are found near 1600 1580, 1500, and 1450 cm^{-1}. The positions of these bands are relatively little affected by the number, orientation, and nature of the substituents. Their intensities vary widely but can be correlated with the electronic interactions between the substituents and the ring [Katritzky et al. (*811–815*)].

Reliable criteria for the nature of the orientation in the ring are found between 5 and 6 μ [Young et al. (*766*)] and between 10 and 15 μ. These bands correspond to the fundamentals and combination bands respectively of the CH out-of-plane deformation [Bomstein (*76*)]. Figs 139 and 140 give a survey of the various possibilities according to the type and extent of alkyl substitution [McMurry and Thornton (*460*)]. The positions of these bands depend to a minor extent on the nature of the substituents, for alkyl groups especially on the degree of branching [Potts (*557*)]. Certain deformation modes of the ring, perpendicular to the plane, may overlap some of these bands.

A survey of the spectra of *polycyclic compounds* is given in a comprehensive paper by Cannon and Sutherland (*111*); this paper also gives references to the early literature. Some examples from this work are given in Fig. 141. The available data do not yet suffice for the derivation of rules which are valid

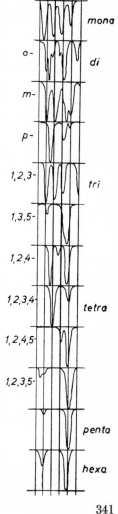

FIG. 139. *Characteristic bands of substituted benzene rings between 5 and 6 μ according to Young* et al. (*766*).

341

in every case. Nevertheless, certain regularities are apparent, e.g. in the position of the bands which are derived from the CH out-of-plane deformation vibrations. From experience with substituted benzenes, these are

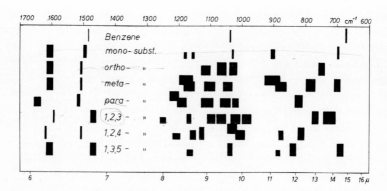

FIG. 140. *Characteristic bands of the substituted benzene rings between 6 and 15 μ according to McMurry and Thornton (350).*

expected to be especially sensitive with respect to the substitution and to occur at *ca.* 650 to 900 cm^{-1}. Table 31 gives a short summary of the bands of a few non-substituted compounds with condensed rings with the limita-

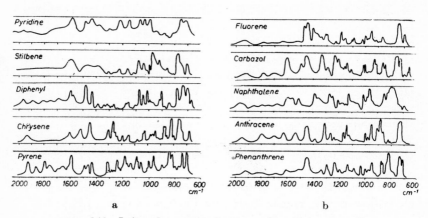

FIG. 141. *Infrared spectra of some polycyclic compounds according to Cannon and Sutherland (111).*

tion that each ring has, at most, two CC bonds in common with another, i.e. a limitation to linear and angular compounds. The band at 810 cm^{-1} is obviously connected with angular annulation. For substituted compounds

342

TABLE 31. *Characteristic Ring Vibrations of some Non-substituted Condensed Ring Systems according to Cannon and Sutherland (111)*

	cm^{-1}		
Benzene		670	
Naphthalene		780	
Anthracene	730		890
Naphthacene	750		900
Phenanthrene	730	810	870
1,2-Benzanthracene	750	810	890
1,2; 5,6-Dibenzanthracene	750	810	890

the situation becomes less clear. Naphthalenes monosubstituted in the 1-position can be distinguished spectroscopically from the corresponding 2-substituted derivatives [Hunsberger (*340*); Seidman (*639*); Pimentel and McClellan (*536*); Mosby (*487*); Ferguson and Werner (*206*); Cencelj and

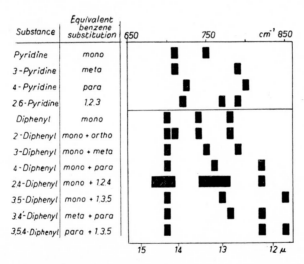

FIG. 142. *Characteristic bands of substituted pyridines and diphenyls [Cannon and Sutherland (111)].*

Hadži (*113*); Luther and co-workers (*438, 439, 440*)]. The former show characteristic bands at 770, 800, 1515, and 1600 cm^{-1}, the latter at 750, 820, 1500, 1600, and 1640 cm^{-1}. Corresponding rules for further types of substitution have been given by Hawkins, Ward and Whiffen (*834*). The position *for anthracenes* and *phenanthrenes* is gradually becoming clear

Results and Applications

[Pacault and Lecomte (520); Orr and Thompson (511); Hunsberger et al. (341); Dannenberg et al. (161); Dannenberg and Naerland (774)].

It has been shown [Laakso et al. (400); Orr and Thompson (511); Thompson et al. (708)] that each ring of *diphenyl compounds* can be treated independently with respect to the CH out-of-plane deformation vibrations. The results from benzenes can be applied to these compounds, although the bands are displaced about 20 cm^{-1} to higher wave numbers and a band at 700 cm^{-1} invariably appears. Thus, diphenyl itself corresponds to a monosubstituted benzene, a 2-substituted diphenyl simultaneously to a mono- and to a ortho-disubstituted benzene, etc. (Fig. 142).

The infrared spectra of heteroaromatic compounds with five- and six-membered rings have been reviewed by Katritzky (810). The vibrations of these compounds may be classified as in Table 31a. Six-membered ring

TABLE 31a. *Spectral Regions (cm^{-1}) associated with Different Classes of Vibrations of Heteroaromatic Compounds*

	Six-membered rings	Five-membered rings
CH stretching	3100–3000	3100–3000
CH in-plane deformation	1250–1000	1250–1000
CH out-of plane deformation	1000– 600	1000– 600
CH out-of-plane (overtones and combinations)	2000–1650	(?)
CC stretching	1600–1300	1600–1250
Ring breathing frequency	*ca.* 1000	1150– 800
CC in-plane bending	700– 600	900– 600
CC out-of plane bending	550	600– 450

compounds (pyridines, pyrimidines, azines, etc.) behave similarly to benzenes with respect to the dependence of the positions and intensities of the bands on the orientation and nature of the substitution [Katritzky (811)]. Five-membered ring compounds (pyrroles, thiophenes, furans, azoles, etc.) have been less studied up to the present.

7. Acetylenes

The spectra of terminal *acetylenes* are characterized by the band corresponding to the CH stretching mode of the C≡CH group which absorbs at especially high wave numbers, 3290–3390 cm^{-1}, or short wavelengths, 3·04–2·95 μ. The C≡C stretching mode is inactive in acetylene itself on symmetry grounds. For monosubstituted acetylenes [Wotiz and Miller (756); Wotiz et al. (757); Sheppard and Simpson (649)], it is found at 2100

344

cm^{-1}, and in disubstituted acetylenes between 2210 and 2280 cm^{-1}. This correlation also applies to cyclic acetylenes [Blomquist *et al.* (*66*); Blomquist and Liu (*65*)]. Conjugation has only a small effect on the position of the C≡C vibration [Allan *et al.* (*11*)]. The CH deformation mode of the C≡CH group lies at rather small wave numbers, e.g. for acetylene itself at 612 (*ia*) and 729 cm^{-1}, and for methyl acetylene at 642 cm^{-1}; generally it is so strong that its overtones are noticeable.

8. *Halogen, Sulphur, Silicon, and Phosphorus Compounds*

Substitution of hydrogen atoms in hydrocarbons by *halogens* considerably alters the infrared spectrum because of the quite different masses and force constants. The corresponding CH vibrations disappear and absorption bands connected with the halogen appear. As expected, the stretching vibrations of the halogen-carbon bond lie at decreasing wave numbers as the atomic weight of the halogen atom increases. With the NaCl prism normally used, only the C—F and the C—Cl bond can be observed, the former at 1000 to 1100 cm^{-1}, and the latter at 650–730 cm^{-1}. For the C—Br bond, the corresponding vibrations lie at 560–650 cm^{-1} and for the C—I bond at 500–600 cm^{-1}. The exact position of these vibrations depends on the environment of the C-atom in question. The frequency is reduced for halogens connected to secondary and tertiary C-atoms compared with those on primary C-atoms; multiple halogen substitution at the same C-atom raises the frequency. Thus, the two stretching vibrations of the CF$_2$ group in polytetrafluoroethylene are found at 1154 and 1213 cm^{-1} [Robinson and Price (*603*)]. The excitation of the stretching vibration of the carbon-halogen bond is usually so strong that the first overtone occurs with notable intensity. The various bending vibrations of these bonds lie at such low wave numbers that they have only recently been reached in the infrared spectrum by the introduction of KBr, CsBr, and CsI prisms. Sometimes they had been found previously, in the form of combination bands, in more readily accessible regions. It may be mentioned that Raman spectroscopy is in many ways more suitable for the investigation of halogen type vibrations. Nevertheless, the recent infrared literature contains so many papers dealing with halogenated hydrocarbons, etc., that only a small selection can be quoted; perfluoroalkyl halides [Hauptschein *et al.* (*312*)]; halogeno-alcohols [Henne and Francis (*317*), Haszeldine (*310*)]; fluorobenzenes [Ferguson *et al.* (*207*)]; chlorobenzenes [Plyler (*544*)]. For the rest, reference should be made to the review literature on infrared spectra.

Many papers have appeared dealing with the infrared spectra of organic *sulphur* compounds [Trotter and Thompson (*717*); Schreiber (*636*); Barnard *et al.* (*37*); Sheppard (*645*); Amstutz *et al.* (*19*); Cymerman and Willis

Results and Applications

(155); Flett (213a); Price and Gillis (561); Detoni and Hadži (170); Mecke et al. (468); Baxter et al. (47)]. This work allows the most important vibrations of sulphur-containing groups to be assigned as follows: SH, 2400 to 2600 and 810 to 930 cm^{-1}; >SO, 1050 cm^{-1}; >SO$_2$, 1150 and 1350 cm^{-1}; C—S, 680 to 690 cm^{-1}; S—S, 430 to 490 cm^{-1}; C=S, 1050 to 1230 cm^{-1} (narrower ranges apply in homologous series).

Available material on the infrared spectra of *silicon* compounds is still more extensive: Wright and Hunter (760); Richards and Thompson (590); Jones et al. (362); Scott and Frisch (637); Clark et al. (119); McBride and Beachell (453); Šimon and McMahon (662); Smith (672); Kakudo et al. (371); Westermark (731). The vibrations apparently cause absorption as follows: the SiH group at 2100 to 2300 cm^{-1}, the SiC bond at 700 to 850 cm^{-1}, and the SiO bond at 1000 to 1100 cm^{-1}.

Data on the infrared spectra of organic *phosphorus* compounds are considerably less numerous: Meyrick and Thompson (473); Daasch and Smith (159); Bellamy and Beecher (52, 53); Bell et al. (51); Harvey and Mayhood (307); Corbridge and Lowe (145); Corbridge (773); From this work it is possible to formulate several rules, although in many cases these are not final. Characteristic bands are found: P—OH, 2500 to 2700 cm^{-1}; P—H, 2300 to 2450 cm^{-1}; phosphinic acids, 1670 to 1750 cm^{-1}; P-phenyl, 1440 to 1000 cm^{-1}; P—O—C (aliphatic), 1000 to 1100 cm^{-1} (however, P—O—CH$_3$, 1190 cm^{-1}, P—O—C$_2$H$_5$, 1160 cm^{-1}); P—O—C (aromatic), 850 to 950 cm^{-1}; P—F, 735 to 890 cm^{-1} (for trivalent phosphorus), 850 to 890 cm^{-1} (for higher valent phosphorus); P—C (aliphatic), 670 to 750 cm^{-1}; P—Cl, 475 to 585 cm^{-1}; P=O, 1200 to 1300 cm^{-1}; P—O—P, 910 to 950 cm^{-1}; pyrophosphates PO—O—PO, 870 to 930 and 670 to 710 cm^{-1}.

2. Substances of High
Molecular Weight

1. Introductory

The preceding sections have shown that infrared spectroscopy is suited to the determination of the structure and constitution, and to the estimation, of numerous small and well-defined molecules. Its application to compounds of high molecular weight, which are often poorly defined, was an obvious extension. In spite of some scepticism, this has led to numerous valuable conclusions, although the investigations in this field will continue for many years. The limits of the application of spectroscopy are quite different for high polymers than for the substances of lower molecular weight [Sutherland (*692*); Thompson *et al.* (*707*); Thompson (*701*)].

The investigation of two large groups of high molecular weight substances has been especially facilitated by infrared spectroscopy. The first group is that of the natural and synthetic polymerization and condensation products, characterized by the periodic occurrence of certain structural elements along a chain or in a surface. High polymer hydrocarbons, rubber, cellulose, polypeptides, plastics, etc., all belong to this group. The second large group is composed of organic compounds of medium to high molecular weight, particularly substances of biochemical and biological interest. Examples of this group are the carbohydrates [Kuhn (*398*)], polysaccharides [Orr and Harris (*512*)], nucleic acids, and amino-acids. As can be seen from the recent literature, more and more attention is being paid to this second group of substances. The aims of the infrared spectroscopic investigation of these high molecular weight substances remain the same, i.e. qualitative and quantitative analysis, and the clarification of structural and constitutional problems. There is a further aim: the attempted correlation of structural properties with physical and technological properties, especially for the natural and synthetic polymers used technically. The determination of the crystalline and amorphous portions of plastics from the infrared spectrum is of especial importance here [Miller and Willis (*479*)]. The spectral investigation of high molecular weight substances and their interpretation depends on the existence of characteristic bands for certain atomic groupings, just as for the low molecular weight compounds. For many of the lower molecular weight compounds it was possible to carry out a vibrational analysis of the molecule, with at most a few alternative possibilities, and thus make a more or less definite assignment of the bands (the number of which is generally relatively small) to definite vibrations or groups. Such a treatment is usually not possible for substances of high molecular weight. As a result of the large number of atoms and the low symmetry, the number of bands observed is large, especially for substances of the second group mentioned above. Nevertheless, this conclusion often does not apply to the true polymerization products, the spectra of which frequently show a considerable simplification compared to those of the corresponding monomer. This manifests itself in the disappearance of the bands corresponding to the groups which directly participate in the polymerization. If the monomer itself possesses a spectrum very rich in bands, as is the case for molecules of somewhat more complicated structure, then this effect is of little importance. High spectral resolution is required in order to record such complicated spectra clearly. Several bands of similar intensity are often found in one region and it is often not possible to decide unambiguously which of these should be assigned to a certain group. The increased interaction between parts of the complex molecule, or between neighbouring molecules, can lead to considerable displacements and alterations in the position and appearance of bands, which makes unambiguous assignments difficult. In addition, high molecular weight substances are often poorly defined, either because they are technical products which differ according to the way in which they are produced, or because they occur in nature as mixtures which can be separated only with difficulty or not at all. A further point is that sample preparation is often particularly difficult for substances of this class. Some of the methods for the preparation of samples already described were first developed in order to overcome difficulties encountered with compounds of high molecular weight.

Bearing in mind these difficulties, the success achieved in the application of infrared spectroscopy to high molecular weight substances is very considerable and encourages further efforts. The investigation of natural and synthetic polymers is already far developed (although certainly not concluded), and relatively few difficulties were encountered with them. By contrast, the investigation of the complicated organic structures of biochemical substances is still in its infancy and has hardly gone beyond recording the spectra, apart from a few conclusions drawn from very well known bands such as those of the OH, NH_2, and CONH groups. The hope that the infrared spectra of long-chain molecules with repeating structural units could give information about their chain length, and therefore the molecular weight, has had to be relinquished. Experience with the long-chain normal paraffins showed that their spectra differed very little above a chain length of about 10 to 14 carbon atoms.

Other techniques and effects must be used in the interpretation of the spectra of high polymers even more than for those of low molecular weight molecules. Examples of techniques which directly affect the spectrum are: selective deuteration, the influence of moderately high temperatures and more especially of low temperatures, and the use of polarized radiation. The last is of particular importance.

2. *Natural and Synthetic High Polymers*

Further discussion is limited to a few selected high molecular weight substances for which numerous reliable infrared data are available. Work principally concerned with purely analytical questions is mentioned only in

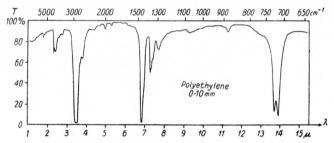

FIG. 143. *Infrared spectrum of polyethylene* [*from the original of the Perkin-Elmer Corp.*].

passing or not at all. For further details, reference should be made to the original papers and reviews of the infrared spectroscopic clarification of structure of natural and synthetic high polymers [Clement (*120*); Brügel (*D, 99*)].

Polyethylene: This thermoplastic material is formed by the polymerization of ethylene. Its structure is essentially a long zigzag chain of CH_2 groups. Corresponding to this simple structure, it shows a very simple infrared spectrum (Fig. 143) with bands at $2 \cdot 4, 3 \cdot 4, 6 \cdot 8, 7 \cdot 2, 7 \cdot 6, 13 \cdot 7,$ and $13 \cdot 9 \mu$, and a few other very weak ones. Detailed investigation has shown that, contrary to expectation, a considerable proportion of methyl groups is present. This proportion varies with the origin; for older products with relatively low molecular weight it can reach 10 per cent with respect to the CH_2 group content. For the present high pressure product, it is still about 3 per cent, whereas the proportion is vanishingly small in low pressure polyethylene [Fox and Martin (*220*); Cross *et al.* (*153*); Richards (*586*); Rugg *et al.* (*614*,

Results and Applications

615)]. Investigations of stretched samples with polarized radiation have shown that these methyl groups do not occur directly attached to the main methylene chain but at the ends of methylene side chains. However, little is known about the exact length of the side chains. From the fact that chains of up to four carbon atoms show characteristic bands in the spectrum and these are not present in the polyethylene spectrum, it has been concluded that the side chains have a somewhat greater length [Elliot *et al.* (*194*); Bryant and Voter (*100*)]. Further bands corresponding with a $0 \cdot 5$ per cent content of unsaturated groups have been found. Oxygen has been detected by infrared spectroscopy in the form of hydroxyl, carbonyl, and ether groups in very small proportions, usually less than $0 \cdot 1$ per cent [Cross *et al.* (*153*)].

The doublet at $13 \cdot 7$ and $13 \cdot 9$ μ is of especial interest in connection with the question of crystalline character. This band is found for all paraffins with more than three methylene groups in the chain, and it has been shown by the investigation of deuterated paraffins to originate in a rocking vibration of the two hydrogen atoms of the CH_2 group [Sheppard and Sutherland (*648*)]. It occurs as a doublet in polyethylene and this was considered to result from vibrations parallel and perpendicular to the axis of the chain. In agreement with this conclusion, a different behaviour was reported for the two components for oriented samples investigated with polarized radiation. However, other authors [Rugg *et al.* (*613*)] have recently reported an almost identical behaviour for the two bands. The component of shorter wavelength (731 cm^{-1}) vanishes when the material is melted [Thompson and Torkington (*705, 706*)], from which the conclusion was drawn that it originates in crystalline regions, the order of which is destroyed on melting. Thus, the intensity of the band is a direct measure of the magnitude of the crystalline portion in polyethylene [Rugg *et al.* (*614, 615*); Stein and Sutherland (*683*); Sandeman and Keller (*624*); Kaiser (*369, 780*)]. This interpretation is supported by observations at low temperatures [King *et al.* (*384*)]. It is known from experience, derived from many substances of known constitution and structure, that the bands of crystalline bodies at very low temperatures are generally considerably sharper and more prominent than, and often somewhat displaced from, those at normal temperatures. The bands of amorphous substances are not influenced by temperature changes. The polyethylene band at 731 cm^{-1} does indeed show a decrease in the half band width (in addition to a small spectral displacement) when the spectrum is taken at $4°$ K, while the other component at 720 cm^{-1} is unaltered. Similarly, a weak band at 1301 cm^{-1} is considered to be a direct measure of the amorphous part of polyethylene [Richards (*586*)].

Infrared spectroscopic investigation of the thermal oxidation and degradation of polyethylene has been very successful for determining its technological properties especially its resistance to temperature [Oakes and Richards (*504*)]. During oxidation, the bands of the oxygen linkages mentioned above increase in intensity and further bands appear. Although it has been stated in the literature that polyethylene which has not been deliberately oxidized possesses no free hydroxyl groups (i.e. not hydrogen-bonded), other spectra show weak but clear indications of the presence of such groups. The hydroxyl groups that are formed by oxidation are shown to be hydrogen-bonded by the spectral position of the corresponding bands. Thus it must be assumed that the hydroxyl groups now appearing (in contrast to those present initially, which are probably end groups) are situated on the chain near to ketonic carbonyl groups and joined to these by means of a hydrogen bond. On stronger oxidation, these bands and those of other oxygen-containing groups become the most intense bands in the spectrum. Studies of the thermal degradation at temperatures above $200°C$, for an initial molecular weight of 13,000 have shown that with decreasing molecular weight an increase takes place in the proportion of olefinic bonds of the types $RCH{=}CH_2$ (909 and 990 cm^{-1}), $RCH{=}CHR'$ (964 cm^{-1}), and $RR'C{=}CH_2$ (887 cm^{-1}). At first, the bands corresponding to all these bonds increase in intensity more or less equally. However, at a molecular weight of about 700, the two latter reach a constant value while the contents of the $RCH{=}CH_2$ bond increase particularly rapidly for degradations to molecular weights below 1000. It is obvious that the normal C—C bonds in the chain are not concerned in the degradation which occurs principally at bonds next to oxygen functions and at points of branching.

The infrared spectrum of chlorinated polyethylenes has also been investigated [Thompson and Torkington (*705, 706*)]. The doublet at 720 and 730 cm^{-1} decreases with increasing content of chlorine and vanishes completely for about 50 per cent chlorine. The bands at 1370 and 1460 cm^{-1} (methylene and methyl groups) decrease in intensity somewhat more slowly; the latter disappears at a chlorine content at which no CH_2 groups free from steric hindrance are left. In their place, new bands appear in increasing proportion. These correspond to the vibrations of chlorine atoms in various environments; rather sudden changes in the spectrum are found at between 40 to 50 per cent and near 74 per cent chlorine. So far it has not been possible to deduce the distribution of chlorine along the chain from the spectrum. However, the spectra of polyvinylchloride and polyvinylidene chloride, in which the chlorine atoms are regularly arranged in the chain, have a different appearance from chlorinated polyethylenes of equal chlorine content.

Polyvinyl alcohol: The infrared spectra (Fig. 144) of this well-crystallized plastic were important for the clarification of the question of how ordered the hydrogen bonds between oxygen atoms in different neighbouring chains

351

were and whether they were all of equal strength. This question was not unambiguously decided by X-ray methods. The two alternatives were, firstly, that hydrogen bonds were completely equivalent, i.e. the OH . . .O bond was about 60° inclined to the chain alternately in the opposite sense and approximately in the plane of the carbon atoms thus bound together [Mooney (*486*)]. The second possibility was that these bonds were statistically distributed along the chain, lay outside the plane mentioned, and were of varying strength [Bunn and Peiser (*103*)]. Infrared investigation with polarized radiation has decided more or less unambiguously in favour of the second alternative [Elliot *et al.* (*196*); Glatt *et al.* (*257*)]. The spectra of simple oriented films show the expected dichroism of the CH stretching

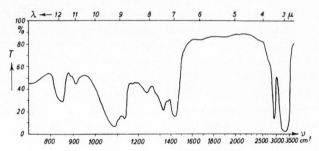

FIG. 144. *Infrared spectrum of polyvinylalcohol* [*Blout and Karplus (68)*].

vibration at 2940 cm^{-1}, while the OH stretching vibration near 3400 cm^{-1} is found to be unoriented. The spectral position of this band indicates strong hydrogen bonding with little or no orientation; this could correspond with either of the alternative structures. The surprisingly broad character of the band provides evidence against the first structure with equivalent and strictly ordered hydrogen bonds. The spectra of samples doubly oriented by hot rolling demonstrate even more clearly the correctness of the second proposal. If the OH . . . O bonds were restricted to the plane of the carbon atoms which they connect, the corresponding band should show strict dichroism if the direction of vibration of the polarized radiation approximated to this plane or coincided with it. This has never been observed.

Polystyrene: Published infrared work concerning this plastic (Fig. 145) is concerned more with the course of the polymerization reaction, and with degradation caused by heat or light, than with the structure itself. Infrared spectral evidence has shown that the vinyl polymerization with benzoyl peroxide as catalyst is controlled principally by the reactivity of the carboxy-radical formed by fission of the peroxide and not by the hydrocarbon radical (Pfann *et al.* (*530*)]. The infrared spectra of polystyrenes which have been subjected to heat or ultraviolet irradiation (Fig. 146) allow conclusions as to the reactions occurring during these processes [Achhammer *et al.* (*1*)]. Although many details await clarification, hydrogen bonds are demonstrated to be present between the hydroxyl groups formed in the course of

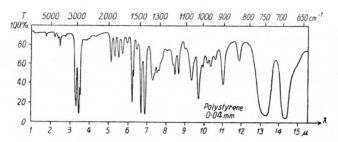

FIG. 145. *Infrared spectrum of polystyrene [from an original of the Perkin-Elmer Corp.].*

the oxidation. These hydroxyl groups can lose water to give ether-type bridges. From a purely technical point of view, infrared spectroscopic measurements of oxidation effects are important as a rapid test of the resistance of polystyrenes to light [Matheson and Boyer (*449*)].

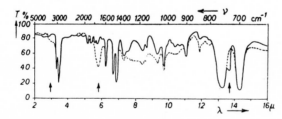

FIG. 146. *Alteration in the polystyrene spectrum after ultraviolet irradiation. Continuous line: non-treated material. Dotted line: material after 200 hours intensive irradiation. The band at 13·7 μ which disappears belonged originally to solvent (toluene) present in the polystyrene [from Achhammer* et al. (*1*)].

Natural rubber: About 15 to 20 years ago, some of the first applications of infrared spectroscopy to high molecular weight compounds were concerned with rubber. Because of defects in the apparatus used at that time, these investigations are no longer of any importance, in spite of some fine

2 A

individual observations. The most important recent work on the structure
and constitution has shown that the two types of natural rubber known as
Hevea rubber (Hevea-, Guayul-, Koksaghyz-rubber) and as Gutta rubber
(Guttapercha, Balata, Chicle) could be differentiated by the infrared spectrum [Richardson and Sacher (*592*); Saunders and Smith (*628*); Sutherland
and Jones (*694*)]. Each of the two types can exist in several modifications
and each of these modifications gives a different spectrum (Fig. 147).
Hevea rubber is known to be amorphous at normal temperatures when
fresh, but it becomes hard and crystalline on long keeping. This crystalliza-

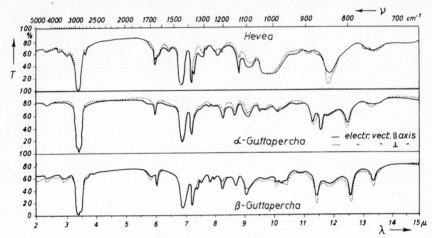

FIG. 147. *Infrared spectra of various types of rubber. Full lines
refer to radiation with the electric vector parallel to the axis of the
rubber molecule, and dotted lines to that with the electric vector
perpendicular to this axis [from Saunders and Smith (628)].*

tion can be accelerated by cooling to -20 to $-30°$C. It manifests itself in
the spectrum principally by alterations in a band at 837 cm^{-1} and also in the
bands at 1137, 1316, and 1367 cm^{-1}. New bands at 870 and 964 cm^{-1} also
occur for the crystalline state. All these phenomena can only be observed
for types of rubber with molecular weights above 100,000. For smaller
molecular weights, the greater thermal motion of the chains apparently
prevents crystallization.

Guttapercha exists at normal temperatures in two crystalline forms,
differentiated as α- and β-guttapercha; above $80°$C, these are converted into
an amorphous modification. Each of the three forms possesses a characteristic spectrum. Films of α-guttapercha are usually obtained from solutions in benzene which have not been heated. β-Guttapercha is obtained by

heating above 80°C and rapid cooling; slow cooling gives back the α-form. Bands at 864, 883, and 802 cm^{-1} are characteristic for α-guttapercha and bands at 876, 796, and 750 cm^{-1} for β-guttapercha.

Chemically, the natural elastomers are all polyisoprenes. It was thus natural to assign the differences between the Hevea and the Gutta type to cis-trans-isomerism. This assumption was confirmed by the spectra, especially those obtained using polarized radiation and oriented samples. In the simple schematic planar polyisoprene model (Fig. 148), the double bonds in the cis form lie parallel to the axis of the chain, but they are inclined to it in the trans form. The crystal constant for β-guttapercha is equal to one isoprene unit, and that for α-guttapercha and Hevea rubber to two isoprene units. Accordingly, a marked π-dichroism of the C=C bond and the corresponding band at 1667 cm^{-1} are to be expected for the cis form. For the trans form, this dichroism should be much less or even go over to σ-dichroism. In fact, Hevea rubber shows π-dichroism and guttapercha a moderate σ-dichroism. According to the length of the crystal constant just mentioned, the bands of β-guttapercha, in addition to those of Hevea rubber and α-guttapercha, are expected to be split. This can be observed, in part as a full splitting and in part as a broadening of the

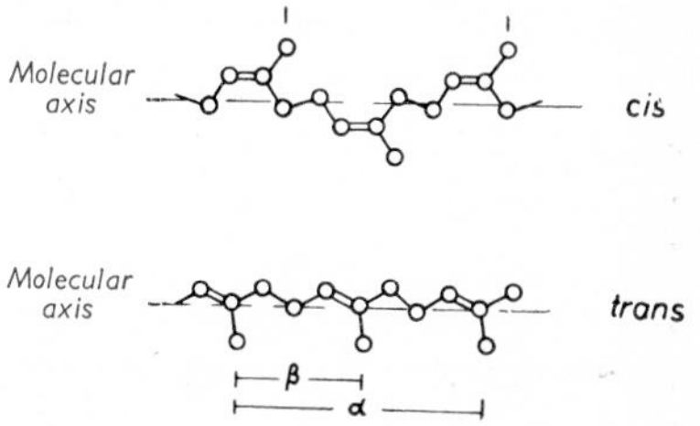

FIG. 148. *Schematic planar polyisoprene model [from Saunders and Smith (628)].*

bands. Complete agreement with expectation is not found, either with respect to this splitting, or with respect to dichroism. This is not surprising considering the simplicity of the planar model taken, which is certainly only a rough approximation. The spectral differences between the two guttapercha forms themselves are probably due to structural differences inside the crystallites. The most marked manifestations of these are to be expected at longer wavelengths than those as yet investigated.

Infrared spectroscopy has been used for the clarification of the technological processes of rubber chemistry as well as for the clarification of structural questions. A new band at 962 cm^{-1} appears during vulcanization with elementary sulphur [Sheppard and Sutherland (646, 647)]. The initial assignment, to one of the possible sulphur-carbon bonds (with fission of a C=C bond), was later demonstrated to be untenable. The new band was shown to be due to new bonds of type RCH=CHR′ and some of type RR′C=CH$_2$ (band at 890 cm^{-1}) which appeared during vulcanization. Because the intensity of the C=C band at 1667 cm^{-1} remains almost constant, it must be

Results and Applications

concluded that the vulcanization process displaces the isoprene double bonds lying in the chain. For vulcanization with S_2Cl_2, the spectral alterations are more marked [Thompson (704)]. New bands appear at 635, 680, and 895 cm^{-1}, and others at 1135 and 1667 cm^{-1} become weaker, especially at the beginning of the process. In this connection, it is interesting that a displacement of the double bond from the main chain into a slide chain according to the scheme

$$\begin{array}{cc} \mathrm{CH_3} & \mathrm{CH_2} \\ | & || \\ \mathrm{C-C=C-C} & \mathrm{C-C-C-C} \end{array}$$

occurs when rubber is chlorinated or reacts in certain other ways [Salomon and v. d. Schee (622); Salomon et al. (623)]. These side-chain vinyl groups are responsible for the known tendency of rubber derivatives to form cross-linkages between the chains. This is very important technologically and does not occur in untreated rubber.

The oxidation of rubber in various ways has often been studied by infrared spectroscopy [Allison and Stanley (13); Cole and Field (134); D'Or and Kössler (175)]. New bands are found at 890, 1175, 1700, 1720, and 3600 cm^{-1}; the last three are especially marked at an advanced state of oxidation. A large increase in the background absorption in the region 1000 to 1300 cm^{-1}, and a decrease in the intensity of the band at 914 and 840 cm^{-1}, are also noticed. The interpretation of these results is by and large clear. The absorption at 3600 cm^{-1} is derived from hydroxyl, that in the 1700 cm^{-1} region from carbonyl and that at 1000 to 1300 from C—O groups. The band at 890 cm^{-1} is assigned to the group RR'C=CH$_2$ and the decrease in intensity mentioned is interpreted as a result of the fission of the double bond and increasing saturation. The fact that two absorption peaks are found in the carbonyl region can be interpreted as the simultaneous formation of aldehyde and ketone groups or alternatively as the effect of conjugation; the colouration of the material observed points to the latter. Carboxyl groups are considered to form only at a late stage of the oxidation. Peroxide groups are certainly concerned in the initiation of the oxidation, but apparently they undergo fission very quickly, because they cannot be observed in the infrared spectrum.

Synthetic rubbers: The same interest has been shown in the infrared spectra of the various types of synthetic rubber as in those of natural rubber [Dinsmore and Smith (174)]. Some very important structural questions have only been satisfactorily answered by infrared spectroscopy. Infrared spectra (Fig. 149) were especially valuable in the conclusions that they gave regarding butadiene polymerization, in particular the relative proportions of 1,2- and

356

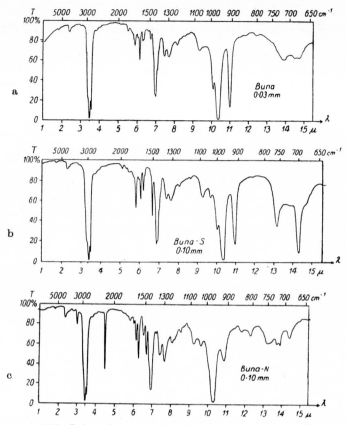

FIG. 149. *Infrared spectra of various types of buna rubber* [*from the original of the Perkin-Elmer-Corp.*].

1,4-addition in the various products and the alteration in the technical properties connected with this ratio. The three possibilities

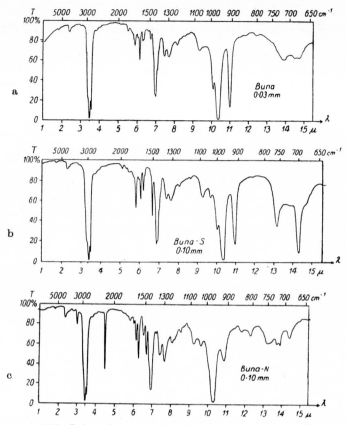

can be differentiated in the spectrum by bands at 996 and 910 cm^{-1} for the 1,2-addition, at 724 cm^{-1} for the 1,4-cis-addition and at 969 cm^{-1} for the 1,4-trans-addition [Field *et al.* (*208*); Hampton (*295*); Hart and Meyer (*303*); Treumann and Wall (*716*); Richardson and Sacher (*592*)].

357

Results and Applications

The characteristic bands of both components are shown in the infrared spectra of the various mixed polymers such as buna-S (butadiene-styrene) and buna-N or perbunan (butadiene-acrylonitrile). For these, phenyl bands at $13 \cdot 2$ and $14 \cdot 3\,\mu$, or the nitrile band at $4 \cdot 5\,\mu$, are shown together with those

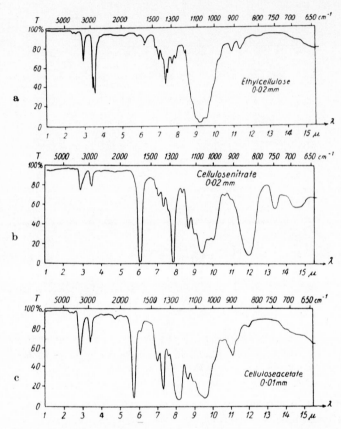

FIG. 150. *The infrared spectra of various cellulose derivatives* [*from the original of the Perkin-Elmer Corp.*].

of butadiene. Obviously, differentiation between 1,2- and 1,4-addition by the bands at $10 \cdot 97$ and $10 \cdot 35\,\mu$ is also possible here.

Cellulose: The infared spectra of various cellulose derivatives are shown in Fig. 150. The marked spectral differences are easily explicable from the differences in chemical structure. Other investigations have been concerned with the study of oxidative degradation, deuteration and the existence of hydrogen bonds and their importance for the formation of crystalline regions.

358

Although a series of papers is already available on this subject, the investigations must still be considered to be in progress (Forziati *et al.* (*216*); Rowen and Plyler (*609*); Rowen *et al.* (*610, 611*)].

Polyamides: The infrared spectroscopic investigation of polyamides is important because of the technical importance of these substances, especially their use as completely synthetic fibres. Such studies are also of great interest because peptide linkages are here incorporated into long hydrocarbon chains. Thus, their investigation is important with reference to the clarification of the structure of the poylpeptides and proteins. The main characteristic of the polyamide spectrum (Fig. 151) is the displacement of the carbonyl band from 1720 to about 1630 cm^{-1} and the appearance of a further strong band near 1560 cm^{-1}. These two phenomena are observed in

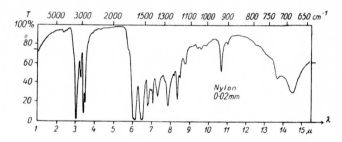

FIG. 151. *Infrared spectrum of a polyamide film* [*from the original of the Perkin-Elmer Corp.*].

the spectra of almost all amides. Their interpretation has been discussed under that of the amide spectra (see page 339). A hydrogen bond-bridge, perpendicular to the molecular axis and between the C=O and to the NH groups of neighbouring chains, is probably an essential structural element. This assumption is confirmed by the spectra. The position of the NH stretching vibration at relatively long wavelengths indicates hydrogen bonding and the observed displacement of the carbonyl band could also originate from this influence. Using polarized radiation, the NH stretching bands are found to be markedly σ-dichroic in oriented samples. Exactly the same conclusions follow from the polyamide spectra in the short-wave infrared region. A band at 6523 cm^{-1}, undoubtedly the first overtone of a NH stretching vibration modified by hydrogen bonding, shows strong σ-dichroism. In melted polyamides this band is no longer observed, but it is replaced by another at 6757 cm^{-1}, where the first overtone of the free NH group is to be expected. This indicates that the bridges between the chains, the origin of the crystalline order, are destroyed on melting.

Results and Applications

Although observations on stretched polyamides gave no final clarification of the mutual orientation of the various linkages, deeper insight was afforded by the use of doubly oriented samples [Elliot *et al.* (*196*)]. The flat plate-like crystallites are thereby oriented in a plane which coincides with the crystallographic 010 plane. According to X-ray results, the C=O bonds lie in this plane and this causes the strong σ-dichroism observed for the corresponding bands. The NH groups, on the other hand, should be inclined to this plane at the same angle as the axis of the CH₂ groups (the CH₂ groups themselves lie in the plane of the C atoms). This angle is found to be 22°, but the infrared spectrum does not give the sign of the angle, because the polyamide crystallites are arranged only in the surface. On the reasonable assumption that the NH groups and the neighbouring C=O groups lie on the same side of the chain, and using the angle of 4° that the latter make with the 010 plane, the angle between the C=O and NH bonds is found to be 18° (Fig. 152). A value of 39° would be expected on the basis of the normal nitrogen valency angle; the large discrepancy is due to the influence of the hydrogen bonding.

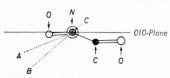

FIG. 152. *Geometry of the polyamide crystallites, according to X-ray and infrared spectroscopic investigations, as viewed along the molecular axis* [*Elliot (196)*].

Polypeptides: As already mentioned, investigations on polypeptides have recently become important in connection with the question of the structure of proteins. Because of the valuable information obtained by infrared spectroscopy on the polyamides, its application to these questions was inevitable [Ambrose and Elliot (*14, 15*); Elliot and Ambrose (*192*); Elliot and Malcolm (*193*); Goldstein and Halford (*265*); Hurd *et al.* (*344*); Fraser (*228*)]. Although these investigations are still proceeding, it has already been found possible to determine from infrared spectra whether a polypeptide exists in the folded (α-form) or extended (β-form) form. It has been shown that the different vibrations of the NH and C=O group show characteristic behaviour with respect to polarized radiation according to the form in which the polypeptide being investigated exists. Table 32 summarizes the results. Although in principle simple infrared measurements with polarized radiation on the NH or C=O fundamental vibration could clarify the position, the bands in question lie in a region of great experimental difficulty because of the water content of biological substances which cannot easily be avoided. Fortunately, bands occur above 4500 cm⁻¹ which allow similar differentiations to be made. The exchange of H₂O for D₂O is also of assistance, although some of the alterations connected in the position and intensity of the bands need to be studied more exactly [Lenormant and Chouteau (*415*)]. Unex-

360

pectedly, the single fibrous protein (collagen) does not fit into the scheme of Table 32; therefore, it is given separately using gelatine as an example.

TABLE 32. *Position and Dichroism of Characteristic Peptide Bands*
[*Ambrose and Elliot (14)*]

| Band | Synthetic polypeptides and α- and β-Proteins | | | | Gelatine | |
| | Position in cm⁻¹ | | Dichroism | | Position cm⁻¹ | Dichroism |
	folded	stretched	folded	stretched		
NH-Deformation	1545	1525	σ	π	1548	π
C=O-Stretching	1660	1640	π	σ	1660	σ
NH-Stretching	3300	3300	π	σ	3335	σ
C=O-Combination	4600	4530	none	π	4590	σ
NH-Combination	4830	4825	σ	π	4865	π

The reason for this discrepancy is probably the existence of a special folded structure for collagen, being different from the α-folded form [Ambrose and Elliot (*14, 15*)].

3. *Condensation Products and Complicated Natural Products*

As already mentioned, the infrared spectroscopic investigation of condensation products, and of compounds of complicated biochemical and biological structure, is still only beginning. However, this is already a fruitful extension of infrared spectroscopy. Often the infrared spectrum is only used to confirm the identity of two substances, e.g. one found naturally and one synthesized; this can be done without any knowledge of the origin of the individual bands. In other work, the infrared spectrum is used as a purely analytical tool based on certain empirical relationships. As an example, the quantitative estimation of unreacted phenol in phenol-formaldehyde synthetic resins by the strong phenol band at $14 \cdot 4 \, \mu$ may be mentioned [Smith *et al.* (*678*)].

The nature, extent, and orientation of bridges formed between recurring structural units such as benzene rings in *condensation products* have already been successfully investigated by means of infrared spectroscopy. In the case of simple phenol- and cresol-formaldehyde resins, information can be obtained as to the orientation of the bridges by means of the very marked influence, already mentioned, of position isomerism in the benzene ring on

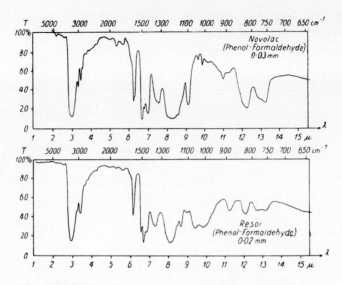

FIG. 153. *Infrared spectrum of a phenol-formaldehyde resin in the Novolac- and Resol-form [from the originals of the Perkin-Elmer Corp.].*

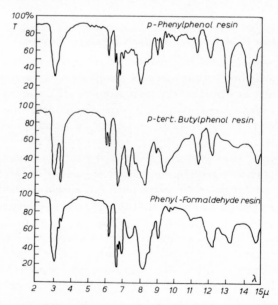

FIG. 154. *Infrared spectra of phenolic resins of different structures [Shreve (499)].*

the spectra. Bands at about 760 and 820 cm^{-1} are usually found for phenol-formaldehyde Novolacs (Fig. 153). The orientation of the substituents is more important than their chemical nature for di- and poly-substitution of the benzene ring. Hence, the first band can be assigned to ortho- or 1,2,3,-substitution, and the second to para- or 1,2,5-substitution [Thompson and Torkington (705); Richards and Thompson (588)]. Conclusions as to the nature of the bridges have also been drawn from the infrared spectrum. The

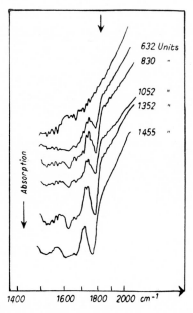

FIG. 155. *Part of the infrared spectrum of penicillin* [*Barnes et al. (43)*].

methylene bridges expected on chemical grounds were indeed found, but also ether bridges, usually in the ortho-position [Bender (54)]. Obviously, other results can be obtained from the spectra, e.g. as to the chemical nature of the components taking part in the condensation [Shreve (651)]. Fig. 154 serves as an example. The presence of a benzene ring is shown by the well-known bands at 6·2 and 6·7 μ among others, that of a phenol by the band at 8·2 μ. The nature of the substituents is indicated by bands at 11·4 and 12·1 μ for alkyl groups, at 13·1 and 14·4 μ for phenyl, etc. In spite of these significant beginnings, most of the work necessary to make available all the possibilities of infrared spectroscopy for the investigation of condensation products still remains to be done. Recently, the infrared spectroscopic investigation of the

363

Results and Applications

condensation products of urea and formaldehydes has received attention [Becher (*49*)].

Extensive investigations of biochemical and biological compounds have already been completed and others are still proceeding. The complexity of the problems and the extent of the investigations, which often involve the spectral determination of numerous compounds, forbids a detailed treatment here. It may be mentioned that, e.g. the investigation of antibiotics owes very much to infrared spectroscopy; conversely, infrared spectroscopic methods and techniques have been much advanced by this work. For example, teams of numerous workers in the U.S.A. investigated the infrared spectra of hundreds of substances in connection with the clarification of the structure of penicillin. This work has been described in comprehensive publications, e.g. (*P, Q*). The use of the infrared spectrum to determine directly biological activity is exemplified by Fig. 155 [Barnes *et al.* (*40*)]. This demonstrates the relationship between the activity (expressed in arbitrary medical units) and the intensity of a band at 1770 cm^{-1}. Once such a relationship has been established, the infrared spectrum, which can be obtained easily and quickly, can be used in this way to control production. Details of the infrared spectra of some important antibiotics have been given by Lacher *et al.* (*401*).

3. Inorganic Compounds

1. *Fundamentals*

The main applications of infrared spectroscopy are to be found in the field of organic chemistry. Nevertheless, the potentialities for inorganic chemistry are considerable. The older infrared literature, up to the end of the second decade of this century, contains a large number of excellent infrared spectroscopic investigations of inorganic compounds. For the most part these are concerned with diatomic polar gases and crystals (simple or complex) which were usually investigated by the reflection method. However, more recent work has also been concerned with inorganic compounds, as can be seen from the compound index of the new Landolt-Börnstein volume (M) on molecular spectra. Some of these have repeated the early investigations with improved methods and instruments and others have dealt with compounds hitherto uninvestigated. Much of the older work was of great importance for the technical development of the infrared region and for the initiation of fundamental theories.

In the range of inorganic compounds investigated by infrared spectroscopy, salts have been prominent, particularly those with complex ions. The reflection method gave useful values for the position of their characteristic vibrations, although many of these are crude compared with what can be achieved now. These values were refined by the absorption method and this was used to investigate the detailed structure of the ions. Besides these fundamental investigations, the application of infrared spectroscopic analytical methods recently appears to have become important for this type of compound. The infrared spectra of the hydrogen halides have been subjected to especially detailed study, because of the simplicity of their spectra and because their simple constitution is well suited to testing and gradually refining the underlying theory.

2. *Normal Vibrations of Ionic Crystals*

The discussion will be concerned principally with ionic crystals, but crystals containing covalent bonds will not be entirely excluded because these give some information about the ionic crystals. Initially the simplest case will be considered, i.e. crystals with mono-atomic ions. In the form of the alkali

365

Results and Applications

halides, these are of great importance to infrared spectroscopy. For these compounds, only so-called lattice vibrations are possible, i.e. movements of the ions relative to one another. Where these are infrared active they cause broad absorption bands in the long-wave infrared region. These bands are so intense that their wings reach into the middle and near infrared regions and are responsible for the absorption observed in these regions at technical thicknesses. If the ions are composed of several atoms, torsion vibrations and so-called internal vibrations occur in addition to the lattice vibrations. For the internal vibrations, the centre of gravity and the positions of the axes of the molecule considered as a whole remain fixed and the vibration takes place in the molecule itself. The vibrations are characteristic of the particular ions involved. The differentiation between lattice and internal vibrations is still possible for molecular crystals, in which the individual molecules are held together by van der Waal's forces or subsidiary valencies in the lattice (e.g. benzene, phosphorus). However, this differentiation is no longer possible for crystals in which the lattice is composed of covalent bonds such as in diamond. Nevertheless, such crystals also show characteristic spectra with sharp prominent bands.

In general, the older work of reflection spectroscopy will not be considered in detail here; for this aspect reference should be made to earlier treatments (M, N, R). Only the alkali halides will be considered in any detail, because this group of substances is of such importance to infrared spectroscopy and because their investigation is an important test of the Born lattice theory and serves as starting point for its extension. As is well known, a pronounced reflection maximum is found on the short-wave side of the normal vibration. From the position of this maximum, the exact frequency of the normal vibration itself can be found using rather complicated formulae. However, the normal vibration frequency is obtained more exactly by absorption measurements on extremely thin layers, the thickness of which should not exceed a few microns, or in some cases even a few tenths of a micron. Such measurements were first carried out by Czerny and Barnes [for literature see (O)] on NaCl. They enable a considerably more stringent test of the lattice theory to be made than the earlier values from reflection measurements. Table 33 compares the observed values of the normal vibrations and those calculated from the Born lattice theory. The agreement is quite satisfactory. However, the reflection measurements surprisingly indicated the presence of a second weaker maximum at shorter wavelength. This observation was confirmed and extended by absorption work. In addition, the course of the extinction coefficient curve deviated considerably from that calculated: values were too small on the shorter wavelength side of the normal vibration and too large on the longer wavelength side. The interpretation and the

366

TABLE 33. *Positions of the Normal Vibrations of Alkali Halides in* μ *[Matossi (O)]*

	Exp.	Calc.		Exp.	Calc.		Exp.	Calc.		Exp.	Calc.
LiF	32·6	—									
NaF	40·6	—	NaCl	61·1	61·6	NaBr	74·7	71·4	NaI	85·5	84·6
			KCl	70·7	74·5	KBr	88·3	88·0	KI	102·0	108·3
			RbCl	84·8	—	RbBr	114·0	—	RbI	129·5	—
			CsCl	102·0	—	CsBr	134·0	—			
TlF	67·5	—	TlCl	117·0	—						

theoretical consequences of these two observations have been shown to be intimately connected with each other [see the detailed discussion with literature references given by Matossi (O)]. It is sufficient to say that these observations have led to the extension of the Born theory by using anharmonic terms.

Recent measurements on complex ions will now be discussed. These measurements have enabled the normal vibrations to be determined, the influence of the positive ion to be studied, and the general applicability to analysis to be tested [Miller and Wilkins (477); Hunt et al. (343); Hunt and Turner (342)]. Fig. 156 gives a summary of the spectral positions of the characteristic bands of the most important complex ions. The inorganic ions possess characteristic bands just as do the atomic groupings and the bond types of organic molecules.† Measurements on ten sulphates are summarized in Fig. 157 which exemplifies the influence of the positive ion. This influence on the exact position of the characteristic sulphate bands at 650 and 1100 cm⁻¹ is noticeable. So far, it has only been possible in one case to demonstrate a regular relation between the band displacement observed and the properties of the positive ion. This was for the 860 cm⁻¹ band of anhydrous carbonates [Hunt et al. (343)]. Certainly, it is expected that the varying charges and sizes of the different positive ions will influence the frequency of the normal vibrations as a result of the variable intermolecular field. In addition, alterations as a result of the symmetry variations are not excluded. Finally, varying degrees of hydration will also effect the characteristic bands. So far, it has not been possible to interpret these effects quantitatively, but this is due to the relatively small amount of data available and also to the deficiencies of present-day theory.

With respect to the application of the characteristic bands for analytical

† The band at $7 \cdot 22$ μ, sometimes mentioned as an inexplicable 'ghost band' in infrared spectra [Launer (406)], has been shown to be a NO_3^- band. It results from the formation of nitrate on NaCl plates placed too near to the Nernst glower where nitric oxide is produced.

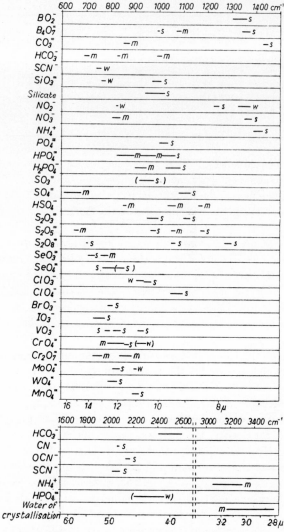

FIG. 156. *Characteristic bands of inorganic ions* [*Miller and Wilkins (477)*].

purposes, Fig. 156 at present appears to give too optimistic an impression. The qualitative analysis of salt mixtures by infrared spectroscopy alone frequently leads to false conclusions. The simultaneous use of several analytical techniques, e.g. emission spectra, infrared spectra and X-ray investigations, may be more favourable. At least in some trial experiments, these

368

have given very encouraging results. Quantitative analysis by infrared spectroscopy has hitherto been hindered by difficulties of sample preparation. Almost all salts must be investigated with the powder or suspension

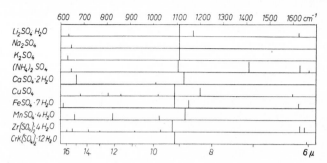

FIG. 157. *Infrared spectra of various sulphates* [*Miller and Wilkins (477)*].

technique, both of which are semi-quantitative at best. It is possible that the further extension of the disc technique, which can be used quantitatively will change this situation markedly.

3. Glasses

Among the infrared spectroscopic investigations devoted to the more complex inorganic systems, the measurements on glasses deserve special mention. Earlier they were a favourite theme of reflection spectroscopy, the application of which gave much valuable information on their structure. Recently various glasses have been the subject of infrared spectroscopic work. Florence *et al.* (*214*) investigated the absorption bands of numerous glasses in the near infrared region and tried to assign them to various atomic groupings, in part of the glass itself, but in part of various other substances. Surface effects were found to be of greater importance than volume effects. Free water, OH^-, CO_3^{--}, and NO_3^- ions, free CO_2, and SiO bonds were frequently found to be the origin of absorption.

Anderson (*20*) investigated barium silicate, boron oxide, and a technical glass, in the sodium chloride range. Both reflection and absorption measurements were used, with special attention to the effect of colouring and etching on structure. The spectra suggest SiO_4 tetrahedra and a Ba octahedron as structural elements. The Ba co-ordination appears to disappear on etching so that a configuration similar to quartz remains.

Absorption measurements of glasses at very high temperatures [Genzel (*247*); von Neuroth (*494, 495*)] are of importance for the technology of glass

Results and Applications

manufacture. By application of the principle of alternating radiation, it was possible to measure the absorption in the near infrared region from room temperature up to 1300° C. Absorption decreased with the temperature up to a transition region (500 to 600° C). Within the transformation interval the behaviour of the glasses varies, sometimes the absorption increases for one part of the spectrum and decreases for another. In the liquid state the absorption again increases with the temperature.

4. Special Effects and Applications

Infrared spectroscopy is an excellent experimental method for the investigation of certain effects connected with the structure of molecules. Only a few of these can be mentioned and the discussion has to be limited to the most important facts without considering the details or the fundamental theory. It must be remembered that infrared spectroscopy is only one of many methods of investigating such effects and its full value is attained only in conjunction with other methods.

1. Hydrogen Bonds

Hydrogen bonding has a large influence on the properties of compounds and has frequently been investigated by infrared spectroscopy [e.g. Davies and Evans (162); Francis (221); Fuson et al. (239); Kuhn (399); Lord and Merrifield (427); Lüttke and Mecke (432); Mecke (461); Rundle and Parasol (616)].

A strongly polar group with 'exchangeable' hydrogen, such as OH or HN as a proton donor, and another group with easily polarized electrons, as a proton acceptor, are required for the formation of hydrogen bonds. Intermolecular and intramolecular hydrogen bonds must be differentiated. The former involve the interaction between two or more different molecules which satisfy the above conditions; these can be molecules of the same kind, or other molecules, e.g. those of the solvent. Intramolecular hydrogen bonds are formed when two groups suitable for their formation are situated at the correct distance within the same molecule. The well-known association of the alcohols and carboxylic acids are examples of intermolecular hydrogen bonds. Intramolecular hydrogen bonding is also known as chelation and is found, e.g. in enolized diketones. The existence of hydrogen bonds causes a displacement of the bands belonging to the donor group. Thus the stretching and deformation vibrations of the hydrogen-bonded OH and NH groups are found at different wavelengths from those for the so-called free OH and NH groups. Intensities are also altered.

Several bands are found, e.g. for the stretching vibration, for intermolecular hydrogen bonds, according to the number of possibilities of association. The intensities of these bands depend on the concentration, i.e. on the degree of association, which is always a chemical equilibrium [Coggeshall and Saier (*128*)]. The band corresponding to non-bonded species is observed at sufficient dilution, i.e. for sufficient mutual separation of the associating molecules. This band decreases in intensity with increasing concentration and the band of the associated species now appears. For phenol, these bands are of quite different appearance (Fig. 158); the associated band is quite broad and diffuse whereas that of the free OH group is sharp and prominent.

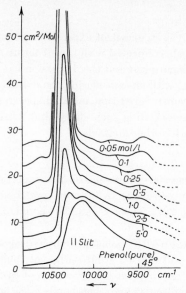

FIG. 158. *Variation of the OH stretching vibration bands of phenol as a result of association* [Mecke (*461*)].

The intensity of the band is a direct measure of the number of associated and non-associated molecules. Provided both bands obey Beer's law, the proportion

$$\alpha = \frac{\epsilon_c}{\epsilon_\infty} \qquad\qquad [\text{IV, 4.1}]$$

of monomeric molecules can be determined at various concentrations by measurements of the extinction coefficient ϵ_c at the concentration c and also ϵ_∞ at infinite dilution. The results are identical whether they are obtained from fundamental vibrations or overtones and agree very well with values found by other methods. Using α and the Duhem-Margules equation

$$d\left[\ln\left(1 - x + \frac{M}{M}.x\right)\right] = xd[\ln \alpha] \qquad\qquad [\text{IV, 4.2}]$$

(where x is the mole fraction), the mean molecular weight $\overline{M}$ of the associated species can be calculated and thus the degree of association

$$f = \frac{\overline{M}}{M} \qquad [\text{IV, 4.3}]$$

Provided the association leads to the polymeric chain molecules, the mean association energy $\overline{W}$ can be calculated from the temperature variation of the bands

$$\overline{W} = RT^2\left[\frac{d \ln \epsilon_c}{dT}\right]_{f=\text{const}} \qquad [\text{IV, 4.4}]$$

Table 34 gives some values of $\overline{W}$ and f determined spectroscopically. Fig. 159 shows the concentration dependence of f for alcohols and phenols in various solvents.

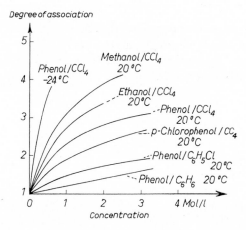

FIG. 159. *Values of the degree of association of some alcohols and phenols determined spectroscopically* [*Mecke (461)*].

Smith and Creitz (675) have recently studied the spectroscopic evidence for the association of alcohols. For the OH stretching vibrations they found the values given in Fig. 160 for the individual cases: $2\cdot74\ \mu$ for the monomer, $2\cdot86$ and $2\cdot76\ \mu$ for the dimer formed by a single bridge, a value between $2\cdot76$ and $2\cdot96\ \mu$ for the dimer with a double bridge, and finally $2\cdot76$, $2\cdot86$, and about $3\ \mu$ for the polymer, according to the extent the OH bond was influenced by the association. These investigations make probable the existence of the dimer with a single bridge that was earlier disputed. The effect of steric hindrance on the association was shown by investigations

373

TABLE 34. *Mean Association Energies $\overline{W}$ and Degrees of Association f of Alcohols and Phenols from Infrared Measurements* [*Mecke (461)*]

Substance	Solvent	$\overline{W}$ kcal/Mol	f at $c = 1$ Mol/l
Methanol	CCl_4	4·72	3·05
	C_6H_6	3·67	—
tert.-Butanol	CCl_4	5·3	2·0
Benzylalcohol	CCl_4	4·60	2·50
	C_6H_6	4·38	—
Phenol	CCl_4	4·35	2·30
	C_6H_6	3·55	1·20
	C_6H_5Cl	3·48	1·50
p-Chlorophenol	CCl_4	3·72	1·90
o-Cresol	CCl_4	3·8	1·7
p-Cresol	CCl_4	4·4	2·3

on branched alcohols to be parallel to the effect of dilution. It was already known for substituted phenols [Coggeshall (*126*)].

The influence of association on the stretching vibration bands is readily interpreted, partly because of their isolated position in the spectrum. The

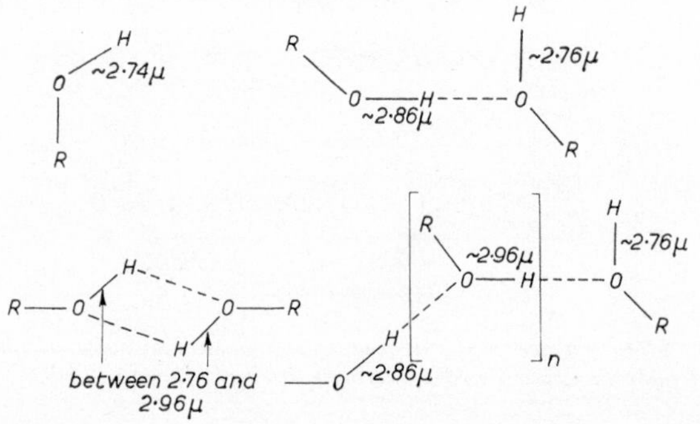

FIG. 160. *Various possibilities for the association of alcohols and their influence on the position of the OH stretching vibrations* [*Smith and Creitz (675)*].

situation for the other bands is less clear, e.g. the OH deformation vibration in the 7 μ region (which contains many bands) or the OH torsional vibration in the 15 μ region. However, specific effects are to be expected here also.

Fig. 159 indicates the influence of the solvent on the degree of association; this is the greater, the more marked the proton acceptor qualities of the solvent. Accordingly, for transition from non-polar to polar solvents, a displacement of the OH band to smaller wave numbers with simultaneous broadening and decrease in intensity is to be expected. As is indeed found for phenol, see Fig. 161 and Table 35. The splitting of the phenol OH bands into two components in benzene, toluene, and cyclohexene is especially noticeable. This phenomenon has been interpreted by Mecke (*461*) as a type of

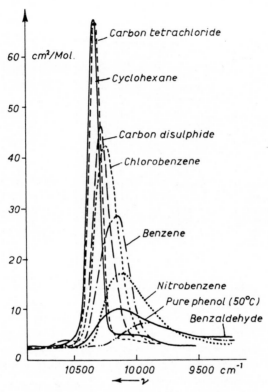

FIG. 161. *Effect of the solvent on the position and intensity of the second overtone of the OH stretching vibration of phenol [Lüttke and Mecke (432)].*

isomerism. For example, the molecular plane of benzene can lie parallel or perpendicular to the OH group. Thus, the polarizability, which depends on the π-electrons, can vary.

Intramolecular hydrogen bonds have been investigated *inter alia* by Barchewitz (*34*) and Mecke and his school (*433, 461, 607*). The substituted phenols will be described as an example. Different results are found with different acceptor group. *o*-Substituted phenols, with halogen, OH, OCH$_3$, CN, or phenyl groups as a substituent, show one and sometimes two bands of

375

Results and Applications

TABLE 35. *Association Effects of Phenol in Various Solvents measured on the Second Overtone of the OH Stretching Vibration [Mecke (461)]*

Solvent	Band at cm^{-1}	Max. Ext.-coeff. cm^2/Mol	Integr. absorption cm/Mol	Half band width cm^{-1}
Cyclohexane	10,340	64·5	8·0	100
Cyclohexene	10,340	18·5	5·6	230
	9,950	7·0	—	—
CCl_4	10,330	64·0	8·2	110
C_2Cl_4	10,325	63·5	7·7	120
C_2HCl_3	10,305	53·0	7·5	120
$CHCl_3$	10,295	49·0	7·7	155
CS_2	10,285	43·5	8·2	160
C_2HCl_5	10,285	54·5	7·7	160
Chlorobenzene	10,250	40·0	9·3	210
Benzene	10,175	27·0	7·8	270
	10,330	10·0	—	—
Toluene	10,140	25·2	9·0	340
	10,330	10·0	—	—
Nitromethane	10,160	11·0	4·5	370
Anisole	10,130	11·6	6·2	430
Nitrobenzene	10,120	14·0	5·5	370
Benzaldehyde	9,920	5·0	2·8	540
Acetophenone	9,870	4·0	2·6	650
Diethylether	9,700	4·0	2·7	700
Acetone	9,600	5·0	3·1	750

varying intensity in the OH stretching region. The weaker of these two bands corresponds to the non-bonded phenol, i.e. to the free OH group. The other band lies at longer wavelength and is usually somewhat broadened (Fig. 162). The splitting has been interpreted as cis-trans-isomerism where only the cis-form is capable of forming a hydrogen bond.† However, if the substituent is a NO_2, CHO, COOR, or COR group only one single, very strong, band is found, much displaced towards longer wavelengths and broad and diffuse (Fig. 162). Thus only the cis-form occurs here and the chelation is especially strong. These two types of substituents are thus sharply differentiated; this is a result of the electron distribution in the donor and acceptor groups, i.e. the size and direction of the mesomeric and inductive effects.

† In the case of the o-halogeno-phenols, repetition of earlier work [Lüttke and Mecke (433)] has shown that the band at the usual wavelength for the OH stretching vibration previously assigned to the trans-form is actually due to contamination by unsubstituted phenol. The trans-configuration therefore does not exist, except for o-iodophenol where a weak but apparently real effect of this type was found [Rossmy et al. (607)].

For reasons of space, details cannot be given here and reference should be made to the discussion in the original paper [Lüttke and Mecke (*433*)].

The investigation of the OH stretching vibration band at 3 μ for crystals with special consideration of the effect of the hydrogen bonding has established an important relationship between the frequency of this vibration and the distance O—O of the two atoms involved in the hydrogen bond [Rundle and Parasol (*616*); Lord and Merrifield (*427*); Hartert and Glemser (*304*)]. In this way the infrared spectrum gives numerical information about the atomic distance by means of very simple measurements. This is very important for the clarification of the structures of crystals containing OH

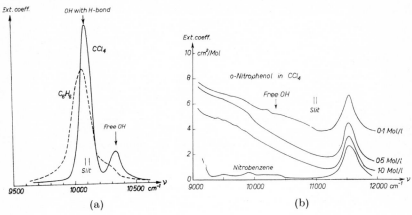

(a) (b)

FIG. 162. *Spectral consequences of chelation in ortho-substituted phenols,* (a) *with halogen,* (b) *with a nitro group as substituent* [*Lüttke and Mecke (432)*].

groups, especially for hydrates and for the whole question of the existence of the so-called hydroxyl bond [Bernal and Megaw (*59*)]. Glemser and Hartert (*258, 259, 305*) have recently used such measurements to clarify the structure of hydroxides, basic salts, and hydrated salts. They reject the hydroxyl bond because of the smooth curve that they found for the relationship just mentioned over the whole of the range of atomic distances in question.

2. Rotational Isomerization

The phenomenon of rotational isomerization was originally studied by means of the Raman effect and this is still the method most favoured. However, a series of papers has recently shown that infrared spectra are suited for the investigation and clarification in many cases. Measurements are

Results and Applications

available for hydrocarbons [Axford and Rank (*28*)], alkyl halides and alcohols [Brown and Sheppard (*92, 94*)], ethane derivatives [Bernstein (*60, 61*); Powling and Bernstein (*559*)], ethylene derivatives [Neu and Gwinn (*493*); Brown and Sheppard (*93*)], chlorinated butadienes [Szasz and Sheppard (*695*)], and butane derivatives [Brown and Sheppard (*91*)].

The purpose of such investigations is the determination of the fundamental vibrations corresponding to the individual isomers and also the quantitative estimation of the energy differences between the isomers, the

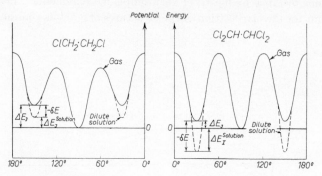

FIG. 163. *Rotational isomerism in ethane derivatives* [*Powling and Bernstein (559)*].

so-called *energy of isomerization* ΔE_J. This is obtained from the ratio K of the concentration of the less stable to the more stable by using the equation

$$K \sim e^{-(\Delta E_J/RT)} \qquad \text{[IV, 4.5]}$$

or an equation derived from this one. The concentration of the individual isomers is obtained from the measurement of the extinctions of the bands unambiguously assigned to them (the applicability of Beer's law is assumed).

The number of isomers depends on the number of minima in the potential energy curve of the molecule in question. For example, there are three for dichloroethane and tetrachloroethane (Fig. 163) [Powling and Bernstein (*559*)]. Two of these, the so-called gauche-forms, are optical isomers and are not spectroscopically distinguishable; the third is the trans-form, which is the only one existing in the crystalline state. The potential energy difference between the minima gives directly the energy of isomerization.

The bands of all the possible isomers are found in the spectrum of the substance in the gaseous and liquid state which shows the phenomenon of rotational isomerism. In order to eliminate them all except one, a spectral investigation has to be carried out at sufficiently low temperatures. For certain substances, cooling alone does not suffice to ensure that the sub-

378

stance really crystallizes and care must be taken that it does not solidify to an amorphous glass. This can easily happen for some alcohols and alkyl halides and Fig. 164 gives an example. Clearly, the spectrum of the glassy modification of the *n*-butylbromide is identical with that of the liquid state (the less stable isomers are frozen in) in contrast to the crystalline substance [Brown and Sheppard (*94*)]. The fundamental vibrations of the most stable isomer can be obtained from the low temperature spectrum in which the number of bands is reduced by the elimination of all the other isomers. Use must be made of the symmetry and activity properties as well as the equally important Raman effect. The assignment of the other isomers existing at

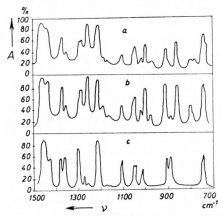

FIG. 164. *Infrared spectrum of* n-*butylbromide in the liquid* (a), *amorphous glass-like* (b), *and crystalline state* (c) [*Brown and Sheppard (94)*].

high temperatures can be attempted based on the analysis of the low temperature isomer. However, care must be taken, as it is possible for bands of the different isomers to coincide. For the determination of the energy of isomerization, intensity measurements on bands which have been assigned unambiguously to the different isomers are necessary for a range of temperatures. From the dependence of K with T, ΔE_J can be found from [IV, 4.5]. Values obtained from measurements on the liquid state often do not agree well with those from measurements on the gaseous state. Special solvent effects are the cause of this discrepancy. According to a suggestion by Powling and Bernstein (*559*), sufficiently dilute solutions should be used. The potential curves for this case are shown by a dotted line in Fig. 163 and it can be seen that the stability relationships can differ compared with the gaseous state as in tetrachloroethane. Obviously, such measurements on

379

Results and Applications

dilute solutions do not give directly the desired energy of isomerization but a value which differs by this by δE. This correction can be calculated using the Onsager theory from the electronic interaction between the molecule in question and its environment (the solvent) provided the dipole moment of the molecule and the dielectric constant of the solvent are known.

3. Induced Absorption

It has already been indicated that absorption of electromagnetic radiation can occur when it interacts with electric quadrupoles and magnetic dipoles. The absorption is usually much weaker than that involving electric dipoles and is therefore usually not observed (see the treatment already given for the gases H_2, O_2 and N_2). A further possibility is the so-called induced absorption which has become of interest recently through a series of good investigations. Induced absorption is the absorption of radiation under the influence of an external field which distorts the symmetry of molecules which are normally transparent. In order to attain a noticeable effect, electrical fields of ca. 10^5 volt/cm are required. Such values are reached when molecules approach each other very closely, e.g. by collisions in compressed gases or liquids. The theoretical treatment has shown [Kastler (373), Mizushima (482), Woodward (753)] that the so-called *induced moments*, i.e. the product of the polarizability of the molecule in question and the field strength governs the induced absorption. The induced moment takes the place of the dipole moment in the normal process of absorption of radiation. The polarizability is similarly all important for the Raman spectrum and the selection and intensity rules of the Raman spectrum apply to induced absorption.

Up to now induced absorption has been observed in H_2, O_2, and N_2, both for the condensed phases and for the strongly compressed gases [Oxholm and Williams (519); van Asselt and Williams (27); Crawford et al. (150); Chisholm et al. (117); Smith et al. (673); Welsh et al. (730); Crawford and Dagg (149); Chisholm and Welsh (116); Hare et al. (299); Allin et al. (12)]. A similar effect is to be expected for the inactive breathing vibration of methane, but this has not yet been observed. The absorption spectra induced experimentally show certain indications of structure which allow further conclusions with regards to the absorbing molecule and the inducting mechanism. Fig. 165 shows the induced absorption spectrum for the fundamental vibration of nitrogen at various temperatures. The main absorption lies at ca. 2350 cm^{-1} for the liquid ($69° A$) with a subsidiary maximum just visible at ca. 2400 cm^{-1}. The spectrum becomes more prominent, with a moderate displacement of the maxima, for the solid β-modification ($43 \cdot 9° A$) Numerous indications of an incompletely resolved fine structure are shown for the solid α-modification ($34 \cdot 4° A$). The results for oxygen are similar. In

380

highly compressed, but not liquified, hydrogen the absorption band has a well-defined fine structure (Fig. 166). At a lower temperature (80° A), five

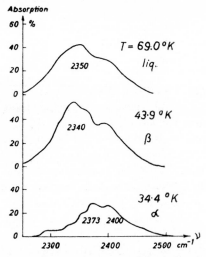

FIG. 165. *Temperature dependence of the induced absorption of liquid nitrogen* [*Smith* et al. (*673*)].

well-resolved absorption maxima are found and some fine structure is still noticeable at room temperature. Similar results are obtained when the

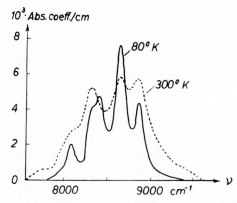

FIG. 166. *Fine structure of the induced absorption of hydrogen for the first overtone* [*Welsh* et al. (*730*)].

compression is carried out by using an added gas rather than by increasing the pressure of the substance being investigated. The fact that the spectra

381

are obtained when the inert gases are added shows that this effect is a true collision-induced absorption and is not due to quadrupole interaction because the inert gases have no quadrupole.

4. Crystals

The structure of molecules in crystals and the arrangement of the atoms in space is usually determined by X-rays. Infrared spectroscopic measurements, especially those with polarized radiation, have been found to be a valuable accessory for the extension and confirmation of the X-ray results particularly relating to the positions of hydrogen atoms. The difficulty of the infrared spectroscopic investigations lies in the preparation of suitable samples. Thin layers of powder, with or without embedding material, usually give no very specific results (because of the statistical distribution of the crystallites over all the possible orientations in space) unless the crystal investigated belongs to the cubic system. Measurements on single crystals at a definite orientation in the optical path are to be recommended. The reflecting microscope is an especially valuable accessory because extremely small crystals can be investigated and they can be oriented in the desired manner with respect to the radiation. However, important investigations can be carried out without the help of this expensive instrument if the position of the direction of growth with respect to the crystallographic axes is known; a definite direction of growth can be ensured, e.g. by crystallizing out between NaCl plates or in the centrifuge.

By the use of polarized radiation at various azimuths, absorption bands belonging to different vibrational modes can be frequently recognized, although in the powder spectrum they all appear close together. Usually conclusions regarding the position of the vibration in the crystal lattice can be drawn from their dichroic behaviour. If the nature and assignment of the vibrations in question are known, information is obtained relating to the positions of the bonds and atoms in the crystal. The vibrations of the crystal lattice are very difficult to obtain in the normal way because they appear in a spectral region which has not yet been experimentally developed. However, lattice vibrations can frequently be observed in the form of their combinations with the molecular vibrations and this allows the determination of the symmetry properties which apply to the crystal lattice as such [Hornig (*330, 331*); Mann and Thompson (*442*); Osberg and Hornig (*513*); Reding and Hornig (*585*); Lax and Burstein (*408*)].

Infrared spectra are also of use for the differentiation and recognition of polymorphic forms of organic and inorganic crystalline substances [Kendall (*378*); Ebert and Gottlieb (*183*)]. Spectra obtained from powders fully suffice for this application, which is useful for structures as complicated as, e.g. copper phthalocyanine. Reflection spectra can also serve well for the structural clarification of crystals including the finer details as Šimon and McMahon (*663*) have shown for the system quartz and cristobalite.

5. *Applications to Thermodynamics*

Vibrational analyses are of great importance for the knowledge of the structure of specific molecules. In addition, they are of great general interest for the theoretical calculation of certain physico-chemical constants. The important calorific and thermodynamic quantities such as specific heat, enthalpy, and entropy, with their direct applications to experimental chemistry and technology, depend on the vibrational frequencies in a more or less complicated manner and on the correct assignment of the frequencies to definite normal vibrations. It is possible [the corresponding formulae and their derivation can be found, e.g. in (*I*)] to calculate these quantities for the gaseous state on the basis of the energy states derived from spectroscopic data and then to compare these values with those determined experimentally. This is a test of the molecular model taken and also (when the model can be assumed) for the correctness of the vibrational assignment. Non-correspondence of such calculated and measured values has frequently directed attention to an unsatisfactory interpretation of a spectrum and has led to the elimination of erroneous assumptions and evaluations.

The relationship between the spectroscopic data and the thermodynamic quantities involves so-called *partition functions*. These are obtained by the following treatment. The number N_n of molecules of a gas in thermal equilibrium which are in a state E_n of the total energy and degeneracy (statistical weight) g_n is given by the Maxwell-Boltzmann law:

$$N_n \sim g_n \cdot e^{-(E_n/kT)} \qquad \text{[IV, 4.6]}$$

where T is the absolute temperature of the gas and k is the Boltzmann constant. The total number of molecules in a given volume is obtained by summation over all the possible states

$$N = \Sigma N_n \sim Q = \Sigma g_n \cdot e^{-(E_n/kT)} \qquad \text{[IV, 4.7]}$$

The quantities Q which occur here are called partition functions. All thermodynamic quantities can be written as a function of Q. Just as the energy of a gas can be divided into a translational part and an internal part which give the total energy by addition, the partition function can be written as the product of two partition functions each of which depends on only one of the types of energy mentioned. The internal energy can be further divided into additive rotational and vibrational parts and correspondingly the special partition function (often written simply as partition function) can be written as a product of a rotational and a vibrational term. Each factor can be calculated from the corresponding energies which can themselves be obtained together with their degrees of degeneracy, from the spectrum.

It should be obvious from these indications that the recognition of the individual rotational and vibrational states, the vibrational species, and the point group to which the molecule belongs, is of especial importance. A characteristic of point groups, the symmetry number, which was given in Table 4 but not further mentioned, is important in this respect. The symmetry number defines the number of indistinguishable positions into which a molecule can be brought by simple rotation. It is not possible to elaborate these relationships here or the influence of the symmetry number on the theoretical thermodynamical calculations based on spectroscopic data. It will only be mentioned that, as a result of the interaction of rotation and vibration discussed earlier, the thermodynamic quantities can often be calculated as functions of the temperature with greater precision from spectroscopic data than by direct experimental determination.

6. Miscellaneous

Infrared spectroscopy is a technique of the broadest possible applicability. The special effects enumerated in the preceding sections certainly do not exhaust its potentialities. In chemical research there is almost *no* problem where it cannot be applied successfully, provided it is possible to adapt the instruments and the preparation of the sample to the problem in question. In this final section a few further topics will be mentioned which have been characterized by numerous investigations of value. One of the most important is the application of infrared spectroscopy to the clarification of *reaction mechanisms* [Gauthier (*244*); Duval and Lecomte (*179*)]. Frequently, infrared spectroscopy has given the correct interpretation, e.g. for the disproportionation of NH_3 [Stedman (*682*)], the N_2O-ethanol reaction [Nightingale *et al.* (*501*)], the ketone-ethanolamine reaction [Daasch (*158*)], the autoxidation of oils etc. [Honn *et al.* (*329*); Khan (*382*)]. Studies of *gas kinetics* have been carried out by infrared spectroscopy [Simard *et al.* (*660*)]. Investigations of *flames* and *explosions* [Benedict *et al.* (*55, 56*); Bullock *et al.* (*102*); Plyler (*546*)] (for which the rapid spectrometers, see page 151, were developed) were especially valuable in connection with the clarification of the combustion processes in rocket motors. Solid state physicists can obtain from infrared spectra the separation between the energy bands in semiconductors and their temperature dependence [Oswald (*515*)]. The investigation of the temperature dependence of the intensity of infrared bands [Slowinski and Claver (*670*); Brown (*89*); Finch and Lippincott (*210*); Hughes *et al.* (*339*); Liddel and Becker (*781*)] opens new possibilities in the structural-theoretical respect. The investigation of chemisorption gives a new impulse to the theory of the activity of catalysts [Pliskin and Eischens (*543*); Eischens (*188*)].

Infrared spectroscopic measurements are also important to meteorologists, astronomers, and astrophysicists. In connection with the radiation economy of the earth, the spectral absorption of the atmosphere has been measured up to very long wavelengths [Gebbie *et al.* (*245*); Kaplan (*372*); Elder and Strong (*189*)]. Further, the determination of traces of gases and their distribution in the levels of the atmosphere is important in connection with the exact composition of the earth's atmosphere [Adel (*6*); Shaw and Claassen (*643*); Benesch *et al.* (*57*); Locke and Herzberg (*424*); Howard *et al.* (*335*)]. There is also the knowledge and evaluation of the infrared part of the spectra of stars, e.g. the sun [Shaw *et al.* (*643, 644*); Mohler (*484*)].

Finally, as yet an individual application of infrared spectroscopy will be mentioned because of its singular character which illustrates well the broad application of the method. This is the identification of *bacterial cultures* by means of their infrared spectra [Stevenson and Bolduan (*685*); Randall and Smith (*569*)].

Literature

The following textbooks, monographs, and reviews are given as general references.

A BARNES, R. B., GORE, R. C., LIDDEL, U., and WILLIAMS, V. Z., *Infrared Spectroscopy, Industrial Applications and Bibliography* (New York, 1944).

B BELLAMY, L. J., *The Infrared Spectra of Complex Molecules* (London, 1954).

C BRÜGEL, W., *Physik und Technik der Ultrarotstrahlung* (Hanover, 1951).

D BRÜGEL, W., Ultrarotspektroskopische Strukturbestimmungen. In HOUWINK, R., *Chemie und Technologie der Kunststoffe*, 3. Aufl., 1. Bd., § 40 (Leipzig, 1954).

E BRÜGEL, W., Ultrarot-Betriebskontrollgeräte. In HENGSTENBERG-STURM-WINKLER, *Messen und Regeln in der chemischen Technik* (Berlin-Göttingen-Heidelberg, 1957).

F COLE, A. R. H., Infrared spectra of natural products. *Fortschr. Chem. org. Naturst.* **13**, 1 (1956).

G CZERNY, M., and RÖDER, H., Fortschritte auf dem Gebiet der Ultrarottechnik. *Ergebn. exakt. Naturw.* **17**, 70 (1938).

H DOBRINER, K., KATZENELLENBOGEN, E. R., and JONES, R. N., *Infrared Absorption Spectra of Steroids, an Atlas* (New York, 1949).

I HERZBERG, G., *Molecular Spectra and Molecular Structure*. Vol. 1: Spectra of Diatomic Molecules. 2. ed. (New York, 1950). Vol. 2: Infrared and Raman Spectra of Polyatomic Molecules. 4. ed. (New York, 1949).

K HUMMEL, D., *Die Identifizierung von Kunststoff- und Lackrohstoffen und daraus hergestellter Erzeugnisse, unter bevorzugter Behandlung infrarotspektroskopischer Methoden* (München, 1957).

L JONES, R. N., and SANDORFY, C., The Application of Infrared and Raman Spectrometry to the Elucidation of Molecular Structure. In WEST, W., *Chemical Applications of Spectroscopy* (New York, 1956).

M LANDOLT-BÖRNSTEIN, *Zahlenwerte und Funktionen aus Physik, Chemie, Astronomie, Geophysik und Technik*. Edited by EUCKEN, A., and HELLWEGE, K. H. Bd. 1: Atom- und Molekülphysik. 2. Teil: Molekeln I (Kerngerüst), Abschnitt Schwingungen und Rotationen, revised by MAIER, W., MECKE, R., KERKHOF, F., SEIDEL, H., and PAJENKAMP, H. (Berlin-Göttingen-Heidelberg, 1951). See also supplement in 3. Teil: Molekeln II, pp. 557–656.

N LECOMTE, J., *Le Rayonnement Infrarouge*. Vol. 1: Applications Biologiques, Physiques et Techniques (Paris, 1948). Vol. 2: La Spectrométrie Infrarouge et ses Applications Physico-Chimiques (Paris, 1949).

O MATOSSI, F., Ergebnisse der Ultrarotforschung. *Ergebn. exakt. Naturw.* **17**, 108 (1938).

P RANDALL, H. M., FOWLER, R. G., FUSON, N., and DANGL, J. R., *Infrared Determination of Organic Structure* (New York, 1949).

Q RASMUSSEN, A. S., Infrared spectroscopy in structure determination and its application to penicillin. *Fortschr. Chem. org. Naturst.* **5**, 331 (1948).

R SCHAEFER, C., and MATOSSI, F., *Das ultrarote Spektrum* (Bd. 10, Struktur der Materie) (Berlin, 1930).

S SUHRMANN, R., and LUTHER, H., Neuere Ergebnisse der Ultrarotspektroskopie, *Fortschr. chem. Forsch.* **2**, 758 (1953).

Literature

T WILLIAMS, V. Z., Infrared instrumentations and techniques. *Rev. sci. Instrum.* **19**, 135 (1948).

U WILSON, E. B., DECIUS, J. C., and CROSS, P. C., *Molecular Vibrations—The Theory of Infrared and Raman Vibrations* (New York, 1955).

Original papers (type of compound investigated is indicated):

1 ACHHAMMER, B. G., REINEY, M. J., and REINHART, F. W. (1951) *J. Res. nat. Bur. Stand.* **47**, 116. Polystyrene degradation.

2 ACQUISTA, N., and PLYLER, E. K. (1952) *J. Res. nat. Bur. Stand.* **49**, 13.

3 ACQUISTA, N., and PLYLER, E. K. (1953) *J. opt. Soc. Amer.* **43**, 977.

4 ADAMS, R. M., and KATZ, J. J. (1956) *J. opt. Soc. Amer.* **46**, 895.

5 ADCOCK, W. A. (1948) *Rev. sci. Instrum.* **19**, 181.

6 ADEL, A. (1951) *Astrophys. J.* **113**, 222. N_2O in the air.

7 AGNEW, J. T., and MCQUISTAN, R. B. (1953) *J. opt. Soc. Amer.* **43**, 999.

8 AGNEW, J. T., FRANKLIN, R. G., and BENN, R. E. (1951) *J. opt. Soc. Amer.* **41**, 76.

9 AHLERS, N. H. E., and FREEDMAN, H. P. (1955) *J. sci. Instrum.* **32**, 61.

10 ALDER, K., SCHÄFER, H. K., ESSER, H., KRIEGER, H., and REUBKE, R. (1955) *Ann. Chem.* **593**, 23. α-Diketones.

11 ALLAN, J. L. H., MEAKINS, G. D., and WHITING, M. C. (1955) *J. chem. Soc.* **1955**, 1874. Conjugated acetylene- and ethylene-compounds.

12 ALLIN, E. J., HARE, W. F. J., and MACDONALD, R. E. (1955) *Phys. Rev.* **98**, 554.

13 ALLISON, A. R., and STANLEY, I. J. (1952) *Analyt. Chem.* **24**, 630.

14 AMBROSE, E. J., and ELLIOT, A. (1951) *Proc. roy. Soc.* **A 206**, 206.

15 AMBROSE, E. J., and ELLIOT, A. (1951) *Proc. roy. Soc.* **A 208**, 75. Proteins.

16 AMBROSE, E. J., ELLIOT, A., and TEMPLE, R. B. (1949) *Proc. roy. Soc.* **A 199**, 183.

17 AMBROSE, E. J., ELLIOT, A., and TEMPLE, R. B. (1951) *Proc. roy. Soc.* **A 206**, 192.

18 AMES, J., and SAMPSON, A. M. D. (1949) *J. sci. Instrum.* **26**, 132.

19 AMSTUTZ, E. D., HUNSBERGER, I. M., and CHESSICK, J. J. (1951) *J. Amer. chem. Soc.* **73**, 1220. Diphenylsulphones, diphenylsulphoxides.

20 ANDERSON, S. (1950) *J. Amer. ceram. Soc.* **32**, 2. Glasses.

21 ANDERSON, D. H., and STEWART, D. W. (1952) *Org. Chem. Bull. Eastmann* **24**, Nr. 4. Spectra of solvents.

22 ANDERSON, D. H., and MILLER, O. E. (1953) *J. opt. Soc. Amer.* **43**, 777.

23 ANDERSON, S., ANDERSON, W. J., and KRAKOWSKI, M. (1950) *Rev. sci. Instrum.* **21**, 574.

24 ARD, J. S. (1953) *Analyt. Chem.* **25**, 1743.

25 ARD, J. S., and FONTAINE, T. D. (1951) *Analyt. Chem.* **23**, 133.

26 VON ARDENNE, M. (1941) *Angew. Chem.* **54**, 144.

27 VON ASSELT, R., and WILLIAMS, D. (1950) *Phys. Rev.* **79**, 1016.

28 AXFORD, D. W. E., and RANK, D. H. (1950) *J. chem. Phys.* **18**, 51.

29 BAIRD, W. S., O'BRYAN, H. M., OGDEN, G., and LEE, D. (1947) *J. opt. Soc. Amer.* **37**, 754.

30 BAK, B., and CHRISTENSEN, D. (1956) *Acta chem. scand.* **10**, 692.

31 BAKER, A. W., and LORD, R. C. (1955) *J. chem. Phys.* **23**, 1636. Cyclopropanes.

32 BAKER, A. W., WRIGHT, N., and OPLER, A. (1953) *Analyt. Chem.* **25**, 1457.

33 BANNING, M. (1947) *J. opt. Soc. Amer.* **37**, 792.

34 BARCHEWITZ, P. (1951) *Compt. Rend.* **232**, 1750.

35 BARCHEWITZ, P., and CHABBAL, R. (1951) *J. Phys. Radium* **12**, 637, 701.

36 BARER, R. (1950) *Disc. Faraday Soc.* **9**, 369.

37 BARNARD, D., FABIAN, J. M., and KOCH, H. P. (1949) *J. Chem. Soc.* **1949**, 2442. Sulphoxides, sulphones.

Literature

38 BARNARD, D., BATEMAN, L., HARDING, A. J., KOCH, H. P., SHEPPARD, N., and SUTHERLAND, G. B. B. M. (1950) *J. chem. Soc.* **1950**, 915. Longchain Monoolefins and derivates.

39 BARNES, R. B., and GORE, R. C. (1949) *Analyt. Chem.* **21**, 7.

40 BARNES, R. B., GORE, R. C., WILLIAMS, E. F., LINSLEY, S. G., and PETERSEN, E. M. (1947) *Analyt. Chem.* **19**, 620. Penicillin.

41 BARNES, R. B., MCDONALD, R. S., WILLIAMS, V. Z., and KINNAIRD, R. F. (1945) *J. appl. Phys.* **16**, 77.

42 BARNES, R. B., MCDONALD, R. S., WILLIAMS, V. Z., and KINNAIRD, R. F. (1946) *J. appl. Phys.* **17**, 532.

43 BARNES, R. P., and PINKNEY, G. E. (1953) *J. Amer. chem. Soc.* **75**, 479. Diketones.

44 BARROW, G. M. (1953) *J. chem. Phys.* **21**, 2008.

45 BARROW, G. M., and SEARLES, S. (1953) *J. Amer. chem. Soc.* **75**, 1175. Cyclic ethers.

46 BARTLESON, J. D., BURK, R. E., and LANKELMA, H. P. (1946) *J. Amer. chem. Soc.* **68**, 2513. Alkylcyclopropanes.

47 BAXTER, J. N., CYMERMAN-CRAIG, J., and WILLIS, J. B. (1955) *J. chem. Soc.* **1955**, 669.

48 BAYLISS, N. S., and BRACKENBRIDGE, C. J. (1955) *Chem. & Ind.* **1955**, 477.

49 BECHER, H. J. (1956) *Chem. Ber.* **89**, 1593, 1951. Urea-formaldehyde condensation products.

50 BECKER, J. A., and BRATTAIN, W. H. (1946) *J. opt. Soc. Amer.* **36**, 354.

51 BELL, J. V., HEISLER, J., TANNENBAUM, H., and GOLDENSON, J. (1954) *J. Amer. chem. Soc.* **76**, 5185. Phosphoryl compounds.

52 BELLAMY, L. J., and BEECHER, L. (1952) *J. chem. Soc.* **1952**, 475, 1701. P-esters, -acids, -amines.

53 BELLAMY, L. J., and BEECHER, L. (1953) *J. chem. Soc.* **1953**, 728. P-compounds in natural products.

54 BENDER, H. L. (1953) *Mod. Plast.* **30**, No. 6, 136.

55 BENEDICT, W. S., HERMAN, R. C., and SILVERMAN, S. (1951) *J. chem. Phys.* **19**, 1325. $(CO + O_2)$-emulsion.

56 BENEDICT, W. S., BULLOCK, B. W., SILVERMAN, S., and GROSSE, A. V. (1953) *J. opt. Soc. Amer.* **43**, 1106.

57 BENESCH, W., MIGEOTTE, M., and NEVEN, L. (1953) *J. opt. Soc. Amer.* **43**, 1119. CO in the air.

58 BERGMANN, G., and KRESZE, G. (1955) *Angew. Chem.* **67**, 685.

59 BERNAL, J. D., and MEGAW, H. D. (1935) *Proc. roy. Soc.* **151**, 384.

60 BERNSTEIN, H. J. (1949) *J. chem. Phys.* **17**, 256.

61 BERNSTEIN, H. J. (1950) *J. chem. Phys.* **18**, 897.

62 BERRY, C. E., and PEMBERTON, J. C. (1946) *Instruments* **19**, 396.

63 BILLINGS, B. H. (1950) *J. Phys. Radium* **11**, 407.

64 BILLINGS, B. H., and PITTMAN, M. A. (1949) *J. opt. Soc. Amer.* **39**, 978.

65 BLOMQUIST, A. T., and LIU, L. H. (1953) *J. Amer. chem. Soc.* **75**, 2153. Cyclic acetylenes.

66 BLOMQUIST, A. T., BURGE, R. E., LIU, L. H., BOHRER, J. C., SUCSY, A. C., and KLEIS, J. (1951) *J. Amer. chem. Soc.* **73**, 5510. Cyclic acetylenes.

67 BLOUT, E. R., and BIRD, G. R. (1951) *J. opt. Soc. Amer.* **41**, 547.

68 BLOUT, E. R., and KARPLUS, R. (1948) *J. Amer. chem. Soc.* **70**, 862.

69 BLOUT, E. R., and LENORMANT, H. (1953) *J. opt. Soc. Amer.* **43**, 1093.

70 BLOUT, E. R., and ABBATE, M. J. (1955) *J. opt. Soc. Amer.* **45**, 1028.

71 BLOUT, E. R., CORLEY, R. S., and SNOW, P. L. (1950) *J. opt. Soc. Amer.* **40**, 415.

72 BLOUT, E. R., BIRD, G. R., and GREY, D. S. (1950) *J. opt. Soc. Amer.* **40**, 304.

73 BLOUT, E. R., PARRISH, M., BIRD, G. R., and ABBATE, M. J. (1952) *J. opt. Soc. Amer.* **42**, 966.

74 BOHN, C. R., FREEMAN, N. K., GWINN, W. D., HOLLENBERG, J. L., and PITZER, K. S. (1953) *J. chem. Phys.* **21**, 719.

75 BOLZ, H. M. (1955) *Jenaer Jahrbuch* **1955**, II. Teil, p. 207.

76 BOMSTEIN, J. (1953) *Analyt. Chem.* **25**, 512. Benzene derivatives.

77 BONINO, G. B., and SCROCCO, E. (1949) *Atti acc. naz. Lincei* **6**, 421. Aldehydes, ketones.

78 BOURGIN, D. G. (1927) *Phys. Rev.* **29**, 794.

79 BOVEY, L. F. H. (1951) *J. opt. Soc. Amer.* **41**, 381.

80 BOVEY, L. F. H. (1951) *J. opt. Soc. Amer.* **41**, 836.

81 BRAITHWAITE, J. G. N. (1955) *J. sci. Instrum.* **32**, 10.

82 BRATTAIN, R. R., RASMUSSEN, R. S., and CRAVATH, A. M. (1943) *J. appl. Phys.* **14**, 418.

83 BRENEMAN, E. J., and WILLIAMS, D. (1953) *Phys. Rev.* **91**, 465. (Short account).

84 BROADLEY, H. R. (1948) *Rev. sci. Instrum.* **19**, 475.

85 BROCKMAN, F. G. (1946) *J. opt. Soc. Amer.* **36**, 32.

86 BROCKMANN, H., and MUSSO, H. (1956) *Chem. Ber.* **89**, 241.

87 BRODERSEN, S. (1954) *J. opt. Soc. Amer.* **44**, 22.

88 BRODERSEN, S. (1956) *J. opt. Soc. Amer.* **46**, 255.

89 BROWN, T. L. (1956) *J. chem. Phys.* **24**, 1281.

90 BROWN, L., and HOLLIDAY, P. (1951) *J. sci. Instrum.* **28**, 27.

91 BROWN, J. K., and SHEPPARD, N. (1952) *Trans. Faraday Soc.* **48**, 128.

92 BROWN, J. K., and SHEPPARD, N. (1950) *Disc. Faraday Soc.* **9**, 144.

93 BROWN, J. K., and SHEPPARD, N. (1951) *J. chem. Phys.* **19**, 976.

94 BROWN, J. K., and SHEPPARD, N. (1955) *Proc. roy. Soc.* A **231**, 555.

95 BROWN, J. K., SHEPPARD, N., and SIMPSON, D. M. (1950) *Disc. Faraday Soc.* **9**, 261. Normal paraffins.

96 BROWN, J. K., SHEPPARD, N., and SIMPSON, D. M. (1953) *Réunion Internationale de la Spectroscopie Moléculaire* (Paris, 1953).

97 BROWNING, R. S., WIBERLEY, S. E., and NACHOD, F. C. (1955) *Analyt. Chem.* **27**, 7.

98 BRÜGEL, W. (1950) *Z. Phys.* **127**, 400.

99 BRÜGEL, W. (1956) *Kunststoffe*, **46**, 47.

100 BRYANT, W. M. D., and VOTER, R. C. (1953) *J. Amer. chem. Soc.* **75**, 6113. Polyethylene.

101 BULLOCK, B. W., and SILVERMAN, S. (1950) *J. opt. Soc. Amer.* **40**, 608.

102 BULLOCK, B. W., FEAZEL, C. E., GLOERSEN, P., and SILVERMAN, S. (1952) *J. chem. Phys.* **20**, 1808. HBr-emission.

103 BUNN, C. W., and PEISER, H. S. (1947) *Nature* **159**, 161.

104 BURCH, C. R. (1947) *Proc. phys. Soc. Lond.* **59**, 41.

105 BURNS, W. G., and GAUNT, J. (1954) Conf. Mol. Spectr. 1954, Inst. Petr., London.

106 BUSWELL, A. M., MAYCOCK, R. L., and RODEBUSH, W. H. (1940) *J. chem. Phys.* **8**, 362.

107 CALIFANO, S., and LÜTTKE, W. (1955) *Z. phys. Chem.* **5**, 240. Oximes.

108 CALIFANO, S., and LÜTTKE, W. (1956) *Z. phys. Chem.* **6**, 83. Oximes.

109 CANDLER, C. (1951) *Nature* **167**, 649.

110 CANNON, C. G. (1955) *Microchim. Acta* **1955**, 555.

111 CANNON, C. G., and SUTHERLAND, G. B. B. M. (1951) *Spectrochim. Acta* **4**, 373. Polycyclic compounds.

112 CARPENTER, G. B., and HALFORD, R. S. (1947) *J. chem. Phys.* **15**, 99.

113 CENCELJ, L., and HADŽI, D. (1955) *Spectrochim. Acta* **7**, 274. Naphthalenes.

114 CHARNEY, E. (1955) *J. opt. Soc. Amer.* **45**, 980.

115 CHILDERS, E., and STRUTHERS, G. W. (1953) *Analyt. Chem.* **25**, 1311.

Literature

116 CHISHOLM, D. A., and WELSH, H. L. (1954) *Can. J. Phys.* **32**, 291.
117 CHISHOLM, D. A., MACDONALD, J. C. F., CRAWFORD, M. F., and WELSH, H. L. (1952) *Phys. Rev.* **88**, 957.
118 CLARK, C. (1950) *Science* **111**, 632.
119 CLARK, H. A., GORDON, A. F., YOUNG, C. W., and HUNTER, M. J. (1951) *J. Amer. chem. Soc.* **73**, 3798. Substituted aryltrimethylsilanes.
120 CLEMENT, P. (1951) *Ind. Plast. Mod.* **3**, No. 5, 33.
121 CLOUD, W. H. (1956) *J. opt. Soc. Amer.* **46**, 899.
122 COATES, V. J. (1951) *Rev. sci. Instrum.* **22**, 853.
123 COATES, V. J., and HAUSDORFF, H. (1955) *J. opt Soc. Amer.* **45**, 425.
124 COATES, V. J., OFFNER, A., and SIEGLER, E. H. (1953) *J. opt. Soc. Amer.* **43**, 984.
125 COGGESHALL, N. D. (1946) *Rev. sci. Instrum.* **17**, 343.
126 COGGESHALL, N. D. (1947) *J. Amer. chem. Soc.* **69**, 1620.
127 COGGESHALL, N. D., and SAIER, E. L. (1947) *J. chem. Phys.* **15**, 65.
128 COGGESHALL, N. D., and SAIER, E. L. (1951) *J. Amer. chem. Soc.* **73**, 5414.
129 COLE, R. (1951) *J. opt. Soc. Amer.* **41**, 38.
130 COLE, A. R. H. (1953) *J. opt. Soc. Amer.* **43**, 807.
131 COLE, A. R. H. (1954) *J. opt. Soc. Amer.* **44**, 741.
132 COLE, A. R. H., and THOMPSON, H. W. (1950) *Trans. Faraday Soc.* **46**, 103. Substituted benzenes.
133 COLE, A. R. H., and JONES, R. N. (1952) *J. opt. Soc. Amer.* **42**, 348.
134 COLE, J. O., and FIELD, J. E. (1947) *Industr. Engng. Chem. (Industr.)* **39**, 174.
135 COLTHUP, N. B. (1947) *Rev. sci. Instrum.* **18**, 64.
136 COLTHUP, N. B. (1950) *J. opt. Soc. Amer.* **40**, 397.
137 COMRIE, L. J. (1949) *J. sci. Instrum.* **21**, 129.
138 CONDON, F. E., and SMITH, D. E. (1947) *J. Amer. chem. Soc.* **69**, 965. Alkylcyclopropanes.
139 CONN, G. K. T., and EATON, G. K. (1954) *J. opt. Soc. Amer.* **44**, 477.
140 CONN, G. K. T., and EATON, G. K. (1954) *J. opt. Soc. Amer.* **44**, 484.
141 CONN, G. K. T., and EATON, G. K. (1954) *J. opt. Soc. Amer.* **44**, 546.
142 CONN, G. K. T., and EATON, G. K. (1954) *J. opt. Soc. Amer.* **44**, 553.
143 COOPER, C. V., and STROUPE, J. D. (1951) *J. opt. Soc. Amer.* **41**, 427.
144 COPE, A. C., and BUMGARDNER, C. L. (1956) *J. Amer. chem. Soc.* **78**, 2812. Cyclic polyolefins.
145 CORBRIDGE, D. E. C., and LOWE, E. J. (1954) *J. chem. Soc.* **1954**, 493, 4555. Phosphorus compounds.
146 CORRSIN, L., FAX, B. J., and LORD, R. C. (1953) *J. chem. Phys.* **21**, 1170. Pyridine.
147 CRAWFORD, B. (1952) *J. chem. Phys.* **20**, 977.
148 CRAWFORD, B. L., and DINSMORE, H. L. (1950) *J. chem. Phys.* **18**, 1682.
149 CRAWFORD, M. F., and DAGG, J. R. (1953) *Phys. Rev.* **91**, 1569.
150 CRAWFORD, M. F., WELSH, H. L., MACDONALD, J. C. F., and LOCKE, J. L. (1950) *Phys. Rev.* **80**, 469.
151 CREITZ, E. C., and SMITH, F. A. (1949) *J. Res. nat. Bur. Stand.* **43**, 365. Olefins.
152 CROSS, L. H., and ROLFE, A. C. (1951) *Trans. Faraday Soc.* **47**, 354.
153 CROSS, L. H., RICHARDS, R. B., and WILLIS, H. A. (1950) *Disc. Faraday Soc.* **9**, 235.
154 CURCIO, J. A., and PETTY, C. C. (1951) *J. opt. Soc. Amer.* **41**, 302.
155 CYMERMAN, J., and WILLIS, J. B. (1951) *J. chem. Soc.* **1951**, 1332. Aromatic disulphides, disulphones, thiosulphonates.
156 CZERNY, M. (1927) *Z. Phys.* **45**, 476.
157 CZERNY, M., KOFINK, W., and LIPPERT, W. (1951) *Ann. Phys.* **8**, 65.
158 DAASCH, L. W. (1951) *J. Amer. chem. Soc.* **73**, 4523.
159 DAASCH, L. W., and SMITH, D. C. (1951) *Analyt. Chem.* **23**, 853. Phosphorus compounds.

390

160 DALY, E. F. (1951) *J. sci. Instrum.* **28**, 308.
161 DANNENBERG, H., SCHIEDT, U., and STEIDLE, W. (1953) *Z. Naturf.* **8b**, 269. Condensed aromatic compounds.
162 DAVIS, M., and EVANS, J. C. (1952) *J. chem. Phys.* **20**, 342.
163 DAVISON, W. H. T. (1951) *J. chem. Soc.* **1951**, 2456. Peroxides, peracids, peresters.
164 DAVISON, W. H. T. (1955) *J. opt. Soc. Amer.* **45**, 227.
165 DECIUS, J. C. (1952) *J. chem. Phys.* **20**, 1039.
166 DEHEER, J. (1952) *J. chem. Phys.* **20**, 637. Allene.
167 DENNISON, D. M. (1928) *Phys. Rev.* **31**, 503.
168 DERFER, J. M., PICKETT, E. E., and BOORD, C. E. (1949) *J. Amer. chem. Soc.* **71**, 2482. Cyclopropanes, cyclobutanes.
169 DERKSEN, W. L., and MONAHAN, T. I. (1952) *J. opt. Soc. Amer.* **42**, 263.
170 DETONI, S., and HADŽI, D. (1955) *J. chem. Soc.* **1955**, 3163. Sulphinic acids.
171 DEW, G. D. (1954) *J. sci. Instrum.* **29**, 277.
172 DEW, G. D., and SAYCE, L. A. (1951) *Proc. roy. Soc.* A **207**, 278.
173 DIMITROFF, J. M., and SWANSON, D. W. (1956) *J. opt. Soc. Amer.* **46**, 555.
174 DINSMORE, H. L., and SMITH, D. C. (1948) *Analyt. Chem.* **20**, 11. Rubber.
175 D'OR, L., and KÖSSLER, J. (1951) *Industr. chim. Belge* **16**, 133.
176 DOWNIE, A. R., MAGOON, M. C., PURCELL, T., and CRAWFORD, B. (1953) *J. opt. Soc. Amer.* **43**, 941.
177 DRAYTON, L. G., and THOMPSON, H. W. (1948) *J. chem. Soc.* **1948**, 1416. Ketene.
178 DUERIG, W. H., and MADOR, I. L. (1952) *Rev. sci. Instrum.* **23**, 421.
179 DUVAL, C., and LECOMTE, I. (1952) *Compt. Rend.* **234**, 2445.
180 DUYCKAERTS, G. (1955) *Spectrochim. Acta* **7**, 25.
181 DUYCKAERTS, G., and PIRLOT, G. (1949) *Bull. Soc. chim. Belg.* **58**, 48.
182 EBERHARDT, W. H. (1950) *J. opt. Soc. Amer.* **40**, 172.
183 EBERT, A. A., and GOTTLIEB, H. B. (1952) *J. Amer. chem. Soc.* **74**, 2806.
184 EDLEN, B. (1953) *J. opt. Soc. Amer.* **43**, 339.
185 EGGERS, D. F., and CRAWFORD, B. L. (1951) *J. chem. Phys.* **19**, 1554.
186 EGGERS, D. F., and LINGREN, W. E. (1956) *Analyt. Chem.* **28**, 1328. Aldehydes.
187 EICHHORN, G., and HETTNER, G. (1948) *Ann. Phys.* **3**, 120.
188 EISCHENS, R. P. (1956) *Z. Elektrochem.* **60**, 782.
189 ELDER, T., and STRONG, J. (1953) *J. Franklin Inst.* **255**, 189. Atmosphere.
190 ELDER, T., and BENESCH, W. (1954) *J. opt. Soc. Amer.* **44**, 279.
191 ELLIOT, A., and AMBROSE, E. J. (1947) *Nature* **159**, 641.
192 ELLIOT, A., and AMBROSE, E. J. (1950) *Disc. Faraday Soc.* **9**, 246. Polypeptides.
193 ELLIOT, A., and MALCOLM, B. R. (1956) *Trans. Faraday Soc.* **52**, 528. Polypeptides.
194 ELLIOT, A., MALCOLM, B. R., and TEMPLE, R. B. (1948) *J. chem. Phys.* **16**, 877. Polyethylene.
195 ELLIOT, A., MALCOLM, B. R., and TEMPLE, R. B. (1948) *J. opt. Soc. Amer.* **38**, 212.
196 ELLIOT, A., MALCOLM, B. R., and TEMPLE, R. B. (1949) *Nature* **163**, 567. Polyvinylalcohol, nylon.
197 ELLIOT, A., MALCOLM, B. R., and TEMPLE, R. B. (1950) *J. sci. Instrum.* **27**, 21.
198 ELLIS, J. W. (1933) *J. opt. Soc. Amer.* **23**, 88.
199 EPSTEIN, M. B., PITZER, K. S., and ROSSINI, F. D. (1949) *J. Res. nat. Bur. Stand.* **42**, 379. Cyclopentene, cyclohexene.
200 EULER, J. (1949) *Z. angew. Phys.* **1**, 569.
201 EVANS, A., HIBBARD, R. R., and POWELL, A. S. (1951) *Analyt. Chem.* **23**, 1604.
202 FABIAN, J., LEGRAND, M., and POIRIER, P. (1956) *Bull. Soc. Chim. Fr.* **1956**, 1499. Imines.
203 FAHR, E., and NEUMANN, W. P. (1955) *Angew. Chem.* **67**, 277.
204 FARMER, V. C. (1955) *Chem. & Ind.* **1955**, 586.

391

Literature

205 FASTIE, W. G., and PFUND, A. H. (1947) *J. opt. Soc. Amer.* **37**, 762.
206 FERGUSON, J., and WERNER, R. L. (1954) *J. chem. Soc.* **1954**, 3645. Naphthalenes.
207 FERGUSON, E. E., HUDSON, R. L., NIELSEN, J. R., and SMITH, D. C. (1953) *J. chem. Phys.* **21**, 1736. Fluorinated benzenes.
208 FIELD, J. E., WOODFORD, D. E., and GEHMAN, S. D. (1946) *J. appl. Phys.* **17**, 386. Synthetic rubber.
209 FIELD, J. E., COLE, J. O., and WOODFORD, D. E. (1950) *J. chem. Phys.* **18**, 1298. Epoxy compounds.
210 FINCH, J. N., and LIPPINCOTT, E. R. (1956) *J. chem. Phys.* **24**, 908.
211 FISCHER, K. A., and BRANDES, G. (1956) *Naturwiss.* **43**, 223.
212 FLETT, M. S. C. (1948) *J. chem. Soc.* **1948**, 1441.
213 FLETT, M. S. C. (1951) *J. chem. Soc.* **1951**, 962. Carboxyl bands.
213a FLETT, M. S. C. (1953) *J. chem. Soc.* **1953**, 347. Thioamides.
214 FLORENCE, J. M., ALLSHOUSE, C. C., GLAZE, F. W., and HAHNER, C. H. (1950) *J. Res. nat. Bur. Stand.* **45**, 121. Glasses.
215 FORD, M. A., WILKINSON, G. R., and PRICE, W. C. (1954) Conf. Mol. Spectr. Inst. Petr., London, 1954.
216 FORZIATI, F. H., ROWEN, J. W., and PLYLER, E. K. (1951) *J. Res. nat. Bur. Stand.* **46**, 288. Cellulose.
217 FOWLER, R. C. (1949) *Rev. sci. Instrum.* **20**, 175.
218 FOX, J. J., and MARTIN, A. E. (1938) *Proc. roy. Soc.* **A 167**, 257.
219 FOX, J. J., and MARTIN, A. E. (1940) *Proc. roy. Soc.* **A 174**, 234.
220 FOX, J. J., and MARTIN, A. E. (1940) *Proc. roy. Soc.* **A 175**, 208.
221 FRANICS, S. A. (1950) *J. chem. Phys.* **18**, 861.
222 FRANICS, S. A. (1951) *J. chem. Phys.* **19**, 505.
223 FRANICS, S. A. (1952) *Analyt. Chem.* **24**, 604. Cyclopentylgroups.
224 FRANICS, S. A. (1953) *Analyt. Chem.* **25**, 1466. Saturated hydrocarbons.
225 FRANCK, B. (1954) *Z. Naturf.* **9b**, 276.
226 FRANK, R. L., and BERRY, R. E. (1950) *J. Amer. chem. Soc.* **72**, 2985. Menthene.
227 FRASER, R. D. B. (1953) *J. opt. Soc. Amer.* **43**, 929.
228 FRASER, R. D. B. (1953) *J. chem. Phys.* **21**, 1511. Proteins.
229 FREEMAN, N. K. (1952) *J. Amer. chem. Soc.* **74**, 2523. Branched chain fatty acids.
230 FREEMAN, N. K. (1953) *J. Amer. chem. Soc.* **75**, 1859. Long chain 2-alkene acids.
231 FRIEDEL, R. A. (1951) *J. Amer. chem. Soc.* **73**, 2881. Cresols, xylenols.
232 FRIEDEL, R. A., and PELIPETZ, M. G. (1955) *J. opt. Soc. Amer.* **45**, 892.
233 FRIEDL, W., and HARTENSTEIN, B. (1955) *J. opt. Soc. Amer.* **45**, 398.
234 FROST, A. A., and TAMRES, M. (1947) *J. chem. Phys.* **15**, 383.
235 FRY, D. L., NUSBAUM, R. E., and RANDALL, H. M. (1946) *J. appl. Phys.* **17**, 150.
236 FUNCK, E. (1956) *Optik* **13**, 524.
237 FUOSS, R. M., and MEAD, D. J. (1945) *Rev. sci. Instrum.* **16**, 223.
238 FUSON, N., JOSIEN, M. L., and SHELTON, E. M. (1954) *J. Amer. chem. Soc.* **76**, 2526. Ketones, quinones.
239 FUSON, N., JOSIEN, M. L., POWELL, R. L., and UTTERBACK, E. (1952) *J. chem. Phys.* **20**, 145.
240 FUSON, N., JOSIEN, M. L., POWELL, R. L., and UTTERBACK, E. (1952) *J. Amer. chem. Soc.* **74**, 1.
241 GARG, S. N. (1953) *J. chem. Phys.* **21**, 1907. Mono-substituted benzenes.
242 GATES, D. M. (1952) *Amer. J. Phys.* **20**, 275.
243 GAUNT, J. (1951) *J. sci. Instrum.* **31**, 315.
244 GAUTHIER, G. (1951) *Compt. Rend.* **233**, 617.
245 GEBBIE, H. A., HARDING, W. R., HILSUM, C., PRYCE, A. W., and ROBERTS, V. (1951) *Proc. roy. Soc.* **A 206**, 87. Atmosphere.
246 GEILING, L. (1951) *Z. angew. Phys.* **3**, 467.
247 GENZEL, L. (1951) *Glastech. Ber.* **24**, 55.

248   GENZEL, L. (1952) *Optik* **9**, 143.
249   GENZEL, L., and NEUROTH, N. (1953) *Z. Phys.* **134**, 127.
250   GENZEL, L., and MÜSER, H. (1953) *Z. Phys.* **134**, 419.
251   GENZEL, L., and ECKHARDT, W. (1954) *Z. Phys.* **139**, 578.
252   GIER, J. T., DUNKLE, R. V., and BEVANS, J. T. (1954) *J. opt. Soc. Amer.* **44**, 558.
253   GIERER, A. (1953) *Z. Naturf.* **8b**, 644, 654. Acetamide derivatives.
254   GIESE, A. T., and FRENCH, C. S. (1955) *J. appl. Spectr.* **9**, 78.
255   GIGUERE, P. A., and HARVEY, K. B. (1956) *Canad. J. Chem.* **34**, 798.
256   GLATT, L., and ELLIS, J. W. (1951) *J. chem. Phys.* **19**, 449.
257   GLATT, L., WEBBER, D. S., SEAMAN, C., and ELLIS, J. W. (1950) *J. chem. Phys.* **18**, 413. Polyvinylalcohol.
258   GLEMSER, O., and HARTERT, E. (1955) *Naturwiss.* **42**, 534.
259   GLEMSER, O., and HARTERT, E. (1956) *Z. anorg. Chem.* **283**, 111.
260   GOERCKE, P. (1951) *Ann. Télécomm.* **6**, 325.
261   GOLAY, M. J. E. (1947) *Rev. sci. Instrum.* **18**, 347, 357.
262   GOLAY, M. J. E. (1949) *Rev. sci. Instrum.* **20**, 816.
263   GOLAY, M. J. E. (1955) *J. opt. Soc. Amer.* **45**, 430.
264   GOLAY, M. J. E. (1956) *J. opt. Soc. Amer.* **46**, 422.
265   GOLDSTEIN, M., and HALFORD, R. S. (1950) *Phys. Rev.* **78**, 336. Proteins.
266   GORDON, R. R., and POWELL, H. (1945) *J. sci. Instrum.* **22**, 12.
267   GORE, R. C. (1950) *Analyt. Chem.* **22**, 7.
268   GORE, R. C. (1951) *Analyt. Chem.* **23**, 7.
269   GORE, R. C. (1952) *Analyt. Chem.* **24**, 8.
270   GORE, R. C. (1954) *Analyt. Chem.* **26**, 11.
271   GORE, R. C. (1956) *Analyt. Chem.* **28**, 577.
272   GORE, R. C., BARNES, R. B., and PETERSEN, E. (1949) *Analyt. Chem.* **21**, 382.
273   GOULDEN, J. D. S., and RANDALL, S. S. (1952) *Nature* **169**, 748, 1108.
274   GRABIEL, C. E., BISGROVE, D. E., and CLAPP, L. B. (1955) *J. Amer. chem. Soc.* **77**, 1293. Nitro-compounds.
275   GREEN, T. M., and HADLEY, L. N. (1955) *J. opt. Soc. Amer.* **45**, 228.
276   GREENLER, R. G. (1954) *J. opt. Soc. Amer.* **45**, 788.
277   GREINACHER, E., LÜTTKE, W., and MECKE, R. (1955) *Z. Elektrochem.* **59**, 23.
278   GREY, D. S. (1951) *J. opt. Soc. Amer.* **41**, 183.
279   GUERTIN, D. L., WIBERLEY, S. E., BAUER, W. H., and GOLDENSON, J. (1956) *Analyt. Chem.* **28**, 1194. Branched chain fatty acids.
280   GUTSCHE, C. D. (1951) *J. Amer. chem. Soc.* **73**, 786. Ketones.
281   GUY, W., and TOWLER, J. H. (1951) *J. sci. Instrum.* **28**, 103, 321.
282   HAAS, C. (1956) *Spectrochim. Acta* **8**, 19.
283   HACSKAYLO, M. (1954) *Analyt. Chem.* **26**, 1410.
284   HADŽI, D., and SHEPPARD, N. (1953) *Proc. roy. Soc.* **A 216**, 247. Dimers of carboxylic acids.
285   HADŽI, D., and SHEPPARD, N. (1951) *J. Amer. chem. Soc.* **73**, 5460. Quinones.
286   HAENDLER, H. M., WHEELER, C. M., and BERNARD, W. J. (1953) *J. opt. Soc. Amer.* **43**, 215.
287   HAGGIS, G. H. (1956) *J. sci. Instrum.* **33**, 491.
288   HALFORD, R. S. (1946) *J. chem. Phys.* **14**, 8.
289   HALFORD, R. S., and SCHAEFFER, O. A. (1946) *J. chem. Phys.* **14**, 141.
290   HALL, R. G. N., and SAYCE, L. A. (1952) *Proc. roy. Soc.* **A 215**, 536.
291   HALL, M. B., and NESTER, R. G. (1952) *J. opt. Soc. Amer.* **42**, 257.
292   HALVERSON, F., and WILLIAMS, V. Z. (1947) *J. chem. Phys.* **15**, 552. Ketene.
293   HAM, N. S., WALSH, A., and WILLIS, J. B. (1952) *J. opt. Soc. Amer.* **42**, 496.
294   HAMMER, C. F., and ROE, H. R. (1953) *Analyt. Chem.* **25**, 668.
295   HAMPTON, R. R. (1949) *Analyt. Chem.* **21**, 923. Synthetic rubber.
296   HAMPTON, R. R., and NEWELL, J. E. (1949) *Analyt. Chem.* **21**, 914. Ester-carbonyl bands.

Literature

297 HANSEN, G., and MOHR, E. (1948) *Spectrochim. Acta* **3**, 584.

298 HARDY, A. C., and DEUCH, E. C. (1948) *J. opt. Soc. Amer.* **38**, 308.

299 HARE, W. F. J., ALLIN, E. J., and WELSH, H. L. (1955). *Phys. Rev.* **99**, 1887.

300 HARMS, D. L. (1953) *Analyt. Chem.* **25**, 1140.

301 HARP, W. R., and RASMUSSEN, R. S. (1947) *J. chem. Phys.* **15**, 778. Ketene.

302 HARRIS, L. (1946) *J. opt. Soc. Amer.* **36**, 597.

303 HART, E. J., and MEYER, A. W. (1949) *J. Amer. chem. Soc.* **71**, 1980. Synthetic rubber.

304 HARTERT, E., and GLEMSER, O. (1953) *Naturwiss.* **40**, 199.

305 HARTERT, E., and GLEMSER, O. (1956) *Z. Elektrochem.* **60**, 746.

306 HARTWELL, E. J., RICHARDS, R. E., and THOMPSON, H. W. (1948) *J. chem. Soc.* **1948**, 1436. Carbonyl bands.

307 HARVEY, R. B., and MAYHOOD, J. E. (1955) *Canad. J. Chem.* **33**, 1552. Phosphorus compounds.

308 HASTINGS, S. H., WATSON, A. T., WILLIAMS, R. B., and ANDERSON, J. A. (1952) *Analyt. Chem.* **24**, 612.

309 HASZELDINE, R. N., ALBERMAN, K. B., and KIPPING, F. B. (1952) *J. chem. Soc.* **1952**, 3284. Cyclobutane, butadienes.

310 HASZELDINE, R. N. (1953) *J. chem. Soc.* **1953**, 1757. Halogen alcohols.

311 HATCH, L. F., D'AMICO, J. J., and RUHNKE, E. V. (1952) *J. Amer. chem. Soc.* **74**, 123.

312 HAUPTSCHEIN, M., STOKES, C. S., KINSMAN, R. L., NODIFF, E. A., and GROSSE, A. V. (1952) *J. Amer. chem. Soc.* **74**, 848, 849, 1347. Perfluoralkyl halides.

313 HAWES, R. C., and GALLAWAY, W. S. (1951) *Science* **113**, 54.

314 HAWES, R. C., CARY, H. H., BECKMAN, A. O., MADSEN, R. G., HARE, G. H., and STICKNEY, M. E. (1949) Symposium on Molecular Structure and Spectroscopy (Ohio State University, 1949).

315 HEDIGER, H. J., and GÜNTHARD, H. H. (1954) *Helv. chim. acta* **37**, 1125.

316 HEIGL, J. J., BELL, M. F., and WHITE, J. U. (1947) *Analyt. Chem.* **19**, 293.

317 HENNE, A. L., and FRANCIS, W. C. (1953) *J. Amer. chem. Soc.* **75**, 991. Fluorinated alcohols.

318 HENRY, R. L. (1948) *J. opt. Soc. Amer.* **38**, 775.

319 HERGERT, H. L., and KURTH, E. F. (1953) *J. Amer. chem. Soc.* **75**, 1622. Flavanones, flavones, chalcones.

320 HERMAN, R. C., and SHAFFER, W. H. (1949) *J. chem. Phys.* **17**, 30. Allene.

321 HERSCHER, L. W. (1949) *Rev. sci. Instrum.* **20**, 833.

322 HERSCHER, L. W., and WRIGHT, N. (1953) *J. opt. Soc. Amer.* **43**, 980.

323 HERZBERG, G., and HERZBERG, L. (1953) *J. opt. Soc. Amer.* **43**, 1037.

324 HERZFELD, N., INGOLD, C. K., and POOLE, H. G. (1946) *J. chem. Soc.* **1946**, 316. Benzene.

325 HIBBARD, R. R., and CLEAVES, A. P. (1949) *Analyt. Chem.* **21**, 486.

326 HOLDEN, R. B., TAYLOR, W. J., and JOHNSTON, H. L. (1950) *J. opt. Soc. Amer.* **40**, 757.

327 HOLLIDAY, P. (1949) *Nature* **163**, 602.

328 HOLZMAN, G. (1950) *Science* **111**, 550.

329 HONN, F. J., BEZMAN, I. I., and DAUBERT, B. F. (1949) *J. Amer. chem. Soc.* **71**, 812.

330 HORNIG, D. F. (1948) *J. chem. Phys.* **16**, 1063. Crystals.

331 HORNIG, D. F. (1950) *Disc. Faraday Soc.* **9**, 115. Crystals.

332 HORNIG, D. F., and O'KEEFE, B. J. (1947) *Rev. sci. Instrum.* **18**, 474.

333 HORNIG, D. F., and MCKEAN, D. C. (1955) *J. phys. Chem.* **59**, 1133.

334 HORNIG, D. F., HYDE, G. E., and ADCOCK, W. A. (1950) *J. opt. Soc. Amer.* **40**, 497.

335 HOWARD, J. N., BURCH, D. E., and WILLIAMS, D. (1956) *J. opt. Soc. Amer.* **46**, 186, 237, 242, 334.

336 HOYER, H. (1956) *Chem. Ber.* **89**, 2677. Isocyanates.

337 HUGHES, H. K. (1952) *Analyt. Chem.* **24**, 1349.

338 HUGHES, R. H., and WILSON, E. B. (1947) *Rev. sci. Instrum.* **18**, 103.

339 HUGHES, R. H., MARTIN, R. J., and COGGESHALL, N. D. (1956) *J. chem. Phys.* **24**, 489.

340 HUNSBERGER, I. M. (1950) *J. Amer. chem. Soc.* **72**, 5626. Naphthalene.

341 HUNSBERGER, I. M., KETCHAM, R., and GUTOWSKY, H. S. (1952) *J. Amer. chem. Soc.* **74**, 4839. Phenanthrene.

342 HUNT, J. M., and TURNER, D. S. (1953) *Analyt. Chem.* **25**, 1169. Minerals.

343 HUNT, J. M., WISHERD, M. P., and BONHAM, L. C. (1950) *Analyt. Chem.* **22**, 1478. Minerals.

344 HURD, C. D., BAUER, L., and KLOTZ, I. M. (1953) *J. Amer. chem. Soc.* **75**, 624. Polypeptides.

345 INGOLD, C. K. (1945) *Trans. Faraday Soc.* **41**, 172. Benzene.

346 INGOLD, C. K., and GARFORTH, F. M. (1946) *Nature* **158**, 163. Benzene.

347 INGRAHAM, L. L., CORSE, J., BAILEY, G. F., and STITT, F. (1952) *J. Amer. chem. Soc.* **74**, 2297. Phenol, catechol.

348 JACQUEZ, J. A., and KUPPENHEIM, H. F. (1956) *J. opt. Soc. Amer.* **46**, 428.

349 JACQUEZ, J. A., MCKEEHAN, W., HUSS, J., DIMITROFF, J. M., and KUPPEN-HEIM, H. F. (1955) *J. opt. Soc. Amer.* **45**, 781, 971.

350 JACQUINOT, P. (1954) *J. opt. Soc. Amer.* **44**, 761.

351 JACQUINOT, P. (1955) *J. opt. Soc. Amer.* **45**, 996.

352 JACQUINOT, P., and CHABBAL, R. (1956) *J. opt. Soc. Amer.* **46**, 556.

353 JAFFE, J. H., and JAFFE, H. (1950) *J. opt. Soc. Amer.* **40**, 53.

354 JAFFE, J. H., RANK, D. H., and WIGGINS, T. A. (1955) *J. opt. Soc. Amer.* **45**, 636.

355 JAMISON, N. C., KOHLER, T. R., and KOPPIUS, O. G. (1951) *Analyt. Chem.* **23**, 551.

356 JANESCHITZ-KRIEGL, H. (1952) *Chem.-Ing.-Tech.* **24**, 158.

357 JENNESS, J. R. (1955) *J. opt. Soc. Amer.* **45**, 671.

358 JONES, L. H. (1956) *J. chem. Phys.* **24**, 1250.

359 JONES, R. C. (1947) *J. opt. Soc. Amer.* **37**, 879.

360 JONES, R. C. (1949) *J. opt. Soc. Amer.* **39**, 327, 344.

361 JONES, R. C. (1953) *J. opt. Soc. Amer.* **43**, 1.

362 JONES, E. A., KIRBY-SMITH, J. S., WOLTZ, P. J. H., and NIELSEN, A. H. (1951) *J. chem. Phys.* **19**, 242. Si-compounds.

363 JONES, R. N., RAMSAY, D. A., KEIR, D. S., and DOBRINER, K. (1952) *J. Amer. chem. Soc.* **74**, 80.

364 JONES, R. N., MCKAY, A. F., and SINCLAIR, R. G. (1952) *J. Amer. chem. Soc.* **74**, 2575. Fatty acids.

365 JOSIEN, M. L., and FUSON, N. (1950) *Compt. Rend.* **231**, 1511. Cyclopropane.

366 JOSIEN, M. L., and FUSON, N. (1951) *J. Amer. chem. Soc.* **73**, 478.

367 JOSIEN, M. L., and FUSON, N. (1952) *Compt. Rend.* **234**, 1680.

368 JOSIEN, M. L., FUSON, N., LEBAS, J. M., and GREGORY, T. M. (1953) *J. chem. Phys.* **21**, 331.

369 KAISER, R. (1956) *Kolloidzschr.* **148**, 168.

370 KAISER, R. (1956) *Z. angew. Phys.* **8**, 429.

371 KAKUDO, M., KASAI, P. N., and WATASE, T. (1953) *J. chem. Phys.* **21**, 1894. Silanols.

372 KAPLAN, L. (1952) *J. Met.* **9**, 139.

373 KASTLER, A. (1950) *Compt. Rend.* **230**, 1596.

374 KAYE, W. (1951) *J. opt. Soc. Amer.* **41**, 277.

375 KAYE, W. (1954) *Spectrochim. Acta* **6**, 257.

376 KAYE, W., and DEVANEY, R. G. (1952) *J. opt. Soc. Amer.* **42**, 567.

377 KAYE, W., CANON, C., and DEVANEY, R. G. (1951) *J. opt. Soc. Amer.* **41**, 658.

378 KENDALL, D. N. (1953) *Analyt. Chem.* **25**, 382.

Literature

379 KENDRICK, E., and SCHURMANN, R. (1954) *J. opt. Soc. Amer.* **44**, 501.
380 KENT, J. W., and BEACH, J. Y. (1947) *Analyt. Chem.* **19**, 290.
381 VON KEUSSLER, V. (1956) *Optik* **13**, 317.
382 KHAN, N. A. (1953) *J. chem. Phys.* **21**, 952. Methyl linolenate.
383 KING, W. H., and PRIESTLEY, W. (1951) *Analyt. Chem.* **23**, 1418.
384 KING, G. W., HAINER. R. M., and MCMAHON, H. O. (1949) *J. appl. Phys.* **20**, 559.
385 KING, G. W., BLANTON, E. H., and FRAWLEY, J. (1954) *J. opt. Soc. Amer.* **44**, 397.
386 KIRBY-SMITH, J. S., and JONES, E. A. (1949) *J. opt. Soc. Amer.* **39**, 780.
387 KIRKLAND, J. J. (1955) *Analyt. Chem.* **27**, 1537.
388 KITSON, R. E. (1953) *Analyt. Chem.* **25**, 1470.
389 KITSON, R. E., and GRIFFITH, N. E. (1952) *Analyt. Chem.* **24**, 334.
390 KIVENSON, G. (1950) *J. opt. Soc. Amer.* **40**, 112.
391 KIVENSON, G., ROTH, A., and RIDER, M. (1949) *J. opt. Soc. Amer.* **39**, 484.
392 KLETZ, T. A., and PRICE, W. C. (1947) *J. chem. Soc.* **1947**, 644. Phenols.
393 KLETZ, T. A., and SUMNER, A. (1948) *J. chem. Soc.* **1948**, 1456. Octenes.
394 KOPPIUS, O. G. (1951) *Analyt. Chem.* **23**, 554.
395 KORNBLUM, N., UNGNADE, H. E., and SMILEY, R. A. (1956) *J. org. Chem.* **21**, 377. Nitrocompounds.
396 KUENTZEL, L. E. (1951) *Analyt. Chem.* **23**, 1413.
397 KUENTZEL, L. E. (1955) *Analyt. Chem.* **27**, 301.
398 KUHN, L. P. (1950) *Analyt. Chem.* **22**, 276. Carbohydrates.
399 KUHN, L. P. (1952) *J. Amer. chem. Soc.* **74**, 2492.
400 LAAKSO, P. V., ORR. S. F. D., and THOMPSON, H. W. (1950) *J. chem. Soc.* **1950**, 221. Polycyclic compounds.
401 LACHER, J. R., BITNER, J. L., and PARK, J. D. (1955) *J. phys. Chem.* **59**, 610. Antibiotics.
402 LACHER, J. R., BITNER, J. L., EMERY, D. J., SEFEL, M. E., and PARK, J. D. (1955) *J. phys. Chem.* **59**, 615. Pyrimidines and purines.
403 LANGEMANN, R. T., and MILLER, T. G. (1951) *J. opt. Soc. Amer.* **41**, 1063.
404 LANE, T. J., SEN, D. N., and QUAGLIANO, J. V. (1954) *J. chem. Phys.* **22**, 1855.
405 LARNAUDIE, M. (1951, 1953) *Compt. Rend.* **232**, 316. Thesis (Paris, 1953). Cyclohexane.
406 LAUNER, P. J. (1952) *J. opt. Soc. Amer.* **42**, 359.
407 LAUNER, P. J., and MCCAULAY, D. A. (1951) *Analyt. Chem.* **23**, 1875. 1,2,3,4-Tetramethylbenzene.
408 LAX, M., and BURSTEIN, E. (1955) *Phys. Rev.* **97**, 39.
409 LEBAS, J. M., and JOSIEN, M. L. (1956) *Bull. Soc. chim. Fr.* **1956**, 53, 57, 62. Mono-substituted benzenes.
410 LECOMTE, J. (1943) *Cah. Phys.* **17**, 1.
411 LECOMTE, J. (1955) *Bull. Soc. chim. Fr.* **1955**, 717. Ketones and aldehydes.
412 LEHRER, E. (1942) *Z. tech. Phys.* **23**, 169.
413 LEJEUNE, R., and DUYCKAERTS, G. (1954) *Spectrochim. Acta* **6**, 194.
414 LENORMANT, H. (1950) *Ann. Chim.* **5**, 459.
415 LENORMANT, H., and CHOUTEAU, J. (1952) *Compt. Rend.* **234**, 2057.
416 LEONARD, N. J., GUTOWSKY, H. S., MIDDLETON, W. J., and PETERSEN, E. M. (1952) *J. Amer. chem. Soc.* **74**, 4070. Esters, ketoesters.
417 LETAW, H., and GROPP, A. H. (1953) *J. chem. Phys.* **21**, 1621.
418 LIEBER, E., LEVERING, D. R., and PATTERSON, L. J. (1951) *Analyt. Chem.* **23**, 1594.
419 LIPPERT, E. (1955) *Angew. Chem.* **67**, 704.
420 LIPPERT, E., and MECKE, R. (1951) *Z. Elektrochem.* **55**, 366.
421 LIPPERT, E., and VOGEL, W. (1956) *Z. phys. Chem.* **6**, 133.
422 LISTON, M. D. (1947) *Rev. sci. Instrum.* **18**, 373.
423 LISTON, M. D. (1947) *J. opt. Soc. Amer.* **37**, 515.
424 LOCKE, J. L., and HERZBERG, G. (1953) *Canad. J. Phys.* **31**, 504.

425 LOOFBOUROW, J. R. (1950) *J. opt. Soc. Amer.* **40**, 317.
426 LORD, R. C., and VENKATESWARLU, P. (1952) *J. chem. Phys.* **20**, 1237. Allene.
427 LORD, R. C., and MERRIFIELD, R. E. (1953) *J. chem. Phys.* **21**, 166.
428 LORD, R. C., and MCCUBBIN, T. K. (1955) *J. opt. Soc. Amer.* **45**, 441.
429 LORD, R. C., and WALKER, R. W. (1954) *J. Amer. chem. Soc.* **76**, 2518. Cyclo-olefins.
430 LORD, R. C., MCDONALD, R. S., and MILLER, F. A. (1952) *J. opt. Soc. Amer.* **42**, 149.
431 LUCK, W. (1951) *Z. Naturf.* **6a**, 191.
432 LÜTTKE, W., and MECKE, R. (1949) *Z. Elektrochem.* **53**, 241.
433 LÜTTKE, W., and MECKE, R. (1951) *Z. physik. Chem.* **196**, 56.
434 LUFT, K. F. (1943) *Z. tech. Phys.* **24**, 97.
435 LUFT, K. F. (1947) *Angew. Chem.* **B 19**, 2.
436 LUFT, K. F. (1954) *Compt. Rend.* **238**, 1651.
437 LUTHER, H., and CZERWONY, G. (1956) *Z. phys. Chem.* **6**, 286.
438 LUTHER, H., and GÜNZLER, H. (1955) *Z. Naturf.* **10b**, 445. Nitronaphthols.
439 LUTHER, H., FELDMANN, K., and HAMPEL. B. (1956) *Z. Elektrochem.* **59**, 1008. Naphthalene.
440 LUTHER, H., BRANDES, G., GÜNZLER, H., and HAMPEL, B. (1956) *Z. Elektrochem.* **59**, 1012. Naphthalene.
441 MAKAS, A. S., and SHURCLIFF, W. A. (1955) *J. opt. Soc. Amer.* **45**, 998.
442 MANN, J., and THOMPSON, H. W. (1952) *Proc. roy. Soc.* **A 211**, 168.
443 MARGOSHES, M., and FASSEL, V. A. (1955) *Spectrochim. Acta* **7**, 14. Benzenes.
444 MARRISON, L. W. (1951) *J. chem. Soc.* **1951**, 1614. Cyclohexane.
445 MARTIN, A. E. (1948) *J. opt. Soc. Amer.* **38**, 70.
446 MARTIN, A. E. (1951) *J. opt. Soc. Amer.* **41**, 56.
447 MARTIN, A. E. (1951) *Trans. Faraday Soc.* **47**, 1182.
448 MARTIN, A. E. (1956) *Ind. Chemist* **32**, 379.
449 MATHESON, L. A., and BOYER, R. F. (1952) *Industr. Engng. Chem. (Industr.)* **44**, 867.
450 MATHIS, R., BOSSON, F., GAUTHIER, G., and LARNAUDIE, M. (1950) *J. Phys. Radium* **11**, 300.
451 MATOSSI, F. (1949) *Phys. Rev.* **76**, 1845.
452 MATOSSI, F., and RAUSCHER, E. (1949) *Z. Phys.* **125**, 418.
453 MCBRIDE, J. J., and BEACHELL, H. C. (1952) *J. Amer. chem. Soc.* **74**, 5247. Methylisocyanosilanes.
454 MCCUBBIN, T. K. (1952) *J. chem. Phys.* **20**, 668.
455 MCCUBBIN, T. K., and SINTON, W. M. (1950) *J. opt. Soc. Amer.* **40**, 537.
456 MCCUBBIN, T. K., and SINTON, W. M. (1952) *J. opt. Soc. Amer.* **42**, 113.
457 MCGOVERN, J. J., and FRIEDEL, R. A. (1947) *J. opt. Soc. Amer.* **37**, 660.
458 MCKINNEY, D. S., and FRIEDEL, R. A. (1948) *J. opt. Soc. Amer.* **38**, 222.
459 MCMAHON, H. O., HAINER, R. M., and KING, G. W. (1949) *J. opt. Soc. Amer.* **39**, 786.
460 MCMURRY, H. L., and THORNTON, V. (1952) *Analyt. Chem.* **24**, 318.
461 MECKE, R. (1951) *Disc. Faraday Soc.* **9**, 161.
462 MECKE, R. (1956) *Tagung Technik d. Ultrarotspektroskopie* (Frieburg, 1956).
463 MECKE, R., and OSWALD, F. (1950) *Z. Phys.* **130**, 445.
464 MECKE, R., and MUTTER, R. (1954) *Z. Elektrochem.* **58**, 1.
465 MECKE, R., and ROSSMY, G. (1955) *Z. Elektrochem.* **59**, 866. Alcohols and phenols.
466 MECKE, R., and MECKE, R. (1956) *Chem. Ber.* **89**, 343. Peptides.
467 MECKE, R., and NOACK, K. (1956) *Angew. Chem.* **68**, 150. Unsaturated ketones.
468 MECKE, R., MECKE, R., and LÜTTRINGHAUS, A. (1955) *Z. Naturf.* **10b**, 367.
469 MEEKS, M. R., WHITTIER, V. E., and YOUNG, C. W. (1951) *Analyt. Chem.* **23**, 792.
470 MELLORS, R. C. (1950) *Science* **112**, 381.

Literature

471 MERTON, T. (1950) *Proc. roy. Soc.* **A 201**, 187.

472 MERTZ, L. (1953) *J. opt. Soc. Amer.* **43**, 220.

473 MEYRICK, C. I., and THOMPSON, H. W. (1950) *J. chem. Soc.* **1950**, 225. Esters of phosphorus oxy-acids.

474 MILAS, N. A., and MAGELI, O. L. (1952) *J. Amer. chem. Soc.* **74**, 1471. Peroxides.

475 MILLER, F. A., and KOCH, S. D. (1948) *J. Amer. chem. Soc.* **70**, 1890. Ketene.

476 MILLER, F. A., and INSKEEP, R. G. (1950) *J. chem. Phys.* **18**, 1519. Cyclopentane.

477 MILLER, F. A., and WILKINS, C. H. (1952) *Analyt. Chem.* **24**, 1253. Inorganic compounds.

478 MILLER, C. H., and THOMPSON, H. W. (1949) *Proc. roy. Soc.* **A 200**, 1. Allene.

479 MILLER, R. G. J., and WILLIS, H. A. (1956) *J. Polym. Sci.* **19**, 485.

480 MILLS, I. M., SCHERER, J. R., CRAWFORD, B., and YOUNGQUIST, M. (1955) *J. opt. Soc. Amer.* **45**, 785.

481 MITZNER, B. M. (1953) *J. opt. Soc. Amer.* **43**, 806.

482 MIZUSHIMA, M. (1950) *Phys. Rev.* **77**, 150.

483 MIZUSHIMA, M. (1953) *J. chem. Phys.* **21**, 1222. Allene.

484 MOHLER, O. C. (1951) *Phys. Rev.* **83**, 464 .

485 MOHLER, N. M., and LOOFBOUROW, J. R. (1952) *Amer. J. Phys.* **20**, 499.

486 MOONEY, R. C. L. (1941) *J. Amer. chem. Soc.* **63**, 2828.

487 MOSBY, W. L. (1952) *J. Amer. chem. Soc.* **74**, 2564. Naphthalene derivatives.

488 MOSS, T. S. (1950) *J. opt. Soc. Amer.* **40**, 603.

489 MÜSER, H. (1951) *Z. Phys.* **129**, 504.

490 MUNDAY, C. W. (1948) *J. sci. Instrum.* **25**, 418.

491 NELSON, R. C. (1949) *J. opt. Soc. Amer.* **39**, 68.

492 NEU, J. T. (1953) *J. opt. Soc. Amer.* **43**, 520.

493 NEU, J. T., and GWINN, W. D. (1950) *J. chem. Phys.* **18**, 1642.

494 NEUROTH, N. (1952) *Glastech. Ber.* **25**, 242.

495 NEUROTH, N. (1953) *Glastech. Ber.* **26**, 66.

496 NEUROTH, N. (1956) *Z. Phys.* **144**, 85.

497 NEWMAN, R., and HALFORD, R. S. (1948) *Rev. sci. Instrum.* **19**, 270.

498 NIELSEN, J. R., and SMITH, D. C. (1943) *Industr. Engng. Chem. (Industr.)* **15**, 609.

499 NIELSEN, J. R., THORNTON, V., and DALE, E. B. (1944) *Rev. Mod. Phys.* **16**, 307.

500 NIELSEN, J. R., CRAWFORD, F. W., and SMITH, D. C. (1947) *J. opt. Soc. Amer.* **37**, 296.

501 NIGHTINGALE, R. E., COWAN, G. R., and CRAWFORD, B. (1953) *J. chem. Phys.* **21**, 1398.

502 NORRIS, K. P., and WILKINS, M. H. F. (1950) *Disc. Faraday Soc.* **9**, 360.

503 NORRIS, K. P., SEEDS, W. E., and WILKINS, M. H. F. (1951) *J. opt. Soc. Amer.* **41**, 111.

504 OAKES, W. G., and RICHARDS, R. B. (1949) *J. chem. Soc.* **1949**, 2929.

505 O'CONNOR, R. T., and GOLDBLATT, L. A. (1954) *Analyt. Chem.* **26**, 1726.

506 OETJEN, R. A., and ROESS, L. C. (1951) *J. opt. Soc. Amer.* **41**, 203.

507 OETJEN, R. A., WARD, W. M., and ROBINSON, J. A. (1946) *J. opt. Soc. Amer.* **36**, 615.

508 OETJEN, R. A., HAYNIE, W. H., WARD, W. M., HANSLER, R. L., SCHAUWECKER, H. E., and BELL, E. E. (1952) *J. opt. Soc. Amer.* **42**, 559.

509 OLDHAM, M. S. (1951) *J. opt. Soc. Amer.* **41**, 673.

510 OPLER, A. (1950) *Analyt. Chem.* **22**, 558.

511 ORR, S. F. D., and THOMPSON, H. W. (1950) *J. chem. Soc.* **1950**, 218. Polycyclic compounds.

512 ORR, S. F. D., and HARRIS, R. J. C. (1952) *Nature* **169**, 544. Polysaccharides.

513 OSBERG, W. E., and HORNIG, D. F. (1952) *J. chem. Phys.* **20**, 1345.

514 OSWALD, F. (1954) *Z. Elektrochem.* **58**, 345.

515 OSWALD, F. (1955) *Z. Naturf.* **10a**, 927.

516 OSWALD, F., and SCHADE, R. (1954) *Z. Naturf.* **9a**, 611.

517 OTTING, W., and STAIGER, G. (1955) *Chem. Ber.* **88**, 828. *o*-Benzoqinones.
518 OVEREND, J., and THOMPSON, H. W. (1953) *J. opt. Soc. Amer.* **43**, 1065. Allene.
519 OXHOLM, M. L., and WILLIAMS, D. (1949) *Phys. Rev.* **76**, 151.
520 PACAULT, A., and LECOMTE, J. (1949) *Compt. Rend.* **228**, 241.
521 PAIGE, E. G. S. (1955) *J. sci. Instrum.* **32**, 150.
522 PALM, A., and WERBIN, H. (1953) *Canad. J. Chem.* **31**, 1004. Oximes.
523 PATTERSON, W. A. (1954) *Analyt. Chem.* **26**, 823. Epoxides.
524 PENNER, S. S., and WEBER, D. (1951) *J. chem. Phys.* **19**, 807, 817.
525 ANON, (1953) *Perkin-Elmer Instr. News* **5**, 5.
526 PERRY, J. W. (1955) *J. opt. Soc. Amer.* **45**, 995.
527 PESTEMER, M. (1951) *Angew. Chem.* **63**, 118.
528 PESTEMER, M. (1955) *Angew. Chem.* **67**, 740.
529 PETERSON, J. H., and DOWNING, J. R. (1951) *J. opt. Soc. Amer.* **41**, 862.
530 PFANN, H. F., WILLIAMS, V. Z., and MARK, H. (1946) *J. Polym. Sci.* **1**, 14.
531 PFUND, A. H. (1947) *J. opt. Soc. Amer.* **37**, 558.
532 PHILPOTTS, A. R., and THAIN, W. (1952) *Analyt. Chem.* **24**, 638. Peroxides.
533 PHILPOTTS, A. R., THAIN, W., and SMITH, P. G. (1951) *Analyt. Chem.* **23**, 268.
534 PIERSON, R. H., and OLSEN, A. L. (1955) *Analyt. Chem.* **27**, 2022.
535 PILSTON, R. G., and WHITE, J. U. (1954) *J. opt. Soc. Amer.* **44**, 572.
536 PIMENTEL, G. C., and MCCLELLAN, A. L. (1952) *J. chem. Phys.* **20**, 270. Naphthalene.
537 PIMENTEL, G. C., MCCLELLAN, A. L., PERSON, W. B., and SCHNEPP, O. (1955) *J. chem. Phys.* **23**, 234. Naphthalene.
538 PIRLOT, G. (1949) *Bull. Soc. chim. Belg.* **58**, 28.
539 PIRLOT, G. (1950) *Bull. Soc. chim. Belg.* **59**, 327.
540 PIRLOT, G. (1950) *Bull. Soc. chim. Belg.* **59**, 352.
541 PITZER, K. S., and SCOTT, D. W. (1943) *J. Amer. chem. Soc.* **65**, 803. Benzene.
542 PLISKIN, W. A., and EISCHENS, R. P. (1955) *J. phys. Chem.* **59**, 1156.
543 PLISKIN, W. A., and EISCHENS, R. P. (1956) *J. chem. Phys.* **24**, 482.
544 PLYLER, E. K. (1950) *Disc. Faraday Soc.* **9**, 100. Chlorobenzene.
545 PLYLER, E. K. (1952) *J. Res. nat. Bur. Stand.* **48**, 281. Methanol, ethanol, propanol.
546 PLYLER, E. K. (1955) *Microchim. Acta* **1955**, 421.
547 PLYLER, E. K., and ACQUISTA, N. (1949) *J. Res. nat. Bur. Stand.* **43**, 37. Cyclo-paraffins.
548 PLYLER, E. K., and ACQUISTA, N. (1953) *J. opt. Soc. Amer.* **43**, 212.
549 PLYLER, E. K., and ACQUISTA, N. (1955) *J. chem. Phys.* **23**, 752.
550 PLYLER, E. K., and ACQUISTA, N. (1956) *J. Res. nat. Bur. Stand.* **56**, 149.
551 PLYLER, E. K., and PETERS, C. W. (1950) *J. Res. nat. Bur. Stand.* **45**, 462.
552 PLYLER, E. K., and BALL, J. J. (1952) *J. opt. Soc. Amer.* **42**, 266.
553 PLYLER, E. K., STAIR, R., and HUMPHREYS, C. J. (1947) *J. Res. nat. Bur. Stand.* **38**, 211. Cyclopentyl, cyclohexyl groups.
554 PLYLER, E. K., GAILAR, N. M., and WIGGINS, T. A. (1952) *J. Res. nat. Bur. Stand.* **48**, 221.
555 PLYLER, E. K., BLAINE, L. R., and TIDWELL, E. D. (1955) *J. Res. nat. Bur. Stand.* **55**, 279.
556 POLO, S. R., and WILSON, M. K. (1955) *J. chem. Phys.* **23**, 2376.
557 POTTS, W. J. (1955) *Analyt. Chem.* **27**, 1027. Benzene.
558 POTTS, W. J., and WRIGHT, N. (1956) *Analyt. Chem.* **28**, 1255.
559 POWLING, J., and BERNSTEIN, H. J. (1951) *J. Amer. chem. Soc.* **73**, 1815.
560 POZEFSKY, A., and COGGESHALL, N. D. (1951) *Analyt. Chem.* **23**, 1611.
561 PRICE, C. C., and GILLIS, R. G. (1953) *J. Amer. chem. Soc.* **75**, 4750. S-compounds.
562 PRICE, W. C., and TETLOW, K. S. (1948) *J. chem. Phys.* **16**, 1157.
563 PRIMAS, H., and GÜNTHARD, H. H. (1954) *Helv. chim. acta* **37**, 360.
564 PRIMAS, H., and GÜNTHARD, H. H. (1955) *Helv. chim. acta* **38**, 1254.

Literature

565 QUINAN, J. R., and WIBERLEY, S. E. (1953) *J. chem. Phys.* **21**, 1896. Alcohols.
566 QUINAN, J. R., and WIBERLEY, S. E. (1954) *Analyt. Chem.* **26**, 1762. Alcohols.
567 RAHBEK, H., and OMAR, M. (1952) *Nature* **169**, 1008.
568 RAMSAY, D. A. (1952) *J. Amer. chem. Soc.* **74**, 72.
569 RANDALL, H. M., and SMITH, D. W. (1953) *J. opt. Soc. Amer.* **43**, 1086. Bacteria.
570 RANDLE, R. R., and WHIFFEN, D. H. (1952) *J. chem. Soc.* **1952**, 4153. Aromatic nitro-compounds.
571 RANDLE, R. R., and WHIFFEN, D. H. (1954) Conf. Mol. Spectr., Inst. Petr., London, 1954. Benzenes.
572 RANK, D. H., and SHEARER, J. N. (1954) *J. opt. Soc. Amer.* **44**, 575.
573 RANK, D. H., and BENNETT, H. E. (1955) *J. opt. Soc. Amer.* **45**, 69.
574 RANK, D. H., RIX, H. D., and WIGGINS, T. A. (1953) *J. opt. Soc. Amer.* **43**, 157, 213.
575 RANK, D. H., SHULL, E. R., and WIGGINS, T. A. (1953) *J. opt. Soc. Amer.* **43**, 214.
576 RANK, D. H., BENNETT, J. M., and BENNETT, H. E. (1956) *J. opt. Soc. Amer.* **46**, 477.
577 RANK, D. H., SHULL, E. R., BENNETT, J. M., and WIGGINS, T. A. (1953) *J. opt. Soc. Amer.* **43**, 952.
578 RAO, K. N. (1950) *J. chem. Phys.* **18**, 213.
579 RASMUSSEN, R. S. (1948) *J. chem. Phys.* **16**, 712. Saturated C_4- and C_5-hydrocarbons.
580 RASMUSSEN, R. S., and BRATTAIN, R. R. (1947) *J. chem. Phys.* **15**, 120, 131. C_2- to C_4-mono-olefine, 2-methyl-2-butene, C_4- and C_5-dienes.
581 RASMUSSEN, R. S., and BRATTAIN, R. R. (1949) *J. Amer. chem. Soc.* **71**, 1073. Esters, lactones, chlorides, and salts of carboxylic acids.
582 RASMUSSEN, R. S., BRATTAIN, R. R., and ZUCCO, P. S. (1947) *J. chem. Phys.* **15**, 135. Octenes.
583 RASMUSSEN, R. S., TUNNICLIFF, D. D., and BRATTAIN, R. R. (1949) *J. Amer. chem. Soc.* **71**, 1068. Ketones.
584 RECART, L. (1951) *Compt. Rend.* **233**, 383. Ketone carbonyl bands.
585 REDING, F. P., and HORNIG, D. F. (1951) *J. chem. Phys.* **19**, 594.
586 RICHARDS, R. B. (1951) *J. appl. Chem.* **1**, 370. Polyethylene.
587 RICHARDS, R. E., and THOMPSON, H. W. (1947) *J. chem. Soc.* **1947**, 1248.
588 RICHARDS, R. E., and THOMPSON, H. W. (1947) *J. chem. Soc.* **1947**, 1260. Phenol resins.
589 RICHARDS, R. E., and THOMPSON, H. W. (1948) *Proc. roy. Soc.* **A 195**, 1.
590 RICHARDS, R. E., and THOMPSON, H. W. (1949) *J. chem. Soc.* **1949**, 124. Silicones.
591 RICHARDS, R. E., and BURTON, W. R. (1949) *Trans. Faraday Soc.* **45**, 874.
592 RICHARDSON, W. S., and SACHER, A. (1953) *J. Polym. Sci.* **10**, 353.
593 RICHARDSON, H. M., FOWLER, R. G., and COFFMAN, M. L. (1953) *J. opt. Soc. Amer.* **43**, 873.
594 RIENTSMA, L. M., and WESTERDIJK, J. B. (1955) *Microchim. Acta* **1955**, 542.
595 ROBERTS, V. (1954) *J. sci. Instrum.* **31**, 226.
596 ROBERTS, V. (1954) *J. sci. Instrum.* **31**, 251.
597 ROBERTS, V. (1955) *J. sci. Instrum.* **32**, 294.
598 ROBERTS, J. D., and CHAMBERS, V. C. (1951) *J. Amer. chem. Soc.* **73**, 5030, 5034. Cycloalkyl halides and acetate.
599 ROBERTS, J. D. and SIMMONS, H. E. (1951) *J. Amer. chem. Soc.* **73**, 5487. Cycloparaffins.
600 ROBINSON, D. Z. (1951) *Analyt. Chem.* **23**, 273.
601 ROBINSON, D. Z. (1952) *Analyt. Chem.* **24**, 619.
602 ROBINSON, T. S. (1952, 1953) *Proc. phys. Soc.* **B 65**, 910 and Thesis (London, 1953).
603 ROBINSON, T. S., and PRICE, W. (1953) *Proc. phys. Soc.* **B 66**, 969. Polytetra-fluoroethylene, ureas.

604 ROCHESTER, J. C. O., and MARTIN, A. E. (1951) *Nature* **168**, 785.

605 RODNEY, W. S. (1955) *J. opt. Soc. Amer.* **45**, 987.

606 ROSS, W. L., and LITTLE, D. E. (1951) *J. opt. Soc. Amer.* **41**, 1006.

607 ROSSMY, G., LÜTTKE, W., and MECKE, R. (1953) *J. chem. Phys.* **21**, 1606.

608 ROUIR, E. V. (1956) *Industr. chim. belge.* **21**, 221.

609 ROWEN, J. W., and PLYLER, E. K. (1950) *J. Res. nat. Bur. Stand.* **44**, 313. Cellulose.

610 ROWEN, J. W., HUNT, C. M., and PLYLER, E. K. (1947) *J. Res. nat. Bur. Stand.* **39**, 133. Cellulose.

611 ROWEN, J. W., FORZIATI, F. H., and REEVES, R. E. (1951) *J. Amer. chem. Soc.* **73**, 4484. Cellulose.

612 RUGG, F. M., CALVERT, W. L., and SMITH, J. J. (1951) *J. opt. Soc. Amer.* **41**, 32.

613 RUGG, F. M., SMITH, J. J., and ATKINSON, J. V. (1952) *J. Polym. Sci.* **9**, 579. Polyethylene.

614 RUGG, F. M., SMITH, J. J., and WARTMAN, L. H. (1953) *J. Polym. Sci.* **11**, 1.

615 RUGG, F. M., SMITH, J. J., and BACON, R. C. (1954) *J. Polym. Sci.* **13**, 535.

616 RUNDLE, R. E., and PARASOL, M. (1952) *J. chem. Phys.* **20**, 1487.

617 RUPERT, C. S. (1952) *J. opt. Soc. Amer.* **42**, 684.

618 RUPERT, C. S., and STRONG, J. (1949) *J. opt. Soc. Amer.* **39**, 1061.

619 RUSSELL, R. A., and THOMPSON, H. W. (1955) *J. chem. Soc.* **1955**, 479.

620 RYASON, R. (1953) *J. opt. Soc. Amer.* **43**, 928.

621 SAIER, E. L., and COGGESHALL, N. D. (1948) *Analyt. Chem.* **20**, 812.

622 SALOMON, G., and VON DER SCHEE, A. C. (1954) *J. Polym. Sci.* **14**, 181.

623 SALOMON, G., VON DER SCHEE, A. C., KETELAAR, J. A. A., and VON EYK, B. J. (1950) *Disc. Faraday Soc.* **9**, 291.

624 SANDEMAN, I., and KELLER, A. (1956) *J. Polym. Sci.* **19**, 401.

625 SANDERS, C. L., and MIDDLETON, E. E. K. (1953) *J. opt. Soc. Amer.* **43**, 58.

626 SANDERSON, J. A. (1947) *J. opt. Soc. Amer.* **37**, 771.

627 SANDS, J. D., and TURNER, G. S. (1952) *Analyt. Chem.* **24**, 791.

628 SAUNDERS, R. A., and SMITH, D. C. (1949) *J. appl. Phys.* **20**, 953. Rubber.

629 SAVITZKY, A., and HALFORD, R. S. (1950) *Rev. sci. Instrum.* **21**, 203.

630 SAVITZKY, A., and ATWOOD, J. C. (1952) *Analyt. Chem.* **24**, 228.

631 SAVITZKY, A., and BRESKY, D. R. (1954) *Industr. Engng. Chem.* (*Industr.*) **46**, 1382.

632 SCHIEDT, U. (1953) *Z. Naturf.* **8b**, 66.

633 SCHIEDT, U., and REINWEIN, H. (1952) *Z. Naturf.* **7b**, 270.

634 SCHLATTER, M. J., and CLARK, R. D. (1953) *J. Amer. chem. Soc.* **75**, 361. Alkylderivatives of toluene and ethylbenzene.

635 SCHURMANN, R., and KENDRICK, E. (1954) *Analyt. Chem.* **26**, 1263.

636 SCHREIBER, K. C. (1949) *Analyt. Chem.* **21**, 1168. Sulphones, sulphides.

637 SCOTT, R. E., and FRISCH, K. C. (1951) *J. Amer. chem. Soc.* **73**, 2599. Unsaturated chlorosilanes.

638 SEARLES, S., TAMRES, M., and BARROW, G. M. (1953) *J. Amer. chem. Soc.* **75**, 71. Esters, lactones.

639 SEIDMAN, J. (1951) *Analyt. Chem.* **23**, 559. Naphthalene.

640 SENSI, P., and GALLO, G. G. (1955) *Gazz. chim. ital.* **85**, 224, 235.

641 SEYFRIED, W. D., and HASTINGS, S. H. (1947) *Industr. Engng. Chem.* (*Anal.*) **19**, 298.

642 SHAW, J. H., and CLAASSEN, H. H. (1951) *Phys. Rev.* **81**, 462.

643 SHAW, J. H., CHAPMAN, R. M., and HOWARD, J. N. (1950) *Phys. Rev.* **79**, 1017.

644 SHAW, J. H., CHAPMAN, R. M., HOWARD, J. N., and OXHOLM, M. L. (1951) *Astrophys. J.* **113**, 268.

645 SHEPPARD, N. (1950) *Trans. Faraday Soc.* **46**, 429. C-S-groups.

646 SHEPPARD, N., and SUTHERLAND, G. B. B. M. (1945) *Trans. Faraday Soc.* **41**, 261.

Literature

647   SHEPPARD, N., and SUTHERLAND, G. B. B. M. (1947) *J. chem. Soc.* **1947**, 1699.
648   SHEPPARD, N., and SUTHERLAND, G. B. B. M. (1947) *Nature* **159**, 739.
649   SHEPPARD, N., and SIMPSON, D. M. (1952) *Quart. Rev.* **6**, 1.
650   SHEPPARD, N., and SIMPSON, D. M. (1953) *Quart. Rev.* **7**, 19.
651   SHREVE, O. D. (1952) *Analyt. Chem.* **24**, 1692.
652   SHREVE, O. D., and HEETHER, M. R. (1950) *Analyt. Chem.* **22**, 836.
653   SHREVE, O. D., HEETHER, M. R., KNIGHT, H. B., and SWERN, D. (1950) *Analyt. Chem.* **22**, 1261, 1498. Unsaturated acids, esters, alcohols, trans-octadecene.
654   SHREVE, O. D., HEETHER, M. R., KNIGHT, H. B., and SWERN, D. (1951) *Analyt. Chem.* **23**, 277. Epoxy compounds.
655   SHREVE, O. D., HEETHER, M. R., KNIGHT, H. B., and SWERN, D. (1951) *Analyt. Chem.* **23**, 282. Peroxides.
656   SIEBERT, W. (1954) *Z. angew. Phys.* **6**, 563.
657   SIEGLER, E. H., and HULEY, J. W. (1956) Conf. Anal. Chem. and Appl. Spectr., Pittsburgh, 1956.
658   SILVERMANN, S. (1948) *J. opt. Soc. Amer.* **38**, 989.
659   SIMARD, G. L., and STEGER, J. (1946) *Rev. sci. Instrum.* **17**, 156.
660   SIMARD, G. L., STEGER, J., MARINER, T., SALLEY, D. J., and WILLIAMS, V. Z. (1948) *J. chem. Phys.* **16**, 836.
661   ŠIMON, I. (1951) *J. opt. Soc. Amer.* **41**, 336.
662   ŠIMON, I., and MCMAHON, H. O. (1952) *J. chem. Phys.* **20**, 905. Alkylsilanes, alkylsiloxanes.
663   ŠIMON, I., and MCMAHON, H. O. (1953) *J. chem. Phys.* **21**, 23. Quartz, cristobalite.
664   SIMPSON, D. M., and SUTHERLAND. G. B. B. M. (1949) *Proc. roy. Soc.* A **199**, 169. Branched paraffins.
665   SIMPSON, O., and SUTHERLAND, G. B. B. M. (1952) *Science* **115**, 1.
666   SINCLAIR, R. G., MCKAY, A. F., and JONES, R. N. (1952) *J. Amer. chem. Soc.* **74**, 2570. Saturated fatty acids and esters.
667   SINCLAIR, R. G., MCKAY, A. F., MYERS, G. S., and JONES, R. N. (1952) *J. Amer. chem. Soc.* **74**, 2578. Unsaturated fatty acids and esters.
668   SIPPEL, A. (1949) *Z. Naturf.* **4a**, 179.
669   SLABEY, V. A. (1952) *J. Amer. chem. Soc.* **74**, 4928, 4930. Cyclic paraffins.
670   SLOWINSKI, E. J., and CLAVER, G. C. (1955) *J. opt. Soc. Amer.* **45**, 396.
671   SMITH, L. G. (1942) *Rev. sci. Instrum.* **13**, 65.
672   SMITH, A. L. (1953) *J. chem. Phys.* **21**, 1997. Methylchlorosilanes.
673   SMITH, A. L., KELLER, W. E., and JOHNSTON, H. L. (1950) *Phys. Rev.* **79**, 728.
674   SMITH, R. A. (1953) *Advanc. Phys.* **2**, 321.
675   SMITH, F. A., and CREITZ, E. C. (1949) *Analyt. Chem.* **21**, 1474.
676   SMITH, F. A., and CREITZ, E. C. (1951) *J. Res. nat. Bur. Stand.* **46**, 145.
677   SMITH, D. C., and MILLER, E. C. (1944) *J. opt. Soc. Amer.* **34**, 130.
678   SMITH, J. J., RUGG, F. M., and BOWMAN, H. M. (1952) *Analyt. Chem.* **24**, 497.
679   SOLOWAY, A. H., and FRIESS, S. L. (1951) *J. Amer. chem. Soc.* **73**, 5000. Substituted acetophenones.
680   STAHL, W. H., and PESSEN, H. (1952) *J. Amer. chem. Soc.* **74**, 5487. Di-alkyl sebacates.
681   STAMM, R. F., and WHALEN, J. J. (1946) *J. opt. Soc. Amer.* **36**, 2.
682   STEDMAN, D. F. (1952) *J. chem. Phys.* **20**, 718.
683   STEIN, R. S., and SUTHERLAND, G. B. B. M. (1955) *J. chem. Phys.* **22**, 1993.
684   STEVENS, S. S. (1954) *Physics Today* **8**, 12.
685   STEVENSON, H. J. R., and BOLDUAN, O. E. A. (1952) *Science* **116**, 111.
686   STEVENSON, H. J. R., and LEVINE, S. (1953) *Rev. sci. Instrum.* **24**, 229.
687   STIMSON, M. M., and O'DONNELL, M. J. (1952) *J. Amer. chem. Soc.* **74**, 1805.
688   STRAUSS, F. B., and THOMPSON, A. E. (1955) *Chem. & Ind.* **1955**, 1402
689   STRONG, F. C. (1952) *Analyt. Chem.* **24**, 338.
690   STUART, A. V. (1953) *J. opt. Soc. Amer.* **43**, 212.

691 STUART, A. V., and SUTHERLAND, G. B. B. M. (1956) *J. chem. Phys.* **24**, 559. Alcohols.

692 SUTHERLAND, G. B. B. M. (1950) *Disc. Faraday Soc.* **9**, 274.

693 SUTHERLAND, G. B. B. M., and WILLIS, H. A. (1945) *Trans. Faraday Soc.* **41**, 181.

694 SUTHERLAND, G. B. B. M., and JONES, A. V. (1950) *Disc. Faraday Soc.* **9**, 281. Polyisoprene.

695 SZASZ, G. J., and SHEPPARD, N. (1953) *Trans. Faraday Soc.* **49**, 358.

696 TANNER, E. M. (1956) *Chem. & Ind.* **1956**, 916.

697 TARTE, P. (1951) *Bull. Soc. chim. Belg.* **60**, 227, 240. Alkyl nitrites.

698 TARTE, P. (1952) *J. chem. Phys.* **20**, 1570. Nitrites.

699 TAYLOR, J. H., RUPERT, C. S., and STRONG, J. (1951) *J. opt. Soc. Amer.* **41**, 626.

700 THOMAS, W. J. O. (1953) *Chem. Ind.* **1953**, 567. Azo-compounds.

701 THOMPSON, H. W. (1947) *J. chem. Soc.* **1947**, 289.

702 THOMPSON, H. W. (1954) Conf. Mol. Spectr., Inst. Petr. (London, 1954).

703 THOMPSON, H. W. (1955) *J. chem. Soc.* **1955**, 4501.

704 THOMPSON, H. W. (1945) *Trans. Faraday Soc.* **41**, 275.

705 THOMPSON, H. W., and TORKINGTON, P. (1945) *Trans. Faraday Soc.* **41**, 246.

706 THOMPSON, H. W., and TORKINGTON, P. (1945) *Proc. roy. Soc.* A **184**, 3, 21.

707 THOMPSON, H. W., NICHOLSON, D. L., and SHORT, L. N. (1950) *Disc. Faraday Soc.* **9**, 222.

708 THOMPSON, H. W., VAGO, E. E., CORFIELD, M. C., and ORR, S. F. D. (1950) *J. chem. Soc.* **1950**, 214. Polycyclic compounds.

709 THORNBURG, W. (1955) *J. opt. Soc. Amer.* **45**, 740.

710 TILTON, L. W., and PLYLER, E. K. (1951) *J. Res. nat. Bur. Stand.* **47**, 25.

711 TOLANSKY, S. (1951) *J. opt. Soc. Amer.* **41**, 425.

712 TOLANSKY, S. (1952) *Nature* **169**, 445.

713 TORKINGTON, P. (1950) *J. opt. Soc. Amer.* **40**, 481.

714 TORKINGTON, P., and THOMPSON, H. W. (1945) *Trans. Faraday Soc.* **41**, 184.

715 TOWLER, J. H., and GUY, W. (1951) *J. sci. Instrum.* **28**, 105.

716 TREUMANN, W. B., and WALL, F. T. (1949) *Analyt. Chem.* **21**, 1161. Synthetic rubber.

717 TROTTER, I. F., and THOMPSON, H. W. (1946) *J. chem. Soc.* **1946**, 481. Thiols, sulphides, disulphides.

718 TSCHAMLER, H., and LEUTNER, R. (1952) *Mh. Chem.* **83**, 1502. Ethers, acetates.

719 TYLER, J. E., and EHRHARDT, S. A. (1953) *Analyt. Chem.* **25**, 390.

720 VAGO, E. E., TANNER, E. M., and BRYANT, K. C. (1949) *J. Inst. Petr.* **35**, 293. Benzene derivatives.

721 VAMPIRI, M. (1954) *Gazz. chim. ital.* **84**, 1087.

722 VANDERBELT, J. M., SCOTT, R. B., and SCHOEB, E. J. (1954) *Appl. Spectr.* **8**, 88.

723 VOETTER, H., and TSCHAMLER, H. (1952) *Mh. Chem.* **83**, 835. Cyclopentane.

724 WALSH, A. (1952) *J. opt. Soc. Amer.* **42**, 94.

725 WALSH, A. (1953) *J. opt. Soc. Amer.* **43**, 215.

726 WALSH, A., and WILLIS, J. B. (1950) *J. chem. Phys.* **18**, 552.

727 WALSH, A., and WILLIS, J. B. (1953) *J. opt. Soc. Amer.* **43**, 989.

728 WARSCHAUER, D. M., and PAUL, W. (1956) *Rev. sci. Instrum.* **27**, 419.

729 WEBER, D., and PENNER, S. S. (1951) *J. chem. Phys.* **19**, 974.

730 WELSH, H. L., CRAWFORD, M. F., MACDONALD, J. C. F., and CHISHOLM, D. A. (1951) *Phys. Rev.* **83**, 1264.

731 WESTERMARK, H. (1955) *Acta chem. scand.* **9**, 947. Alkylsilanes.

732 WHEATLEY, P. J., VINCENT, E. R., ROTENBERG. D. L., and COWAN. G. R. (1951) *J. opt. Soc. Amer.* **41**, 665.

733 WHIFFEN, D. H., and THOMPSON, H. W. (1946) *J. chem. Soc.* **1946**, 1005. Diketene.

734 WHITCOMB, S. E., NIELSEN, H. H., and THOMAS, L. H. (1940) *J. chem. Phys.* **8**, 143.

735 WHITE, J. U. (1942) *J. opt. Soc. Amer.* **32**, 285.

2D*

Literature

736 WHITE, J. U. (1947) *J. opt. Soc. Amer.* **37**, 713.

737 WHITE, J. U. (1950) *Rev. sci. Instrum.* **21**, 629.

738 WHITE, J. U., and HOWARD, J. N. (1956) *J. opt. Soc. Amer.* **46**, 662.

739 WHITE, J. U., and LISTON, M. D. (1950) *J. opt. Soc. Amer.* **40**, 29, 36, 93.

740 WHITE, J. U., LISTON, M. D., and SIMARD, R. G. (1949) *Analyt. Chem.* **21**, 1156.

741 WIBERLEY, S. E., and BUNCE, S. C. (1952) *Analyt. Chem.* **24**, 623. Cyclopropane derivatives.

742 WILKINS, M. H. F. (1950) *Disc. Faraday Soc.* **9**, 363.

743 WILLIAMS, V. Z. (1951) *Science* **113**, 51.

744 WILLIAMS, H. R., and MOSHER, H. S. (1955) *Analyt. Chem.* **27**, 517. Alkyl hydroperoxides.

745 WILLIS, J. B. (1951) *Aust. J. sci. Res.* A **4**, 172.

746 WILLITS, C. H., DECIUS, J. C., DILLE, K. L., and CHRISTENSEN, B. E. (1955) *J. Amer. chem. Soc.* **77**, 2569. Purines.

747 WILSON, E. B., and WELLS, A. J. (1946) *J. chem. Phys.* **14**, 578.

748 WINSTON, H., and HALFORD, R. S. (1949) *J. chem. Phys.* **17**, 607.

749 WOLLMAN, S. H. (1949) *Rev. sci. Instrum.* **20**, 220.

750 WOOD, D. L. (1950) *Rev. sci. Instrum.* **21**, 764.

751 WOOD, D. L. (1955) *Rev. sci. Instrum.* **26**, 787.

752 WOODHULL, E. H., SIEGLER, E. H., and SOBCOV, H. (1954) *Industr. Engng. Chem. (Industr.)* **46**, 1396.

753 WOODWARD, L. A. (1950) *Nature* **165**, 198.

754 WORMSER, E. M. (1953) *J. opt. Soc. Amer.* **43**, 15.

755 WOTIZ, J. H. (1951) *J. Amer. chem. Soc.* **73**, 693.

756 WOTIZ, J. H., and MILLER, F. A. (1949) *J. Amer. chem. Soc.* **71**, 3441. Acetylenes.

757 WOTIZ, J. H., MILLER, F. A., and PALCHAK, R. J. (1950) *J. Amer. chem. Soc.* **72**, 5055. Propargyl alcohol, propargyl bromide.

758 WRIGHT, N. (1941) *Industr. Engng. Chem. (Anal.)* **13**, 1.

759 WRIGHT, N. (1948) *J. opt. Soc. Amer.* **38**, 69.

760 WRIGHT, N., and HUNTER, M. J. (1947) *J. Amer. chem. Soc.* **69**, 803. Methyl-polysiloxanes.

761 YASUMI, M. (1955) *Bull. chem. Soc. Japan* **28**, 489.

762 YATES, K. P., and BUHL, R. F. (1955) *J. opt. Soc. Amer.* **45**, 192.

763 YATES, P., ARDAO, M. I., and FIESER, L. F. (1956) *J. Amer. chem. Soc.* **78**, 650.

764 YOSHINAGA, H., and OETJEN, R. A. (1955) *J. opt. Soc. Amer.* **45**, 1085.

765 YOUNG, A. S. (1955) *J. sci. Instrum.* **32**, 142.

766 YOUNG, C. W., DUVALL, R. B., and WRIGHT, N. (1951) *Analyt. Chem.* **23**, 709. Benzene and derivatives.

767 ZBINDEN, R., BALDINGER, E., and GANZ, E. (1949) *Helv. phys. acta* **22**, 411.

768 ZEISS, H. H., and TSUTSUI, M. (1953) *J. Amer. chem. Soc.* **75**, 897. Alcohols.

769 ZERWEKH, C. E. (1949) *Rev. sci. Instrum.* **20**, 371.

770 ZUIDEMA, G. D., COOK, P. L., and ZYL, G. V. (1953) *J. Amer. chem. Soc.* **75**, 294. Epoxy compounds.

771 BERGMANN, G., and KAISER, H. (1960) *Z. InstrumKde.* **68**, 201.

772 COLLIER, G. L., and SINGLETON, F. (1956) *J. appl. Chem.* **6**, 495.

773 CORBRIDGE, D. E. C. (1956) *J. appl. Chem.* **6**, 456. Phosphorus compounds.

774 DANNENBERG, H., and NAERLAND, A. (1957) *Z. Naturf.* **12b**, 1. Anthracene derivatives.

775 FLAIG, W., and SALFELD, J. C. (1956) *Naturwiss.* **43**, 467. Benzoquinones.

776 GREENLER, R. G. (1957) *J. opt. Soc. Amer.* **47**, 130.

777 HEXTER, R. M., and DOWS, D. A. (1956). *J. chem. Phys.* **25**, 504.

778 JAFFE, J. H., and KIMEL, S. (1956) *J. chem. Phys.* **25**, 374.

779 JONES, L. H., and CHAMBERLAIN, M. M. (1956) *J. chem. Phys.* **25**, 365.

780 KAISER, R. (1956) *Kolloidzschr.* **149**, 84.

781 LIDDEL, U., and BECKER, E. D. (1956) *J. chem. Phys.* **25**, 173.

782 LÜTTKE, W. (1957) *Z. Elektrochem.* **61**, 302. Nitroso-compounds.

783 RENNDORF, R. (1957) *J. opt. Soc. Amer.* **47**, 176.

784 PINCHAS, S. (1957) *Analyt. Chem.* **29**, 334. Aldehydes.

785 RANK, D. H., GUENTHER, A. H., SHEARER, J. N., and WIGGINS, T. A. (1957) *J. opt. Soc. Amer.* **47**, 144.

786 RODNEY, W. S., and MALITSON, I. H. (1956) *J. opt. Soc. Amer.* **46**, 956.

787 SPEDDING, H., and WHIFFEN, D. H. (1956) *Proc. roy. Soc.* **A 238**, 245.

788 STERNGLANZ, H. (1956) *Appl. Spectr.* **10**, 77.

789 WENZEL, F., SCHIEDT, U., and BREUSCH, F. L. (1957) *Z. Naturf.* **12b**, 71. Fatty acids.

790 WIBERLEY, S. E., SPRAGUE, J. W., and CAMPBELL, J. E. (1957) *Analyt. Chem.* **29**, 210.

791 BAKER, A. W. (1957) *J. phys. Chem.* **61**, 450.

792 BAKER, A. W., CRANE, R. A., OPLER, A., and DIETHELM, C. A. (1956) *Appl. Spectr.* **10**, 83.

793 BIRD, G. R., and SHURCLIFF, W. A. (1959) *J. opt. Soc. Amer.* **49**, 235.

794 BLACK, E. D., MARGERUM, J. D., and WYMAN, G. M. (1957) *Analyt. Chem.* **29**, 169.

795 BRACKETT, F. S. (1957) *J. opt. Soc. Amer.* **47**, 636.

796 BRADLEY, K. B., and POTTS, W. J. (1958) *Appl. Spectr.* **12**, 77.

797 DRECHSLER, L. (1957) *Naturwiss.* **44**, 533.

798 DRECHSLER, L. (1958) *Jenaer Jahrbuch* **1958** I, p. 55.

799 EDWARDS, D. F., and BRUEMMER, M. J. (1959) *J. opt. Soc. Amer.* **49**, 860.

800 FORD, M. A., PRICE, W. C., SEEDS, W. E., and WILKINSON, G. R. (1958) *J. opt. Soc. Amer.* **48**, 249.

801 FRASER, R. D. B. (1958) *J. opt. Soc. Amer.* **48**, 1017.

802 GATEHOUSE, B. M. (1957) *Chem. & Ind.* **1957**, p. 1352.

803 GORDON, J. M. (1958) *J. opt. Soc. Amer.* **48**, 583.

804 GORE, R. C. (1958) *Analyt. Chem.* **30**, 570.

805 HARRICK, N. J. (1959) *J. opt. Soc. Amer.* **49**, 379.

806 HEAVENS, O. S., RING, J., and SMITH, S. D. (1957) *Spectrochim. Acta* **10**, 179.

807 HERSHENSON, H. M. (1959) *Infrared Absorption Spectra, Index for 1945–1957* (New York and London, 1959).

808 JONES, R. N., FAURE, P. K., and ZAHARIAS, W. (1959) *Rev. univ. Min.* **15**, 417.

809 JONES, R. N., and NADEAU, A. (1958) *Spectrochim. Acta* **12**, 183.

810 KATRITZKY, A. R. (1959) *Quart. Rev.* **13**, 353.

811 KATRITZKY, A. R. (1958) *J. chem. Soc.* **1958**, 4162.

812 KATRITZKY, A. R., and JONES, R. A. (1959) *J. chem. Soc.* **1959**, 3670.

813 KATRITZKY, A. R., and LAGOWSKI, J. M. (1958) *J. chem. Soc.* **1958**, 4155.

814 KATRITZKY, A. R., and SIMMONS, P. (1959) *J. chem. Soc.* **1959**, 2051.

815 KATRITZKY, A. R. (1959) *J. chem. Soc.* **1959**, 2058.

816 MEAKINS, G. D., and NELSON, K. A. (1957) *Chem. & Ind.* **1957**, 1415.

817 MELOCHE, V. W., and KALBUS, G. E. (1958) *J. Inorg. Nucl. Chem.* **6**, 104.

818 MITZNER, B. M. (1957) *J. opt. Soc. Amer.* **47**, 328.

819 OLSEN, A. L., JOHNSON, D. J., and PIERSON, R. H. (1956) *J. opt. Soc. Amer.* **46**, 354.

820 PLYLER, E. K., BLAINE, L. R., and NOWAK, M. (1957) *J. Res. nat. Bur. Stand.* **58**, 195.

821 POTTS, W. J., and WRIGHT, N. (1956) *Analyt. Chem.* **28**, 1255.

822 RANK, D. H., GUENTHER, A. H., SHEARER, J. N., and WIGGINS, T. A. (1957) *J. opt. Soc. Amer.* **47**, 144.

823 RAO, K. N., RYAN, L. R., and NIELSEN, H. H. (1959) *J. opt. Soc. Amer.* **49**, 216.

824 RAO, K. N., COBURN, T. J., GARING, J. S., ROSSMANN, K., and NIELSEN, H. H. (1959) *J. opt. Soc. Amer.* **49**, 221.

Literature

825 SCHWARZ, H. P., CHILOS, R. C., DREISBACH, L., MASTRANGELO, S. V., and
     KLESCHIK, A. (1958) *Appl. Spectr.* **12,** 35.
826 SMITH, S. D., and HEAVENS, O. S. (1957) *J. sci. Instrum.* **34,** 492.
827 STITT, F., and BAILEY, G. F. (1957) *Analyt. Chem.* **29,** 1557.
828 WASHBURN, W. H. (1956) *Appl. Spectr.* **2,** No. 1, Nov. 1956.
829 WASHBURN, W. H., and MAHONEY, M. J. (1958) *Analyt. Chem.* **30,** 1053.
830 WHITE, J. U., and ALPERT, N. L. (1958) *J. opt. Soc. Amer.* **48,** 460.
831 WHITE, J. U., WEINER, S., ALPERT, N. L., and WARD, W. M. (1958) *Analyt.
     Chem.* **30,** 1694.
832 ZENCHELSKY, S. T., and SHOWELL, J. S. (1957) *Analyt. Chem.* **29,** 167.
833 PEMSLER, J. P. (1957) *Rev. sci. Instrum.* **28,** 274.
834 HAWKINS, J. G., WARD, E. R., and WHIFFEN, D. H. (1957) *Spectrochim. Acta* **10,**
     105.

Index

407

Index

Index

Index

412

Index

Index

Index

418